Camera Image Quality Benchmarking

Color Appearance Models (3rd Edition)
Mark D. Fairchild

2.5D Printing: Bridging the Gap between 2D and 3D Applications
Carinna Parraman and Maria V. Ortiz Segovia

Published in Association with the Society for Imaging Science and Technology

Camera Image Quality Benchmarking

Jonathan B. Phillips
Google Inc., USA

Henrik Eliasson
Eclipse Optics AB, Sweden

With contributions on video image quality by Hugh Denman

This edition first published 2018
© 2018 John Wiley and Sons Ltd

The right of Jonathan B. Phillips and Henrik Eliasson to be identified as the authors of this work / the editorial material in this work has been asserted in accordance with law.

Registered Offices
John Wiley & Sons, Inc., 111 River Street, Hoboken, NJ 07030, USA
John Wiley & Sons Ltd, The Atrium, Southern Gate, Chichester, West Sussex, PO19 8SQ, UK

Editorial Office
The Atrium, Southern Gate, Chichester, West Sussex, PO19 8SQ, UK

For details of our global editorial offices, customer services, and more information about Wiley products visit us at www.wiley.com.

Wiley also publishes its books in a variety of electronic formats and by print-on-demand. Some content that appears in standard print versions of this book may not be available in other formats.

Limit of Liability/Disclaimer of Warranty
While the publisher and authors have used their best efforts in preparing this work, they make no representations or warranties with respect to the accuracy or completeness of the contents of this work and specifically disclaim all warranties, including without limitation any implied warranties of merchantability or fitness for a particular purpose. No warranty may be created or extended by sales representatives, written sales materials or promotional statements for this work. The fact that an organization, website, or product is referred to in this work as a citation and/or potential source of further information does not mean that the publisher and authors endorse the information or services the organization, website, or product may provide or recommendations it may make. This work is sold with the understanding that the publisher is not engaged in rendering professional services. The advice and strategies contained herein may not be suitable for your situation. You should consult with a specialist where appropriate. Further, readers should be aware that websites listed in this work may have changed or disappeared between when this work was written and when it is read. Neither the publisher nor authors shall be liable for any loss of profit or any other commercial damages, including but not limited to special, incidental, consequential, or other damages.

The content of the book, including the views and opinions expressed, are solely attributable to the Authors and do not reflect any official policy or practice of Kodak, its subsidiaries or affiliates. Kodak's permission should not imply any approval or verification by Kodak of the factual and opinion contents of the book.

Library of Congress Cataloging-in-Publication Data

Names: Phillips, Jonathan B., 1970- author. | Eliasson, Henrik, author. |
    Denman, Hugh, 1978- contributor.
Title: Camera image quality benchmarking / by Jonathan B. Phillips, Henrik
    Eliasson ; with contributions on video image quality by Hugh Denman.
Description: Hoboken, NJ : John Wiley & Sons, 2017. | Includes
    bibliographical references and index. |
Identifiers: LCCN 2017024315 (print) | LCCN 2017041277 (ebook) | ISBN
    9781119054528 (pdf) | ISBN 9781119054511 (epub) | ISBN 9781119054498
    (cloth)
Subjects: LCSH: Image processing. | Imaging systems–Image quality.
Classification: LCC TA1637 (ebook) | LCC TA1637 .P483 2017 (print) | DDC
    771.3–dc23
LC record available at https://lccn.loc.gov/2017024315

Cover Design: Wiley
Cover Image: © jericho667/Getty Images

Set in 10/12pt WarnockPro by SPi Global, Chennai, India

Contents

About the Authors

Source: Courtesy of Weinberg-Clark Photography

Jonathan B. Phillips is Staff Image Scientist at Google where his responsibilities include overseeing the approach to defining, measuring, and developing image quality for consumer hardware. His involvement in the imaging industry spans more than 25 years, including an image scientist position at NVIDIA and two decades at Eastman Kodak Company where he was Principal Scientist of Imaging Standards. His focus has been on photographic quality, with an emphasis on psychophysical testing for both product development and fundamental perceptual studies. His broad experience has included image quality work with capture, display, and print technologies. He received the 2011 I3A Achievement Award for his work on camera phone image quality and headed up the 2012 revision of ISO 20462 – Psychophysical experimental methods for estimating image quality – Part 3: Quality ruler method. He is a United States delegate to the ISO Technical Committee 42/Working Group 18 on photography and a longstanding member of the IEEE CPIQ (Camera Phone Image Quality) initiative. With sponsorship from Kodak, Jonathan's graduate work was in color science in the Munsell Color Science Lab and the Center for Imaging Science at Rochester Institute of Technology. His undergraduate studies in chemistry and music were at Wheaton College (IL).

Henrik Eliasson received his Masters and PhD degrees in physics from Göteborg University. His thesis work focused on relaxation processes in polymers around the glass transition temperature. He has been working in the optics and imaging industry for the past 17 years, first as a consultant designing optical measurement systems and between 2003 and 2012 as a camera systems engineer at Sony Ericsson/Sony Mobile Communications. There he engineered the camera systems in many of the successful products made by the company. He was also deeply involved in the image quality improvement work and in building up the camera labs as well as designing and implementing new image quality assessment methods. In this role he was the company representative in the CPIQ (Camera Phone Image Quality) initiative, where he was a key contributor in developing many of the image quality metrics.

He has also been a Swedish delegate to the ISO Technical Committee 42/Working Group 18 on photography. His experience and expertise in imaging covers many different areas, such as color science, optical measurements, image sensor characterization and measurements, as well as algorithm development and image systems simulations and visualization. He also has a keen interest in photography in general, providing many of the photographs found in this book. Currently, he is working at Eclipse Optics in Sweden as an image sensor and image analysis specialist. Dr. Eliasson is a Senior Member of SPIE.

Series Preface

At the turn of the century, a cellular phone with an on-board camera did not exist and the film camera market had just hit its historic peak. In 1999, digital cameras were introduced, and in the early 2000s cameras were first integrated into mobile phones. By 2015, more than 25% of the world's population were using smartphones. With this explosion of "pocket supercomputers", the need to understand and evaluate the quality of the pictures captured by digital cameras has increased markedly, and a resource for *Camera Image Quality Benchmarking* has become essential. In this so-named book, part of the *Wiley-IS&T Series in Imaging Science and Technology*, Jonathan Phillips and Henrik Eliasson provide information on image quality metrics, how they were developed, why they are needed, and how they are used.

This book provides the framework for understanding the visual quality of digitally captured images. It defines image quality and its attributes, and sketches a detailed perspective on the qualitative and quantitative approaches to the evaluation of captured images. There are many existing sources for learning about the subjective and objective procedures for evaluating image quality; however, this book goes many steps further. It provides the reader with an understanding of the important elements of the camera itself as well as of the physiology and physicality of the human visual system. This awareness of both the human and machine capture systems provides the background needed to understand why the accepted metrics were developed. The book also elucidates why measuring perceived quality has been such an intractable problem to solve.

Additionally, a key contribution of *Camera Image Quality Benchmarking* is that it provides detailed information on how to set up a lab for efficiently conducting this work. This means describing the testing, including how to select image content and observers when needed. This information is invaluable to those who aim to understand the capabilities of camera prototypes and to evaluate finished products.

The authors have been engaged in the development of camera-captured image quality measurements for many years. Their complementary backgrounds provide them with a somewhat different emphasis. Mr. Jonathan Phillips has generally been focused on subjective and applied aspects of image quality evaluation. He has played an important role in image capture evaluation in industry. As a seasoned Image Scientist, currently at Google, and previously at NVIDIA and Kodak, he has been deeply engaged in the development and evaluation of image quality measurements and the use of these to foster improved capture products. Additionally, he has been a key member of the IEEE Camera Phone Image Quality (CPIQ) initiative and the ISO Technical Committee 42 on photography. Thus, he has been instrumental in the development of international standards for quantifying photographic quality. The research focus of Mr. Phillips' graduate work in Color Science at the Rochester Institute of Technology was on perceptual image quality. His undergraduate studies were in chemistry and music at Wheaton College (IL). His accomplishments include the 2011 Achievement Award from International Imaging Industry Association for his contributions to the CPIQ image quality test metrics. His academic and industrial backgrounds serve as a solid foundation for making this valuable contribution to the **Wiley-IS&T Series in Imaging Science and Technology**.

Partnering Mr. Phillips' attention to the subjective and applied aspects, Dr. Henrik Eliasson has generally been focused on the objective and theoretical side of image quality measurement. Dr. Eliasson completed his graduate work in Physics at Göteborg University. Since then, he has designed optical measurement systems and, more recently, engineered camera systems for Sony Ericsson/Sony Mobile Communications. His work at Sony Ericsson also involved establishing the camera labs as well as designing and implementing improved image quality evaluation techniques. Currently, he is working as a consultant at Eclipse Optics in Sweden, with a focus on image sensor technology and image analysis. His publications cover a breadth of imaging and image quality topics including optics simulation, white balancing assessment, and image sensor crosstalk characterization. He, like Mr. Phillips, has played an important role in the CPIQ (Camera Phone Image Quality) initiative. He has served as a Swedish delegate in the ISO Technical Committee 42/Working Group 18 on photography. Dr. Eliasson is a Senior Member of SPIE. Together, the two authors bring significant experience in and understanding of the world of **Camera Image Quality Benchmarking**.

As cameras become ubiquitous for everything from selfies to "shelfies" (inventory management), and from surveillance to purveyance (automated point of sale), image quality assessment needs to become increasingly automated, so the right information is disambiguated and the unnecessary images are discarded. It is hard to imagine a world without digital image capture, and yet we have only begun. This book is sure to maintain its relevance in a world where automated capture, categorization, and archival imaging become increasingly critical.

Susan P. Farnand
Steven J. Simske

Preface

The seed for the content of this book started in 2011 when Nicolas Touchard of DxO Labs in France, being a participant in the Camera Phone Image Quality (CPIQ) initiative just like us, contacted us about a short course they wanted to teach on objective and subjective camera image quality benchmarking at the then SPIE/IS&T Electronic Imaging conference (now IS&T International Symposium on Electronic Imaging). Nicolas proposed to have us join some of his colleagues, Harvey (Hervé) Hornung, Frédéric Cao and Frédéric Guichard, to plan this new course. In 2012, we launched our short course on camera image quality benchmarking with a set of nearly 400 slides, with the DxO team being the major contributor. Over a period of several years, the course was supplemented and revised, with particular attention to adding video image quality to our initial focus of still imaging. Five individuals were involved with the class instruction over time: apart from Hervé and the two of us (Henrik and Jonathan), also Nicolas Touchard and Hugh Denman. When John Wiley & Sons Ltd asked Jonathan in 2014 about converting our course slides to a book, he contacted each of the course contributors with the same inquiry. Finally, the two of us decided we were up to the challenge. As we began the writing, we realized we needed to convince Hugh, who was at YouTube at the time, to join our efforts as a contributing author on the topic of video image quality.

We have been involved in image quality measurements for many years, in the mobile industry as well as more generally. Our backgrounds are slightly different: while Jonathan has mainly been focusing his efforts on the subjective and pragmatic side of image quality assessment, Henrik has been looking more at objective image quality metrics and the theory of such. Thus, the book has been naturally divided between us, with Jonathan responsible for Chapters 1, 5, 8, and 9, and Henrik for Chapters 2, 3, 4, 6, and 7. We need to mention here also the contribution from Hugh Denman, who has been responsible for the video-related content in Chapters 1 and 3 through 8.

We have met regularly, nearly every weekend via webcam, for the past several years as we have been collaborating on the book. To increase our productivity, we have even had "Writers Workshops" in our respective countries, with Henrik having spent time in San Jose with Jonathan, and Jonathan traveling to spend time in both Sweden with Henrik and Ireland with Hugh. Those workshops as well as weekly meetings have enabled us to assemble this book in a cohesive fashion, including both still and video image quality benchmarking material. Our photography in the following chapters includes images of scenes from our respective countries among others, which we have captured throughout the writing process and are excited to share with the readers of the book.

It has been very interesting to follow the evolution of the camera from a specialized piece of equipment that one carried mostly during vacations and other important events to a ubiquitous component in a mobile communications device one always has in one's pocket or carryall. 15 years ago, it would have been hard to believe that the image quality provided by an average mobile phone camera today would actually be better in most cases compared with the compact cameras then available. Still, there are areas in which the mobile phone camera is lacking, like low-light photography and zooming capabilities. At the same time, the basic principle of photography and video capture has not changed significantly: you point the camera toward the subject you wish to capture and press the shutter button. The camera then records the image or video and one can view it, for example, directly on screen or as a printout at a later time. What has changed tremendously is in the way we use the photos and videos taken, as well as the subjects captured. From being "frozen memories" helping us to recall important moments in our lives and brought out to be viewed at specific occasions, photos are now to a large extent consumed immediately and used to augment social interactions. This has led to an enormous increase in the quantity of pictures taken.

Even with this exponentially increasing volume of new photos taken every day, which inevitably leads to a shorter attention span with regard to viewing the images, the quality of the images is still important. This is not least evident from the advertising made by the large mobile phone companies, where camera image quality has a prominent place. Therefore, the quantification of image quality has not become less important, but rather more so.

One often hears the claim that image quality cannot be measured due to reasons such as its subjective nature, or the complexities involved. What we have realized over the years is that image quality measurements are indeed hard to perform, but not prohibitively so. This book will demonstrate this point through theoretical reasoning but also through examples. Furthermore, even with the development of new technologies, many of the traditional concepts and techniques remain valid today. They do need continuous development, but the effort spent on learning the important theories and methods is not wasted, and will instead help substantially in understanding many of the new technologies and how to develop metrics to cope with their particular quirks and intricacies. It is therefore our hope that this book will give the reader and benchmarking practitioner a good start in tackling the challenges that lie ahead in image quality camera benchmarking and characterization.

In order to provide even more useful and accurate content, we sought out notable authorities and peers in the field to both review and provide dialog on chapters related to their respective expertise. Of note, we appreciate and thank the following: Kjell Brunnström, Mark D. Fairchild, Harvey (Hervé) Hornung, Paul Hubel, Elaine W. Jin, Kenneth A. Parulski, Andreas von Sneidern, Nicolas Touchard, and Mats Wernersson.

We want to give a very special thanks to our contributing author, Hugh Denman, for providing invaluable knowledge and insights to the video-related sections spread throughout the book.

We have been given very valuable help by the Wiley staff, Ashmita Thomas Rajaprathapan and Teresa Netzler. We would also like to thank our commissioning editor, Alexandra Jackson, who helped us through the initial stages of the book proposal process and launch of writing.

We also want to thank DxO Labs for providing images as well as video examples for the electronic version of the book. Nicolas Touchard of DxO has been invaluable in supporting us, and also for providing so much extremely valuable feedback at the beginning of this project. We are grateful to the team at Imatest for allowing us to use images of their test charts and equipment, and to Dietmar Wüller at Image Engineering for many fruitful discussions and helpful comments as well as graciously letting us use images of their equipment.

Jonathan B. Phillips
Henrik Eliasson

Of course there are many others to thank along the way. For me, I first want to thank the outstanding colleagues from Kodak who have been my inspiration in so many ways. Starting out my professional career in Kodak's Image Science Career Development Program was a superb introduction to the world of imaging science. When contemplating this book project, I sought out the advice from instructors in that program who have been excellent advisors over the years: Brian W. Keelan and Edward J. Giorgianni. I thank my former Kodak managers, Kenneth A. Parulski and Brian E. Mittelstaedt, who provided the opportunities for me to branch into mobile imaging and the image quality standards efforts in ISO Technical Committee 42/Working Group 18 on photography and the IEEE CPIQ (Camera Phone Image Quality) initiative. I also thank NVIDIA and Google for continuing the sponsorship of my participation in ISO and CPIQ with support from Margaret Belska, Boyd A. Fowler, and Vint Cerf. In many ways, much of the content of this book is built on the dedicated expert efforts of the members and delegates of these image quality standards bodies. Additionally, fundamental to the writing of this book are the faculty and staff of the Munsell Color Science Lab and Center for Imaging Science at Rochester Institute of Technology, who were my instructors and guides as I spent over eight years there for my graduate work. Finally, I thank the many observers who have participated in countless subjective studies, without which we would not have the understanding of camera image quality benchmarking presented in this book.

Jonathan B. Phillips
San Jose, California

Writing this book would never have been possible if I hadn't started to become involved in imaging standards activities some ten years ago through the CPIQ initiative. For me, this has been a great learning experience, and I have enjoyed tremendously being a part of this community where I have met so many outstanding and great personalities over the years. I am thankful to my former employer and my managers there, Martin Ek, Pontus Nelderup, Fredrik Lönn, Per Hiselius, and Thomas Nilsson, for giving me the opportunity to participate in these activities. I would also like to thank my father, Lars Eliasson, for reviewing parts of the book and giving valuable feedback. Last, but in no way least, my wife, Monica, has played a big role in making this possible, not only by putting up with all the evenings I have spent by the computer, but also for giving enthusiastic support and encouragement as well as providing valuable insights and suggestions.

Henrik Eliasson
Blentarp, Sweden

List of Abbreviations

3A	Auto exposure, auto white balance, and autofocus
3AFC	Three-Alternative Forced Choice
ACF	Autocorrelation Function
ACR	Absolute Category Rating
ADC	Analog to Digital Converter
APS	Active Pixel Sensor
AWB	Auto White Balance
AWGN	Additive White Gaussian Noise
BRDF	Bidirectional Reflectance Distribution Function
BSI	Backside Illumination
CAM	Color Appearance Model
CAT	Chromatic Adaptation Transformation
CCD	Charge Coupled Device
CCM	Color Correction Matrix
CCT	Correlated Color Temperature
CDS	Correlated Double Sampling
CI	Confidence Interval
CIE	International Commission on Illumination
CMF	Color Matching Function
CMOS	Complementary Metal Oxide Semiconductor
CPIQ	Camera Phone Image Quality
CPU	Central Processing Unit
CRA	Chief Ray Angle
CRI	Color Rendering Index
CRT	Cathode Ray Tube
CSF	Contrast Sensitivity Function
CTF	Contrast Transfer Function
CWF	Cool White Fluorescent
DCR	Degradation Category Rating
DCT	Discrete Cosine Transform
DFD	Displaced Frame Difference
DFT	Discrete Fourier Transform
DIS	Digital Image Stabilization
DMOS	Difference Mean Opinion Score or Degradation Mean Opinion Score

DMX	Digital Multiplex
DOF	Depth of Field
DRS	Digital Reference Stimuli
DSC	Digital Still Camera
DSCQS	Double Stimulus Continuous Quality Scale
DSIS	Double Stimulus Impairment Scale
DSLR	Digital Single Lens Reflex
DVD	Digital Versatile Disc
EI	Exposure Index
EIS	Electronic Image Stabilization
Exif	Exchangeable Image File Format
FFT	Fast Fourier Transform
FOV	Field of View
FPN	Fixed Pattern Noise
fps	Frames per second
FR	Full Reference
FSI	Frontside Illumination
GOP	Group Of Pictures
GSM	Global System for Mobile Communications
HDR	High Dynamic Range
HEVC	High Efficiency Video Coding
HFR	High Frame Rate
HR	Hidden Reference
HRC	Hypothetical Reference Circuit
HVS	Human Visual System
IEEE	Institute of Electrical and Electronics Engineers
IHIF	Integrated Hyperbolic Increment Function
IR	Infrared Radiation
ISO	International Organization for Standardization
ISP	Image Signal Processor
ITU	International Telecommunication Union
JND	Just Noticeable Difference
JPEG	Joint Photographic Experts Group
LCA	Lateral Chromatic Aberration
LCD	Liquid Crystal Display
LED	Light Emitting Diode
LSB	Least Significant Bit
LSF	Line Spread Function
LVT	Light Valve Technology
MOS	Mean Opinion Score, Metal Oxide Semiconductor
MP	Megapixels
MPEG	Moving Picture Experts Group
MSE	Mean Squared Error
MTF	Modulation Transfer Function
NPS	Noise Power Spectrum
NR	No Reference
OECF	Opto-Electronic Conversion Function

OIS	Optical Image Stabilization
OLPF	Optical Lowpass Filter
OM	Objective Metric
OTF	Optical Transfer Function
PDF	Probability Density Function
ppi	Pixels per inch
PSD	Power Spectral Density
PSF	Point Spread Function
PSNR	Peak Signal to Noise Ratio
PTC	Photon Transfer Curve
PTF	Phase Transfer Function
PVS	Processed Video Sequence
PWM	Pulse Width Modulation
QE	Quantum Efficiency
QL	Quality Loss
REI	Recommended Exposure Index
RIP	Raster Image Processor
RMS	Root Mean Square
RR	Reduced Reference
SAMVIQ	Subjective Assessment of Multimedia Video Quality
SCS	System Coverage Space
SDSCE	Simultaneous Double Stimulus for Continuous Evaluation
SNR	Signal to Noise Ratio
SOS	Standard Output Sensitivity
SQF	Subjective Quality Factor
SQS	Standard Quality Scale
SRS	Standard Reference Stimuli
SSCQE	Single Stimulus Continuous Quality Evaluation
SSIM	Structural Similarity
SSMR	Single Stimulus with Multiple Repetition
UCS	Uniform Chromaticity Scale
UHD	Ultra High Definition
VCX	Valued Camera eXperience
VDP	Visible Differences Predictor
VGA	Video Graphics Array
VGI	Veiling Glare Index
VIQET	Video Quality Experts Group Image Quality Evaluation Tool
VQEG	Video Quality Experts Group
VQM	Video Quality Model

About the Companion Website

This book is accompanied by a companion website:

www.wiley.com/go/benchak

The website includes:

- Videos
- Animated GIFS

Scan this QR code to visit the companion website

1

Introduction

Camera imaging technology has evolved from a time-consuming, multi-step chemical analog process to that of a nearly instantaneous digital process with a plethora of image sharing possibilities. Once only a single-purpose device, a camera is now most commonly part of a multifunctional device, for example, a mobile phone. As digital single lens reflex (DSLR) cameras become more sophisticated and advanced, so also mobile imaging in products such as smartphones and tablet computers continues to surge forward in technological capability. In addition, advances in image processing allow for localized automatic enhancements that were not possible in the past. New feature algorithms and the advent of computational photography, for example, sophisticated noise reduction algorithms and post-capture depth processing, continue to flood the market. This necessitates an ever expanding list of fundamental image quality metrics in order to assess and compare the state of imaging systems. There are standards available that describe image quality measurement techniques, but few if any describe how to perform a complete characterization and benchmarking of cameras that consider combined aspects of image quality. This book aims to describe a methodology for doing this for both still and video imaging applications by providing (1) a discourse and discussions on image quality and its evaluation (including practical aspects of setting up a laboratory to do so) and (2) benchmarking approaches, considerations, and example data.

To be most useful and relevant, benchmarking metrics for image quality should provide consistent, reproducible, and perceptually correlated results. Furthermore, they should also be standardized in order to be meaningful to the international community. These needs have led to initiatives such as CPIQ (Camera Phone Image Quality), originally managed by the I3A (International Image Industry Association) but now run as part of standards development within the IEEE (Institute of Electrical and Electronics Engineers). The overall goal of this specific CPIQ work is to develop an image quality rating system that can be applied to camera phones and that describes the quality delivered in a better way than just a megapixel number. In order to accomplish this, metrics that are well-correlated with the subjective experience of image quality have been developed. Other imaging standards development includes the metrics by Working Group 18 of Technical Committee 42 of the International Organization for Standardization (ISO) and the International Telecommunication Union (ITU). Theses standards bodies have provided, and continue to develop, both objective and subjective image quality metrics. In this context, objective metrics are defined measurements for which the methodology and results are independent of human perception, while subjective metrics are defined measurements using human observers to quantify human response.

In following chapters, the science behind these metrics will be described in detail and provide groundwork for exemplary benchmarking approaches.

1.1   Image Content and Image Quality

Before delving into the specifics related to objective and subjective image quality camera benchmarking, exploration of the essence of photography provides justification, motivation, and inspiration for the task. As the initial purpose for photography was to generate a permanent reproduction of a moment in time (or a series of moments in time for motion imaging), an understanding of what constitutes the quality of objects in a scene will necessitate what to measure to determine the level of image quality of that permanent reproduction. The more a photograph or video represents the elements of a physical scene, the higher the possible attainment of perceived quality can become.

The efforts to create the first permanent photograph succeeded in the mid-1820s when Nicéphore Niépce captured an image of the view from his dormer window—a commonplace scene with buildings, a tree, and some sky. The image, produced by a heliographic technique, is difficult to interpret when observing the developed chemicals in the original state on a pewter plate (see Figure 1.1). In fact, the enhancement of this "raw" image, analogous to the image processing step in a digital image rendering, produces a scene with more recognizable content (see Figure 1.2). But, even though key elements are still discernible, the image is blurry, noisy, and monochrome. The minimal sharpness and graininess of the image prevent discernment of the actual textures in the scene, leaving the basic shapes and densities as cues for object recognition. Of note is the fact that the west and east facing walls of his home, seen on the sides of the image, are simultaneously illuminated by sunlight. This is related to the fact that the exposure was eight hours

Figure 1.1 Image of first permanent photograph circa 1826 by N. Niépce on its original pewter plate. *Source*: Courtesy of Gernsheim Collection, Harry Ransom Center, The University of Texas at Austin.

Figure 1.2 Enhanced version of first permanent photograph circa 1826 by N. Niépce. *Source*: Courtesy of Gernsheim Collection, Harry Ransom Center, The University of Texas at Austin.

in length, during which the sun's position moved across the sky and exposed opposing facades (Gernsheim and Gernsheim, 1969). Needless to say, the monochrome image is void of any chromatic information.

That we can recognize objects in the rustic, historic Niépce print is a comment on the fundamentals of perception. Simple visual cues can convey object information, lighting, and depth. For example, a series of abstract lines can be used to depict a viola as shown in Figure 1.3. However, the addition of color and shading increases the perceived realism of the musical instrument, as shown in the center image. A high quality photograph of a viola contains even more information, such as albedo and mesostructure of the object which constitute the fundamental elements of texture, as shown on the right. Imaging that aims for realism contains the fundamental, low level characteristics of color, shape, texture, depth, luminance range, and motion. Faithful reproduction of these physical properties results in an accurate, realistic image of scenes and objects. These properties will be described in general in the following sections and expanded upon in much greater detail in later chapters of the book, which define image quality attributes and their accompanying objective and subjective metrics.

1.1.1 Color

Color is the visual perception of the physical properties of an object when illuminated by light or when self-luminous. On a basic level, color can describe hues such as orange, blue, green, and yellow. We refer to objects such as yellow canaries, red apples, blue sky, and green leaves. These colors are examples of those within the visible wavelength spectrum of 380 nm to 720 nm for the human visual system (HVS). However, color is more complex than perception of primary hues: color includes the perception of lightness and brightness, which allows one to discriminate between red and light red (i.e., pink), for example, or to determine which side of a uniformly colored house is facing

Figure 1.3 Three renditions of a viola. Left: line sketch; middle: colored clip art (Papapishu, 2007); right: photograph. Each shows different aspects of object representation. *Source*: Papapishu, https://openclipart.org/detail/4802/violin. CC0 1.0

Figure 1.4 Example illustrating simultaneous contrast. The center squares are identical in hue, chroma, and lightness. However, they appear different when surrounded by backgrounds with different colors.

the sun based on the brightness of the walls. These are relative terms related to the contextualized perception of the physical properties of reflected, transmitted, or emitted light, including consideration of the most luminous object in the scene. Color perception is also impacted by the surrounding colors—even if two colors have the same hue, they can appear as different hues if surrounded by different colors. Figure 1.4 shows an example of this phenomenon called *simultaneous contrast*. Note in this example that the center squares are identical. However, the surrounding color changes the appearance of the squares such that they do not look like the same color despite the fact that they are measurably the same.

There are other aspects of the HVS that can influence our perception of color. Our ability to adapt to the color cast of our surroundings is very strong. This *chromatic adaptation* allows us to discount the color of the illumination and judge color in reference to the scene itself rather than absolute colorimetry. When we are outside during sunlight hours, we adapt to the bright daylight conditions. In a similar manner, we adapt to indoor conditions with artificial illumination and are still able to perceive differences in color. Perceptually, we can discern colors such as red, green, blue, and yellow under either condition. However, if we were to measure the spectral radiance of a physical object under two strongly varying illuminant conditions, the measurements would be

Figure 1.5 Example illustrating chromatic adaptation and differences between absolute and relative colorimetry. The fruit basket in the original photo clearly exhibits varying hues. A cyan bias is added to the original photo to generate the middle photo. With chromatic adaptation, this photo with the cyan cast will have perceptible hue differences as well, allowing the observer to note a yellowish hue to the bananas relative to the other fruit colors. However, the bottom photo illustrates that replacing the bananas in the original photo with the cyan-cast bananas (the identical physical color of the bananas in the middle cyan-cast photo) results in a noticeably different appearance. Here, the bananas have an appearance of an unripe green state because chromatic adaptation does not occur. *Source*: Adapted from Fairchild 2013.

substantially different. An example is presented in Fairchild (2013) in which a fruit basket that is well-balanced for daylight exhibits distinct hue differences among the fruit. This is illustrated in the top photo in Figure 1.5. Relative to other fruit in the basket, apples on the right look red, oranges look orange, bananas look yellow, and so on. A cyan cast can be added to the photo such that its overall appearance is distinctly different from the original photo. However, with some time to adapt to the new simulated illumination conditions as presented in the middle photo, chromatic adaptation should occur, after which the fruit will once again begin to exhibit expected relative color such as the bananas appearing to have a yellowish appearance and the apples on the right having a reddish appearance. If, however, the bananas (only) in the original photo are replaced with those having the cyan cast, the chromatic adaptation does not take place; the bananas take on an unripe green appearance relative to the other fruit colors. So, too, the physical spectral reflectance is distinctly different for the bananas in the original and cyan-cast versions, though interpreted differently in the middle and bottom examples.

At times, due to the adaptive nature of the HVS, we can perceive color that is not physically present in a stimulus. A physiological example is part of our viewing experience every day, though we don't usually make note of the phenomenon. The signal of light detection in the eye travels to the brain via the optic nerve. This region is a *blind spot* in our vision because there are no light sensors present in the retina in this position. However, the HVS compensates for the two occlusions (one from each eye) and fills in the regions with signals similar to the surrounding are such that the occlusions of the optic nerves are not normally noticed. This filling in phenomenon encompasses both color and texture. In fact, the HVS is even adaptable to the level of filling in blindspots with highly detailed patterns such as text (though experimental observers could not actually read the letters in the filled-in region) (Ramachandran and Gregory, 1991)! Therefore, it should not be surprising that there are conditions that can result in the HVS filling in information as the signal to the eye is processed even if a blindspot is not present. As such, there can be a perception of a color even when there is no physical stimulus of a hue. An example of such a phenomenon is the *watercolor illusion* in which the HVS detects a faint color filling in shapes which have an inner thin chromatic border of the perceived hue surrounded with an adjacent darker border of a different hue. The filled region's hue is lighter than the inner border, however. Figure 1.6 shows shapes with undulating borders, which typically instill stronger filling in than linear borders. As should be seen due to the illusion, the regions within the shapes have an apparent watercolor-like orange or green tint whereas the regions outside of the shapes do not have this faint hue. However, the inside of the shapes are not orange or green; all regions on either side of the undulating borders are physically the same and would have the same colorimetric values if measured, that is, the value of the white background of the page.

An object has many physical properties that contribute to its color, including its reflectance, transmittance or emittance, its angular dependency, and its translucency. Thus, quantifying color has complexity beyond characterizing the spectral nature of

Figure 1.6 With a thin chromatic border bounded by a darker chromatic border, the internal region is perceived by the HVS to have a faint, light hue similar to the inner chromatic border even though the region has no hue other than the white background on the rest of the page. The regions within the shapes fill in with an orange or green tint due to the nature of the undulating borders and the hue of the inner border.

the color-defining element, such as a chromophore, dye, or pigment. Suppose we have a satin bed sheet and a broadcloth cotton bed sheet which are spectrally matching in hue, that is, having the same dye. However, we are able to discern a material difference because the satin sheet looks shiny and the broadcloth looks dull in nature. This difference in material appearance is because the satin has a woven mesostructure with very thin threads that generates a smooth, shiny surface when illuminated whereas the surface of the broadcloth is more diffuse due to thicker thread, lower thread count and a different weave, thus lacking the degree of shininess of satin. Yet, the color of the satin and broadcloth have matching color from a spectral standpoint. Another example of the complexity of color is the challenge of matching tooth color with a dental implant. Because teeth are translucent, the appearance of the whiteness is dependent on the lighting characteristics in the environment. Similar to placing a flashlight beam near the surface of marble, light can pass through a tooth as well as illuminate it. Thus, the challenge in matching a tooth appearance includes both a lightness and whiteness match as well as opaqueness. If a dental implant has a different opaqueness from the actual damaged tooth, there will be lighting environments in which the implant and tooth will not match even if the physical surface reflections of the white are identical.

Color measurements using colorimetry take into account the spectral properties of the illuminant, the spectral properties of the object, and the HVS. However, colorimetry has fundamental limitations when applied to the plethora of illuminants, objects, and people in the real world. In order to generate equations to estimate first-order color perception, data of (only) 17 color-normal observers were combined to generate the 1931 *standard observer* (Berns, 2000). That it was necessary to have more than one observer to make a standard observer is indicative of the inter-observer variability that exists in color perception. More recent works have confirmed that while this *observer metamerism* does exist, the 1931 standard observer remains a reasonable estimate of the typical color-normal observer (Alfvin and Fairchild, 1997; Shaw and Fairchild, 2002). In addition, inter-observer variability has been noted to be up to eight times greater than the differences inherent in the comparisons between the 1931 standard observer and five newer alternatives (Shaw and Fairchild, 2002). Thus, colorimetric quantification of colors incorporating the 1931 standard observer may predict color accuracy to a certain match level though an individual observer may not perceive the level as such. This becomes especially important considering the quality of colors in a scene that are captured by a camera and then observed on display or in printed material—the source of the colors of the scene, the display, and the printed material are composed of fundamentally differing spectral properties, but are assumed to have similar color for a high quality camera. In fact, color engineering could indeed have generated colors in a camera capturing system that match for the 1931 standard observer, but that matching approach does not guarantee that each individual observer will perceive a match or that the colorimetric match will provide the same impression of the original scene in the observer's mind.

Colorimetric equations are fundamental in quantifying the objective color image quality aspects of a camera. Measurements such as color accuracy, color uniformity, and color saturation metrics described later in the book utilize CIELAB colorimetric units to quantify color-related aspects of image quality. If, for example, the *color gamut* is wide, then more colors are reproducible in the image.

Quantifying the color performance, for example, color gamut, provides insight into an important facet of image quality of a camera system. However, as noted in previous examples, the appearance of color is more complex than the physical measurement of color alone, even when accounting for aspects of the HVS. Higher orders of color measurement include *color appearance models*, which account for the color surround and viewing conditions, among other complex aspects. Color appearance phenomena described in the examples above should point to the importance of understanding that sole objective measurements of color patches do not always correspond to the actual perception of the color in a photo. Challenges in measuring and benchmarking color will be discussed in more detail in further chapters.

1.1.2 Shape

A fundamental characteristic of object recognition in a scene is the identification of basic geometric structure. Biederman (1987) proposed a recognition-by-components theory in which objects are identified in a bottom-up approach where simple components are first assessed and then assembled into perception of a total object. These simple components were termed geometrical ions (or geons) with a total of 36 volumetric shapes identified, for example, cone, cylinder, horn, and lemon. Figure 1.7 has four examples showing how geons combine to form visually related, but functionally different, common objects. For example, in the center right a mug is depicted, whereas in the far right the same geons are combined to form a pail.

The vertices between neighboring geons are very important in distinguishing the overall object recognition: occlusions that overlap the vertices confuse recognition, whereas occlusions along geon segments can be filled in successfully (though this may require time to process perceptually). Biederman provides an example of the difference between these two scenarios (Biederman, 1987). Figure 1.8 contains an object with occluded vertices and a companion image in which only segments are occluded. The latter image on the right can be recognizable as the geons that comprise a flashlight whereas the former object is not readily discernible.

This bottom-up approach described above differs from Gestalt theory, which is fundamentally a top-down approach. "The whole is greater than the sum of the parts" is a generalization of the Gestalt concept by which perception starts with object recognition rather than an assimilation of parts. An example that bridges bottom-up and top-down theories is shown in Figure 1.9 (Carraher and Thurston, 1977). Top-down theorists point out that a Dalmatian emerges out of the scene upon study of the seemingly random

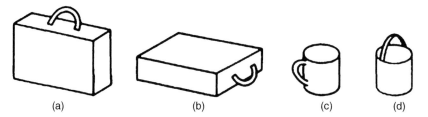

(a)	(b)	(c)	(d)

Figure 1.7 Examples showing how geons combine to form various objects. Far left: briefcase; center left: drawer; center right: mug; far right: pail. *Source*: Biederman 1987. Reproduced with permission of APA.

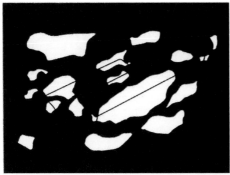

Figure 1.8 An example of an occluded object. Left: the vertices are occluded, making discernment of the object difficult. Right: only segments are occluded. In this right image, the object is more recognizable as a flashlight. *Source*: Biederman 1987. Reproduced with permission of APA.

Figure 1.9 An image associated with top-down processing in order to recognize the shape of a Dalmatian exploring the melting snow. *Source*: Republished with permission of John Wiley & Sons Inc, from Optical Illusions and the Visual Arts, Carraher and Thurston, Van Nostrand Reinhold Company, 7th printing, 1977; permission conveyed through Copyright Clearance Center, Inc.

collection of black blobs, while more recent research points to bottom-up processing for observers who found other objects in this scene such as an elephant or a jogger stretching out (van Tonder and Ejima, 2000). Regardless of the standpoint of bottom-up or top-down processing, shape is an important element of faithful scene reproduction.

Therefore, the spatially related aspects of an image will impact the perceived quality of the camera performance as pertaining to shape reproduction. Objective camera image quality metrics that are critical to shape quality include the spatial frequency response (SFR), resolution, bit depth, and geometric distortion. For example, a sharper image should increase the ability of the observer to see edges and, thus, shape and form in the image. Greater quality of shape and form, in turn, provides better camera image quality.

Figure 1.10 Influence of texture on appearance of fake versus real fruit. The fruits on the left in the top panoramic photo are all fake while the fruits on the right are real. Closer inspection of the pear surfaces can be seen in the bottom pair of images. The fake pear is on the left and the real pear is on the right. The texture appearance of the fake pear is composed of red paint drops.

1.1.3 Texture

Variations in apparent surface properties are abundant in both natural and synthetic physical objects. The HVS is adept at distinguishing these texture properties of objects. For example, in the field of mineralogy, an extensive vocabulary has been defined to describe the visual appearance of rock material (Adelson, 2001). These terms include words such as greasy, vitreous (glassy), dull, dendritic, granular, porous, scaly, and felted. While some of these terms such as greasy and scaly may conjure up specific visual differences, many of the mineralogists' terms refer to subtle changes in surface properties. This highlights the sophistication of the HVS as well as the importance of being able to generate realistic representations of objects in imaging systems. Appearance of material properties has been the focus of ongoing research and publications in the fields of perceptual psychology and computer graphics (Adelson, 2001; Landy, 2007; Motoyoshi *et al.*, 2007; Dorsey *et al.*, 2008; Rushmeier, 2008). Related to food appearance, there are fake products on the market that mimic real food. The top panoramic image in Figure 1.10 contains both fake and real fruits. Material properties that might provide clues as to which is which include texture and glossiness—attributes needing closer inspection. The bottom pair of images shows a crop of the fake pear surface on the left and the real pear surface on the right. In fact, the fake pear does have texture, but it is made with red paint drops whereas the real pear on the right has naturally occurring darker spots and even some surface scratches present in the lower right. As arranged in the panoramic photo at the top, the fake fruits are all on the left. This example shows that the appearance of material properties, for example, texture of fruits, influences the perception and interpretation of objects.

In photographic images, texture enhances object recognition. With changes in texture, an object can transform from appearing pitted and rough to appearing very smooth and

Figure 1.11 Left: the original image; right: after applying a sigma filter similar to one that would be used to reduce image noise (See Chapter 4 for more information on sigma filters.). Note the loss of texture in the hair, skin, and clothing, which lowers overall quality even though edges of the face, eyes, and teeth remain mostly intact.

shiny. Texture elements can also provide contextual information such as the direction of wind across a body of water. Many objects contain important texture elements such as foliage, hair, and clothing. Loss of texture in these elements can degrade overall image quality. As texture decreases, objects can begin to appear waxy and melted as well as becoming blurry. Figure 1.11 shows an example in which the original image on the left has been filtered on the right to simulate an image processing algorithm that reduces image noise (though in this particular example, the original image does not suffer from noise in order to accentuate the filtering result for demonstration). As can be seen, the filtering reduces the quality of the image because of blurring of the hair, skin, and clothing. Thus, objective image quality metrics that quantify texture reproduction are important for camera benchmarking.

1.1.4 Depth

Depth is an important aspect of relating to objects in the physical world. In a three-dimensional (3D) environment, an observer is able to distinguish objects in part by discerning the physical differences in depth. For example, an observer can tell which objects in a room may be within reach compared to objects that are in the distance due in part to binocular disparity of the left and right eyes. However, two-dimensional (2D) images are able to convey a sense of depth despite the lack of a physical third dimension. Several visual cues provide depth information in conventional pictorial images (Coren *et al.*, 2004):

- Interposition (object occlusion)
- Shading (variations in amount of light reflected from object surfaces)

- Aerial or atmospheric perspective (systematic differences in contrast and color when viewed from significant distances)
- Retinal and familiar size (size-distance relation based on angular subtense and previous knowledge of objects)
- Linear perspective (convergence of parallel lines)
- Texture gradient (systematic changes in size and shape as distance to viewer changes)
- Height in the plane (position relative to the horizon)

An additional visual cue for depth, specific to video imaging, is relative motion (motion parallax). When present, all of these visual cues are processed by the HVS in order to interpret the relationship between objects and illumination in the scene. However, because these are cues related to pictorial images, they are fundamentally monocular in nature. Thus, binocular aspects of the HVS, for example, convergence and binocular disparity, are not utilized to determine depth in these cases. In addition, the monocular function of lens accommodation for pictorial viewing is defined by physical distance to the picture, not by the various distances to objects that may be depicted in the scene. Thus, accommodation does not serve as a depth cue in the two-dimensional image scenario. However, realistic imaging is still able to convey depth and dimensionality with pictorial information void of 3D.

These visual cues for depth are dependent on camera image quality—images with elements such as sharp edges, high bit depth, and good color reproduction provide quality that is able to represent depth more fully even in a 2D scene. For example, an image that is blurry, low in bit depth, and monochrome has noticeably less perceptual depth to the objects in the scene compared to an image with high sharpness, sufficient bit depth, and color. Figure 1.12 contains a photograph pair demonstrating this comparison. The top image is monochrome, limited in bit depth, and noticeably blurry. In this image, the source of the surface modulations is non-obvious and the visual interpretation of the curvature and interposition has ambiguity. However, in the bottom version, the color and increased sharpness enable the viewer to better interpret the depth within the scene, including the structure of the sugar granules on the surface of the striped candy.

1.1.5 Luminance Range

Without illumination or self-luminance, scene content would not be discernible: light is a fundamental aspect of perception and imaging. Scene content contains objects that are illuminated or self-luminated by photons. The quantity of photons and surface reflectance or transmittance properties determine the luminance levels within a scene. For example, an object illuminated by candlelight will have a very small number of incident photons compared to the quantity when illuminated by sunlight. Color and surface properties determine the reflectance levels of the illuminated object. Shiny, metallic surfaces reflect a large percentage of incident light as do white, glossy objects. Dull, black objects and occlusions inhibit photon travel, resulting in low reflectance.

The HVS adapts to both light and dark conditions, expanding to an optimal range for a given environment (Fairchild, 2013). Yet, adaptation is not complete—this can be ascertained in one's cognition of being in a moonlit environment versus a daylight environment. Thus, images should be able to represent both a form of absolute luminance and luminance range. If the camera's exposure of a scene is not sufficient, the image will look too dark compared to an ideal representation of the scene or what the observer

Figure 1.12 Top: monochrome candy ribbons with low sharpness and bit depth, bottom: colorful candy ribbons with substantial sharpness. Note that the bottom image is more able to convey a sense of depth versus the top image.

recalls; the absolute luminance is not optimal. At worst, the image might be completely dark with indiscernible scene content. Conversely, if the camera's exposure of a scene is too high, the image will look too light at best and completely washed out at worst. For either case, the image quality can vary widely when observing an exposure series for a given scene.

Similarly, the lower the dynamic range of the rendered image, the more limited the image will be regarding representation of luminance range in the scene. As such, renderings with low dynamic range can lower the quality of scenes with high dynamic range. For example, glossy objects have high dynamic range when illuminated with direct light. Research has shown that rendering glossy objects with more dynamic range increases observer perception of glossiness (Phillips *et al.*, 2009). Thus, as an example, an image with lower dynamic range will have more limitations in representing the attribute of glossiness of an object compared to an image with higher dynamic range.

Figure 1.13 contains a tetraptych of images demonstrating variations in luminance levels and dynamic range. The first three images show an exposure series that shows how changes in the absolute luminance levels emphasize and reveal highlights and shadows in the scene: the underexposure by 2 f-stops of the camera allows one to see details in the shale gorge wall and sunlit trees in the background while the overexposure by 2 f-stops

Figure 1.13 Variations in luminance levels and dynamic range for an example scene. (a) Underexposed by 2 f-stops. (b) Normal exposure. (c) Overexposed by 2 f-stops. (d) Normal exposure with localized tonemapping.

allows one to the see details in the clothing on the models. The final image has localized tone mapping applied to the scene, which results in a rendition with optimized dynamic range in which more highlight and shadow details are apparent; this scene has optimized exposure and dynamic range, which in turn results in higher image quality.

Contributed by Hugh Denman

Motion within a scene is extremely informative for distinguishing and recognizing objects. Perception of motion allows us to determine critical aspects such as the velocity, that is, speed and direction, and the dimensionality of moving objects around us as well as depth in the scene. In this way, motion provides salient information to assess our environment. In fact, motion is of such critical importance that it is encoded by the HVS as a first-order, low level visual percept, similar to edge and texture perception. This contrasts with the naïve supposition that visual perception supplies a continuous stream of "images" of the scene, and that higher-level processes infer motion by comparing successive images (Sekuler *et al.*, 1990; Nakayama, 1985).

   Broadly speaking, there are three sorts of stimuli which give rise to the perception of motion. The first may be termed "actual motion," generated by moving elements in the scene or by the motion of the observer. The second is the well-known "apparent motion" effect. If a static stimulus is presented in a succession of spatial locations, with an interval of less than 100 ms between presentations, the stimulus will be perceived to move continuously, rather than being perceived to disappear and reappear in different locations (which, incidentally, is the percept if the interval increases above 100 ms). For example, a row of lights, each of which is briefly lit in succession in an otherwise dark scene, creates the impression of a single moving light if the delay between each light's blink is less than 100 ms. This effect has long been exploited in visual entertainments, from flip-book animations to the kinetoscope and the cinema. While the term "persistence of vision" continues as a description of the apparent motion phenomenon, this dates to an early misconception of the eye as a sort of camera in which a retinal after-image is retained between stimuli (Anderson and Anderson, 1993). The more neutral term "beta movement" is preferred—Max Wertheimer coined this in the founding monograph of Gestalt psychology, "*Experimentelle Studien über das Sehen von Bewegung*" (Wertheimer, 1912).

   The same monograph describes the third sort of motion stimulus, the "phi phenomenon," in which the subject perceives motion without perceiving anything move. Consider a pair of stimuli, each depicting the same object (a small disc is often used), with a small spatial distance between the object positions. If these stimuli are presented in continuous alternation, with a very short switching interval (less than 30 ms), a flickering image of the object is perceived in both locations simultaneously—and a perception of motion between the object locations is also induced. This motion has no contour: the motion percept is not affected by the shape of the stimulus object (Steinman *et al.*, 2000). Because the motion percept is not associated with any moving object, Wertheimer termed this "pure" motion perception, and concluded that motion perception is "as primary as any other sensory phenomenon".

   It is now known that there is an area within the visual cortex, designated MT or V5, which encodes an explicit representation of perceived motion in terms of direction and speed. This area is also concerned with the somewhat related task of depth perception. The motion percepts arising here can be experienced "out-of-context" through various *motion aftereffect* visual illusions (Anstis, 2015). For example, if one stares at a waterfall for a few minutes and then looks away, a perception of upward motion is superimposed on the scene—this is due to neuronal adaptation within area MT. There are accounts of patients with damage to this area who experience *akinetopsia*: the inability to perceive motion. Temporary disruptions to motion perception, arising from migraines or consumption of hallucinogens, are termed episodes of *cinematographic vision*.

Motion perception has been extensively studied using *random-dot cinematograms*. These consist of a series of images, each depicting a field of dots. Most of the dots appear in a different, random position in each image, but a subset of dots are made to move from image to image. A sample random-dot cinematogram is shown in Figure 1.14a and b. When presented statically, side-by-side, these two images should appear to contain entirely random dot fields. However, if presented in alternating superposition, one pair of dots will be seen to correspond (via motion) from frame to frame, while the others are appearing and disappearing. In Figure 1.14c and d, the pair of dots that have apparent motion have been highlighted in red.

The principal psychophysical parameter determined by these experiments is the maximum displacement at which motion can be detected, denoted d_{max}. Dots that are displaced by more than this amount are not perceived as moving, but rather as disappearing and reappearing in a new location. That this sort of correspondence can be readily established in temporal succession, but not in spatial (i.e., side-by-side) presentation, is due to the specialized motion perception machinery of the HVS.

d_{max} increases with eccentricity (i.e., toward the periphery of vision), from about 9 minutes of arc at the center of the visual field, to about 90 minutes of arc at 10 degrees off-axis. Thus, motion perception is an attribute of vision whose performance improves off-axis, unlike most others such as color perception and acuity. d_{max} can also be increased by low-pass filtering the stimulus (for example, introducing a blur by squinting). This suggests that the presence of high spatial frequencies can prevent the perception of motion. Random-dot cinematograms have also been used to investigate *motion metamerism*: a pair of stimuli in which the dots follow distinct motion trajectories can induce indistinguishable motion percepts, if certain statistics of the motions are identical.

Stimuli giving rise to apparent motion effects, such as random-dot cinematograms as shown in Figure 1.14, can be ambiguous regarding the underlying, continuous motion paths. For example, a pair of dots displaced by the same distance from one image to the next could have traveled in parallel, or could have crossed paths *en route* to the new positions. This is shown in Figure 1.14e and f. Such ambiguities are resolved at a low level: there is no perception of ambiguous motion, nor a conscious choice of motion hypothesis. Thus, the machinery of motion perception consists not only of correspondence matching apparatus, but also apparently the imposition of constraining assumptions such as parsimony, inertia and rigidity of objects (Ramachandran and Anstis, 1986; Gepshtein and Kubovy, 2007).

Camera systems rely on the apparent motion effect to capture convincing video—the frame rate must be high enough to induce the motion percept. As mentioned above, this beta movement effect requires playback rates of about 10 frames per second, or higher. For example, cinema has traditionally used 24 frames per second (fps) for playback rate. In addition, for realistic motion presentation, the capture rate must match the intended playback rate. Thus, regarding benchmarking image quality, the camera frame rate capture and playback directly impact the visual quality.

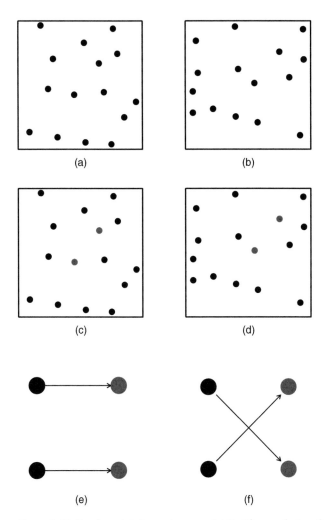

Figure 1.14 Random-dot cinematograms. **(a)** First frame of a two-frame cinematogram. **(b)** Second frame of a two-frame cinematogram. **(c)** The same frame shown in (a), with moving dots shown in red. **(d)** The same frame shown in (b), with moving dots shown in red. **(e)** A plausible motion hypothesis for a two-frame cinematogram in which the dots move from the positions in black to those in red. **(f)** Another plausible motion hypothesis for a two-frame cinematogram in which the dots move from the positions in black to those in red.

As well as the frame rate, the exposure time per frame (shutter speed) affects the perception of motion. Longer exposure times introduce *motion blur*, which increases the perceived smoothness of motion but reduces the visual detail in each frame. A lack of motion blur at lower frame rates (24–30 fps) can result in motion *judder*: jerky movement of objects in the scene. In cinema, the shutter speed is varied according to the motion content and directorial intent. Consumer cameras typically choose frame rate and frame exposure time automatically, to enable correct exposure according to the light level. This excludes the possibility of manipulating the quality of motion capture by manipulating these parameters.

It is clear that once the essential technical requirements for motion capture are met, the motion capture performance of a camera is highly dependent on the nature of the motion itself to be captured. Tests for the motion capture performance of cameras are not yet highly developed, and standardized motion test targets are only beginning to appear. However, several metrics for digital video quality are available and these can be utilized to assess the motion capture performance of a camera. These are discussed in later chapters.

1.2   Benchmarking

Now that we have explored these six key aspects of the essence of imaging—color, shape, texture, depth, luminance range, and motion—we can begin to explore the task of image quality benchmarking. Photographic technology has evolved immensely since Nicéphore Niépce captured the first permanent photograph with his camera obscura in the late 1820s. As mentioned at the beginning of this chapter, cameras have been primarily single-purpose devices over the past centuries: for capturing still images and/or video—though of varying complexity and capability. More recently, mobile phone cameras have evolved from low-resolution, low quality gadgets into fully-fledged photographic and videographic tools, dwarfing placement of traditional cameras in the marketplace. Because of this revolutionary development, the imaging industry has been revitalized regarding the necessity of being able to specify and characterize image quality in a reliable and consistent way, and in a way that also correlates with human vision.

The process of objective and subjective camera image quality benchmarking varies both in breadth and depth, depending on the intent of the benchmarking. A key component of benchmarking is determining what questions need to be answered—if one can envision the type of information needed from the process, then the steps to obtain the benchmarking will become clearer. For example, is the benchmarking intended to compare isolated components of the system such as the sensor or the optics? Then, objective metrics and specialized equipment for characterizing these components can be utilized. If, however, the intention follows the main topic of this book—that of camera system benchmarking—then the integrated behavior of the components needs to be incorporated into the analysis. Typically, this means quantification of key image quality aspects as listed below:

- Exposure and tone
- Dynamic range
- Color
- Shading
- Geometric distortion
- Stray light
- Sharpness and resolution
- Texture blur
- Noise
- Color fringing
- Image defects

The integrated measurement of these aspects of image quality provides a means of predicting how a consumer will perceive photographs from a particular camera. Subsequently, benchmarking is possible when comparing results from multiple cameras. Objective and subjective metrics used to quantify these attributes are described in detail in following chapters. Note that many of the attributes listed above are dependent on the spatial scale. For example, sharpness and resolution of an image will be perceived differently for a given photograph, depending on how close the image is viewed or how much magnification is applied to generate the size of the viewed image, whereas the quality of the exposure and tone of the image typically remains constant under these conditions. Thus, benchmarking should incorporate the use case of the still image or video clip in order to provide more meaningful and appropriate results.

In order to expand on this use case topic, let us return again to the early days of photography. The process to view early photography captured on glass plates was commonly one in which a photograph was generated by a contact print method—a print made by shining light through the glass negative placed in direct contact with the light sensitive emulsion layer of the paper. For this situation, no magnification of the image in the negative was applied to the viewed image: the size of the objects in the negative was the same as the size of the objects in the print. As glass plates transitioned to film made of flexible cellulose support, the print-making process commonly held to that of contact prints. However, as film evolved, film machinery configurations and standards led to the size of 35 mm film for motion pictures (Fullerton and Söderbergh-Widding, 2000; Dickson, 1933), which then became popularized by Leica for still photography. In order to make photographic prints from this film format, the film was no longer placed directly on the photosensitive paper, but was instead projected onto the paper from a distance by means of a lens. For this situation, the photograph became a magnified version of the image in the negative because the print size could be several times larger than the original image. As such, the image quality aspects of the photograph could differ significantly from what was directly measured in the film image. For example, to print a traditional 4R 4 × 6 inch (10 × 15 cm) print, the 35 mm negative is magnified approximately three times in height; to print a traditional 5R 5 × 7 inch (13 × 18 cm) print, the 35 mm negative is magnified approximately four times in height, and so on. Thus, small changes in spatially related image quality properties of the negative become increasingly more important as the source of the image becomes smaller and the size of the photographic output becomes larger, magnifying the aspects of these scale-dependent image quality attributes.

Considering how this aspect of magnification relates to the state of benchmarking digital camera image quality, suppose that one captures a digital image using a mobile phone. Early phones had cameras with digital sensor resolutions of 640 × 480 pixels, or 0.3 megapixels (MP), that is, not a lot of information compared to current camera phones with sensors that strongly exceed this resolution. Given that the resolution of the phone displays coincident with these VGA (Video Graphics Array) sensors were even *less*, the process of displaying a photo actually required a *downsampling* of the image. This reduction in pixel resolution in essence increased the perceived image quality of the scale-dependent attributes, such as sharpness and resolution, from what the sensor captured. Thus, it is not a surprise that people were disappointed with the quality of 4R 4 × 6 inch prints when they first tried printing photos from their early camera phones because the typical print assumed 1800 × 1200 pixels (2 MP) minimum resolution for baseline image quality and their cameras were only capturing 0.3 MP images.

Even with the current advancement of resolutions of camera phone displays and sensors, most often the screen resolution on phones is significantly lower than that of the sensor such that the display on a phone limits the consumer from viewing the native image quality of the camera. For example, a phone with a 4 MP display for a 16 MP camera would have to downsample the camera image height by approximately two times to show on the display for the typical use case of observing camera phone image quality. If, however, the consumer were to magnify the image by zooming in on the phone display, then the perceived quality for this use case would be closer to that of the native camera resolution. Another use case example is to observe the quality of an image on a computer display. Depending on the resolution, physical size, and viewing distance for a given monitor (among other conditions), the perceived image quality of a photo would vary. For example, 4k UHD (ultra high definition) monitors are 3840 × 2160 pixels (8 MP). For sensors with smaller resolution than 4k UHD, the image would have to be magnified. However, for sensors with a resolution of 8 MP or higher, the impact of resolution on image quality would be similar.

Thus, the current limitations of image quality are not really about the megapixel resolution of the sensor, at least for most use cases of cameras with 8 MP sensors or higher. Often, other performance factors of the camera system such as the pixel size, the full well capacity of the image sensor, optics, and the image signal processing (ISP) pipeline are the limiting factors above the sensor resolution. However, the use case(s) for the benchmarking will dictate how important the magnification (digital zoom) aspect is for comparing cameras.

Suppose you want a general benchmark comparison of how consumer cameras compare. One way to approach the task is to generate the image quality assessment for each camera given the type of scene content and application categories that are important to the consumer. The concept of *photospace*, based on the probability distribution of subject illumination level and subject-to-camera distance in photos taken by consumers, has been used to define the scene content categories that are important to include in development analysis related to benchmarking (Rice and Faulkner, 1983). For example, scene content such as a macro photo of a check for bank deposit, a photo of friends in a dimly lit bar, a typical indoor portrait, an indoor stadium sports event, and a daylight landscape photo are all common and important scenes for the typical consumer, representing various illumination levels and subject-to-camera distances. As such, these examples of photospace would provide salient scene categories to include in a comparative assessment.

Modes of viewing photos or videos of these scenes include applications such as viewing on the display of the source camera or camera phone, viewing on a tablet computer, viewing on a UHD monitor or television, or enlarging the photo to hang on a wall as artwork. A simple matrix example adapted from concepts by I3A CPIQ is shown in Figure 1.15 with image quality assessment of the various combinations of scene content and application use cases (Touchard, 2010). Various means can be used to populate the matrix such as an image quality scale value or simplified assessment such as symbols or colors conveying the benchmark assessment. From this type of assessment, a general benchmark comparison can be made between cameras for given combinations of scene content and application use cases.

Figure 1.15 This example benchmark matrix shows the image quality assessment of various scene content and application categories for a consumer camera. Note how the quality varies as these categories change. *Source*: Adapted from Touchard (2010).

	Phone Display	Tablet Display	4k TV	Enlarged Wall Art
Macro	4	4	4	2
Dim Bar	3	3	2	2
Indoor Portrait	5	5	5	4
Sports	4	4	4	3
Landscape	5	5	5	4

What one should notice in the example benchmark matrix in Figure 1.15 is that the uses cases include different aspects of scene content such as illumination level, distance from the camera to the subject, and motion of the subject as well as application of the image such as magnification of the viewed output. The key image quality aspects listed earlier in the chapter should be quantified for each use case in the matrix. Most commonly, this involves quantifying the behavior of a camera based on specific individual image quality metrics for these aspects of color, shape, texture, depth, luminance range, and motion.

Many image quality metrics exist for defining objective image quality. More recently, objective metrics have begun to expand into the realm of subjective evaluation, resulting in perceptually correlated image quality metrics. For example, ISO 15739, written by the Technical Committee 42 of the ISO, incorporates a noise metric extension that predicts the subjective impact of a noise pattern (ISO, 2013). In addition, the image quality metrics by the CPIQ working group of the IEEE Standards Association contain equations to predict the subjective quality loss to a photograph for a given metric value (IEEE, 2017). However, as technology continues to evolve, image quality attributes also continue to migrate, necessitating new and revised means of quantification. These challenges become continually important as new hardware and software aspects introduce more and more spatially localized characteristics into the images and video frames.

Because image quality is in essence a subjective matter, quantifying subjective image quality is just as important as quantifying objective image quality. Systematic science, as established in the field of psychophysics, can be used to measure and quantify what observers perceive about image quality. Chapter 5 will define and discuss this type of subjective evaluation as it relates to image quality metrics. Further discussion on subjective evaluation will continue in subsequent chapters. As noted above, a set of objective measurements can only address the image quality attributes being measured, which makes it possible that a benchmarking approach is not comprehensive. Thus, subjective evaluation should be incorporated into any comprehensive image quality benchmarking approach, either by ensuring that objective metrics contain perceptually correlated metrics or by including subjective image quality metrics themselves into the benchmarking formula. The remainder of this book will spell out the reasoning and details behind this premise.

1.3 Book Content

The following section provides summaries of the remaining book content, providing the reader with a concise explanation of the aim of each chapter.

Chapter 2: Defining Image Quality

Chapter 2 will provide a broad overview of image quality as well as the necessary definitions of the key terms that will be used through the book. First, image quality itself will be defined; then we will define its attributes and how they are categorized. We will also define the difference between global and local attributes. A section will be devoted to defining the specifics and differences of objective versus subjective image quality assessment methods.

Chapter 3: Image Quality Attributes

In Chapter 3, we will describe image quality attributes in more detail. When attempting to quantify image quality using objective measurements, one usually divides the overall impression into several separate "nesses"—sharpness, graininess, colorfulness, and so on—examples of attributes of image quality (Engeldrum, 2000). Each of these attributes has their own distinct signature. Starting out from the categorization of local, global, or video-specific, this chapter will describe each of the attributes in detail, providing many example images and figures. The chapter will conclude with a discussion about measurable attributes versus unmeasurable artifacts for still images as well as video.

Chapter 4: The Camera

The fourth chapter will first describe the different hardware and software components that constitute a digital camera and its architecture. In particular, we will describe how digital camera components (the lens, the image sensor and the image signal processing (ISP)) all contribute to the performance and image quality of a camera. We will establish the connection between each component and the image quality attributes described in the previous chapter. Finally, for each component, we will detail the key parameters that influence image quality (e.g., aperture of the lens, etc.).

Chapter 5: Subjective Image Quality Assessment

Many psychophysical methods exist for quantifying subjective image quality with human observers. Chapter 5 will review key psychometric techniques, such as category scaling, forced-choice comparisons, acceptability ratings, and mean opinion score (MOS), and will emphasize the strengths and weaknesses of each methodology. The review will also explore the similarities and differences between still and video subjective evaluation techniques and how these are able to quantify important perceptual aspects of the human visual system's assessment of image quality. Particular focus will be on the anchor scale method and how that can be used to quantify overall image quality in just noticeable differences (JNDs) for still images in such a way that JNDs of various attributes can be combined to predict image quality of the camera.

Chapter 6: Objective Image Quality Assessment

Objective image quality metrics are by definition independent of human perception. Even so, by carefully choosing the methodology, it is possible to provide objective metrics that can be well-correlated with human vision. The content of Chapter 6 will provide an overview of existing metrics connected to the image quality attributes discussed in Chapter 3. We will describe in detail the "best" metrics for each of the attributes and also discuss pros and cons if there is more than one metric from which to choose for a given attribute. For instance, sharpness can be measured using resolution bars, sinusoidal Siemens stars, or slanted edges. Each of these methods will provide different results in some situations and it is important to understand the underlying reasons for the discrepancies. Moreover, practical issues, such as the choice of correct white point in color measurements, will be addressed in order to minimize the confusion which often arises because of the complexities.

Chapter 7: Perceptually Correlated Image Quality Metrics

In order for objective image quality metrics to be more meaningful to benchmarking, they need to be well-correlated with perception. Two approaches to accomplishing this are typically used, either through methods involving models of the human visual system, or by employing more empirical methods where some known aspect(s) of the human visual system can be taken into account, for example, correlations of adjacent pixels, and so on. Furthermore, some methods may be dependent on comparing the result to some known reference, while other methods may not. In Chapter 7, a large number of such methods will be discussed, including concepts such as mean square error (MSE) and peak signal to noise ratio (PSNR). We will also discuss methods to correlate the results of measurements on sharpness and noise to how these attributes are subjectively experienced. We will introduce the concept of contrast sensitivity functions (spatial and temporal), opponent color spaces, and so on, but also metrics mostly used in video quality assessment, such as the structural similarity index (SSIM) and similar. The importance of viewing conditions will also be stressed.

Chapter 8: Measurement Protocols—Building Up a Lab

When it comes to performing accurate and repeatable measurements, it is absolutely critical to establish and define the so-called protocols. The protocols provide a full description of the testing conditions that are required when performing image quality measurements. Chapter 8 will successively go over the protocols to be applied for objective and then subjective measurements. We will show how protocols are specific to each of the individual image quality attributes or parameters being measured. Discussion will include how protocols, such as those specifying lighting conditions, can vary as test equipment technology evolves.

Chapter 9: The Camera Benchmark Process

The first step to building a camera benchmark is to determine the key image quality attributes to be measured; then a method must be established to weight and combine them to obtain a global scale so that one can benchmark all cameras against each other. In Chapter 9, we will show how a comprehensive camera benchmark should combine subjective and objective image quality assessment methodologies, and how some can substitute some others when correlation is established. We will describe the ideal benchmark and will show that, given the intrinsic subjectiveness of image quality, various approaches nearing the ideal might reach different conclusions. The chapter will also describe a number of existing camera benchmarking systems and will point to the ones that are the most advanced. Example benchmarking data will be shared for a collection of cameras, highlighting how various individual metrics can sway results. Finally, we will detail the possible evolution to move even closer to the ideal benchmark and highlight the technologies that remain to be developed to achieve this goal.

Chapter 10: Conclusion

The concluding chapter will restate the value and importance of a benchmarking approach that includes perceptually correlated image quality metrics. The section will also highlight future computational photography and hardware technologies that will be entering the mainstream consumer electronics market and how they impact the future of image quality metrics. Discussion will cover the challenges of benchmarking systems for the continually evolving camera imaging technology, image processing, and usage models.

Summary of this Chapter

- The more a photograph represents the elements of a physical scene, the higher the possible attainment of perceived quality can become.
- Key aspects of the essence of photography are color, shape, texture, depth, luminance range, and motion.
- Objective image quality evaluation involves making measurements, for which the results as well as methodology are independent of human perception.
- Subjective image quality evaluation is fundamentally a measurement quantifying human perception.
- Image quality is fundamentally a perceptual matter—it should include the perspective of an observer. Therefore, quantifying the subjective component is just as important as quantifying the objective component for the purpose of benchmarking image quality of cameras.
- To be most useful and relevant, benchmarking metrics for image quality should provide consistent, reproducible, and perceptually correlated results.
- Image quality is use case dependent: that which is deemed acceptable for one specific case may be unacceptable in other cases.
- The conditions under which a particular image or video is *captured* are important to define and understand when evaluating image quality. For example, camera performance under bright levels of illumination will almost certainly yield better image quality compared to capturing under dim levels of illumination.
- The conditions under which a particular image or video is *viewed* are important to define and understand when evaluating image quality. For example, viewing an image

on a mobile phone screen will almost certainly yield a different impression compared to a large format print of the image made on a high quality printer and hung on a wall.

- A set of objective measurements can only address the image quality attributes being measured, which makes it possible that a benchmarking approach is not comprehensive.
- Objective image quality metrics become more meaningful when the visual correlation is defined.
- Comprehensive benchmarks incorporate both objective and subjective image quality evaluation.

References

Adelson, E. (2001) On seeing stuff: The perception of materials by humans and machines. *Proc. SPIE*, **4299**, 1–12.

Alfvin, R.L. and Fairchild, M.D. (1997) Observer variability in metameric color matches using color reproduction media. *Color Res. Appl.*, **22**, 174–188.

Anderson, J. and Anderson, B. (1993) The myth of persistence of vision revisited. *J. Film Video*, **45**, 3–12.

Anstis, S. (2015) Seeing isn't believing: How motion illusions trick the visual system, and what they can teach us about how our eyes and brains evolved. *The Scientist*, **29** (6).

Berns, R.S. (2000) *Principles of Color Technology*, John Wiley & Sons, Inc., New York, NY, USA, 3rd edn.

Biederman, I. (1987) Recognition-by-components: A theory of human image understanding. *Psychol. Rev.*, **94** (2), 115–147.

Carraher, R.G. and Thurston, J.B. (1977) *Optical Illusions and the Visual Arts*, Van Nostrand Reinhold Company, New York, NY, USA.

Coren, S., Ward, L., and Enns, J. (2004) *Sensation and perception*, John Wiley & Sons, Inc., Hoboken, NJ, USA, sixth edn.

Dickson, W.K.L. (1933) A brief history of the kinetograph, the kinetoscope and the kineto-phonograph. *J. Soc. Motion Pic. Eng.*, **21**, 435–455.

Dorsey, J., Rushmeier, H., and Sillion, F. (2008) *Digital Modeling of Material Appearance*, Morgan Kaufmann Publishers, Burlington, MA, USA.

Engeldrum, P.G. (2000) *Psychometric Scaling: A Toolkit for Imaging Systems Development*, Imcotek Press, Winchester, MA, USA.

Fairchild, M.D. (2013) *Color Appearance Models*, John Wiley & Sons Ltd, Chichester, UK, 3rd edn.

Fullerton, J. and Söderbergh-Widding, A. (eds) (2000) *Moving Images: From Edison to the Webcam*, vol 5, John Libbey & Company Pty Ltd, Sydney, Australia.

Gepshtein, S. and Kubovy, M. (2007) The lawful perception of apparent motion. *J. Vision*, **7**, 1–15.

Gernsheim, H. and Gernsheim, A. (1969) *The history of photography: from the camera obscura to the beginning of the modern era*, McGraw-Hill, New York, NY, USA.

IEEE (2017) IEEE 1858-2016, IEEE Standard for Camera Phone Image Quality. IEEE.

ISO (2013) ISO 15739:2013 Photography – Electronic Still Picture Imaging – Noise Measurements. ISO.

Landy, M.S. (2007) Visual perception: A gloss on surface properties. *Nature*, **447**, 158–159.

Motoyoshi, I., Nishida, S., Sharan, L., and Adelson, E.H. (2007) Image statistics and the perception of surface qualities. *Nature*, **447**, 206–209.

Nakayama, K. (1985) Biological image motion processing: A review. *Vision. Res.*, **25**, 625–660.

Papapishu (2007) https://openclipart.org/detail/4802/violin. (accessed 29 May 2017).

Phillips, J.B., Ferwerda, J.A., and Luka, S. (2009) Effects of image dynamic range on apparent surface gloss, in *Proceedings of the IS&T 17th Color and Imaging Conference*, Society for Imaging Science and Technology, Albuquerque, New Mexico, USA, pp. 193–197.

Ramachandran, V.S. and Anstis, S.M. (1986) The perception of apparent motion. *Scientific American*, **254** (6), 102–109.

Ramachandran, V.S. and Gregory, R.L. (1991) Perceptual filling in of artificially induced scotomas in human vision. *Nature*, **350**, 669–702.

Rice, T.M. and Faulkner, T.W. (1983) The use of photographic space in the development of the disc photographic system. *J. Appl. Photogr. Eng.*, **9**, 52–57.

Rushmeier, H. (2008) The perception of simulated materials, in *ACM SIGGRAPH 2008 classes*, ACM, New York, NY, USA, SIGGRAPH '08, pp. 7:1–7:12.

Sekuler, R., Anstis, S., Braddick, O.J., Brandt, T., Movshon, J.A., and Orban, G. (1990) The perception of motion., in *Visual perception: The neurophysiological foundations*, (eds L. Spillman and J. Werner), Academic Press, London, pp. 205–230.

Shaw, M.Q. and Fairchild, M.D. (2002) Evaluating the 1931 CIE color-matching functions. *Color Res. Appl.*, **27**, 316–329.

Steinman, R.M., Pizlo, Z., and Pizlo, F.J. (2000) Phi is not beta, and why Wertheimer's discovery launched the gestalt revolution. *Vision Res.*, **40**, 2257–2264.

Touchard, N. (2010) I3A - Camera Phone Image Quality, in *Image Sensors Europe*, image-sensors.com.

van Tonder, G. and Ejima, Y. (2000) Bottom-up clues in target finding: Why a Dalmatian may be mistaken for an elephant. *Perception*, **29**, 149–157.

Wertheimer, M. (1912) Experimentelle studien über das sehen von bewegung. *Zeitschrift für Psychologie und Physiologie der Sinnesorgane*, **61**, 161–265.

2

Defining Image Quality

Before going into the details of image quality evaluation, this chapter provides a broad overview as well as the necessary definitions of the key terms that will be used throughout the book. The concept of image quality itself, as applicable within the scope of this book, is defined, together with its attributes and how they may be categorized. Specifically, we provide two main categories of image quality attributes: local and global. A section is also devoted to defining the specifics and differences of objective versus subjective image quality assessment methods and how they can be combined. This chapter includes definitions of image quality for both still and video capture.

2.1　What is Image Quality?

In modern society we are surrounded by photographic images, still as well as moving. Anyone who has ever looked at an image in print, on a computer screen, in a newspaper, and so on, has some idea about what constitutes good or bad image quality. At the same time, the systematic assessment of image quality by means of measurements has proven to be a non-trivial task. The reasons for this are many, and some examples are given in the following discussion. An image may be pleasing aesthetically even though the technical quality is low. Based on personal preferences, one image may be more pleasing than another depending on the subject matter. Therefore, the emotional response to a photograph may be so strong that it overwhelms the impression of technical excellence. The technical requirements in, for example, surveillance or medical imaging, may lead to images that are far from aesthetically pleasing, but that will provide maximum detectability of objects in the depicted scene. On the other hand, an aesthetically pleasing photograph may not necessarily depict the scene in a realistic way, and may in fact introduce artifacts that make the image deviate considerably from the scene. Therefore, in order to have a fair chance of succeeding in providing a reliable methodology for the quantification of image quality, one needs to start with carefully defining what is meant by "image quality." This will certainly be dependent on the uses of the images being produced, as well as the observers of the images. As a consequence, the context must be taken into consideration, and depending on the particular use case at hand, the most appropriate of a set of different definitions must be chosen.

In the context of this work, image quality assessment for consumer imaging, the camera is mainly used to capture a moment, without too much consideration of aesthetic

Camera Image Quality Benchmarking, First Edition. Jonathan B. Phillips and Henrik Eliasson.
© 2018 John Wiley & Sons Ltd. Published 2018 by John Wiley & Sons Ltd.
Companion website: www.wiley.com/go/benchak

or artistic aspects. With this in mind, it is interesting to reflect upon how the use of the camera has shifted from being a "memory collector," producing prints for the photo album, to capturing brief moments meant to be shared or consumed instantaneously and often only over a limited period of time. The trend in recent years of taking "selfies" also shows how the photographic subject is changing to some extent from having been other persons than the photographer to the photographer her/himself. Another example is the widespread use of the phone camera as an aid to remembering, for example, text in signs, price tags, business cards, or other such information that was previously written down. These changes may potentially have an effect on the requirements on the camera with respect to image quality. As an example, a special "document" setting may be found in some cameras, intended to make images with text more readable. In such a mode, contrast and sharpness may be enhanced to very high levels and the color saturation reduced in order to decrease the noise level.

The requirements on image quality may shift depending on the intended use of the camera, even in the more general, "photographic," use case. This can be illustrated through another example. Distortion is the modification of shape and perspective due to lens aberrations and geometric considerations. For a camera used mainly for landscape or architecture photography, optical distortion is a very important image quality parameter, since it will modify the shape of straight lines, found in, for example, buildings and the horizon. For portrait photography, on the other hand, optical distortion is usually not as important as in the former case, since the subject in this case will help to mask the effect of the distortion. Therefore, the requirements on optical distortion are usually much more severe for landscape photography compared to portraits. It should be observed that the distortion discussed here is lens distortion (to be explained in detail in Chapters 3 and 4), and not the geometric effect due to a short distance between camera and subject which is inevitable if a wide angle lens is used and that is often mistaken for lens distortion.

Going back to the "document mode" case above, where the camera is used solely for capturing black text on a white background, the image quality definition might be narrowed down considerably. In this case, we only need to take into consideration the readability of the captured text, ignoring all potential side effects due to increased contrast and sharpness as well as possibly decreased color saturation and other settings. In the distortion example, on the other hand, there is no need to change the actual definition of image quality, even though the requirements are changing.

In the literature, image quality is often defined in broad terms, for instance, according to Engeldrum (2000), image quality is "...the integrated set of perceptions of the overall degree of the excellence of the image." This is indeed a quite general definition that does not make any distinction between imaging for consumers and other fields with quite disparate requirements. It also does not take into account the differences in judgment due to personal or emotional preferences. A somewhat more specific definition is provided by Keelan (2002): "The quality of an image is defined to be an impression of its merit or excellence, as perceived by an observer neither associated with the act of photography, nor closely involved with the subject matter depicted." This definition is still rather broad, and may be narrowed down further in order to make it more appropriate for our particular use case, that is, consumer imaging. The definition might also take into account the purpose of defining image quality within the scope of this book, namely camera image quality benchmarking. However, there is a danger in putting too much emphasis on this, since it might mean ending up with a definition that is adapted to the metrics, when it really should be the other way around.

For consumer images, where artistic considerations are generally not of the highest importance, it is tempting to include a phrase in the image quality definition that requires an image of high quality to be close in appearance to the original scene. However, such a definition is clearly not appropriate since an image may be modified substantially with respect to some attribute(s), but still appear more pleasing and be perceived to have higher merit than an image more true to the original scene. Wang and Bovik (2006) point out this aspect of image quality, giving an example with a contrast-enhanced image that in the context of a simple image difference metric would be determined to be degraded compared to the unmodified image, while perceptually there would be no difference between them. As they also point out, at least one study has indicated that there is no obvious correlation between image *fidelity* and image *quality* (Silverstein and Farrell, 1996). Therefore, our image quality definition should not require that the image must be *identical* to a two-dimensional projection of the scene.

Another important point to make is the difference between personal preferences. Certainly, there will be a spread between individuals in what is considered the "best" image. However, what is also relevant is the difference in preferences between photographic professionals and non-professionals. For instance, an engineer working with tuning image quality algorithms or otherwise involved in camera development may have completely different preferences compared to a naïve observer (Persson, 2014).

With these considerations, we can put together an image quality definition that is relevant for image quality benchmarking of consumer cameras. We define image quality as:

> *the perceived quality of an image under a particular viewing condition as determined by the settings and properties of the input and output imaging systems that ultimately influences a person's value judgment of the image.*

2.2 Image Quality Attributes

In Chapter 1, the components of image understanding and perception were discussed. The impression of an image can be related to a few characteristics of the scene, viz., color, shape, texture, depth, luminance range, and (in the case of video) motion. The way these properties are rendered in the image will, as pointed out in the previous chapter, be directly related to how the quality of the image is perceived. This limited number of characteristics implies that it may be possible to classify elements of image quality into a restricted set of *attributes*, which makes it possible to break down the task of measuring image quality into smaller, more manageable pieces. In fact, with the above definition of image quality in mind, if asking a person his/her opinion of the quality of a certain image, the answer may be "blurry," "good colors," "too grainy," and so on. This is actually a description of the image quality attributes we call sharpness, color, and noise. It would seem that the number of attributes available is large enough to make any classification more or less impossible. However, as already implied, one finds that to a large extent the quality of an image can be attributed to a handful of characteristics. A few of the most obvious ones are sharpness, graininess/noisiness, color, lightness, and tonal reproduction/contrast, as well as geometrical attributes like optical distortion.

A very important observation regarding the image quality attributes is that even though all other attributes may be indicating a high quality image, a poor rating of one single attribute will completely overshadow the impression of the other attributes, leading to a negative overall impression. For instance, an image may exhibit very low levels of noise, have perfect sharpness, no optical distortion, and so on, but the white balance may be completely off, so that white and neutral objects appear green. Obviously, such an image will be rated as being of low quality, regardless of the other attributes. A technique to deal with this property will be discussed in the following chapters.

According to Keelan (2002), image quality attributes may be classified into four broad categories: personal, aesthetic, artifactual, or preferential. A personal attribute reflects some aspect that is only relevant to a person directly involved with the image, for example, the photographer herself. This could be a particular person being present in the image, or a particular place that evokes some pleasant memory. An aesthetic attribute, on the other hand, may also be relevant to someone not directly involved in the picture taking process. This could be reflected in, for example, the composition of the image. An artifactual attribute is more directly related to the technical quality of the image, and if a change in the attribute is above the threshold of detectability, the larger the difference becomes, the more objectionable the image quality will become due to the attribute. For instance, sharpness can be regarded as an artifactual attribute. The blurrier an image becomes, all other attributes kept constant, the lower the image quality will be perceived. Preferential attributes are also related to the technical quality of the image, but in contrast to artifactual attributes, there exists an optimum in perceived quality with changes in the attribute. An example is color saturation. As the saturation is increased, the perceived image quality will increase up to a certain point after which it will start to decrease again. Furthermore, the optimal level may be different between observers to some extent. This can be illustrated by comparing the color rendition between cameras from different camera manufacturers, which can show clearly discernible differences, partly due to differences in the hardware and software, but most likely in part also due to the preferences of the design engineers.

Another classification of image quality attributes may be made in terms of *global* and *local*. A global attribute is visible at all scales, and independent of viewing distance. The opposite is the case for a local attribute. Figure 2.1 illustrates this. The original image to the left has been degraded in two of its attributes: color saturation and sharpness, shown in the middle and to the right. The differences between the original image and the desaturated copy are clearly seen on both scales. However, between the original and blurred image the difference should be clearly seen in the larger images, but to a much lesser extent at the smaller scale in the bottom images. The classification into global and local attributes will become very important when implementing perceptually correlated measurement methods, since it means that the viewing conditions must be taken into consideration in the metrics.

In order for image quality attributes to be useful for image quality assessment, they must be *measurable*, and it must be possible to relate the measurement results to human perception. For the common attributes, for example, sharpness, noise, color, distortion, and shading, there do exist measurement methods, as will be shown and elaborated in much more detail in later chapters. However, there are attributes that are very difficult, if not impossible, to measure. In this category, image compression and other processing artifacts are found. Part of what makes such artifacts difficult to measure is that they are

Figure 2.1 Illustration of the difference between global and local attributes. The images in the top and bottom rows have been manipulated in the same way and only differ in size. The leftmost image is the original, unmodified image. In the middle, the original image was modified by reducing the color saturation. To the right, the original image was blurred. In the top row, these modifications are easily seen. In the bottom row, the difference between the leftmost and middle images is as clearly seen as in the top row. The difference between the leftmost and rightmost images is, however, not as obvious. Therefore, color rendition represents a global attribute, while sharpness can be categorized as a local attribute.

very dependent on the particular (and often proprietary) image processing algorithms, making any classification very difficult. Therefore, there are situations when it will not be possible to obtain a quantitative assessment of all relevant attributes. In this case, the only possible way forward may be to perform a subjective image quality experiment in order to obtain a quantification of overall image quality. In the following section, a comparison of objective and subjective image quality assessment methods is made.

2.3 Subjective and Objective Image Quality Assessment

As image quality is so intimately linked with perception, and therefore dependent on the judgment of human observers, the most accurate assessment of the quality of an image needs to be made by gathering opinions from persons watching the image. Since these opinions can vary substantially between observers, because of individual preferences as well as variations in the tolerance to boredom and other such individual properties, a large number of judgments must be collected and analyzed using statistical methods. In order to produce meaningful results, the experiment must be carefully designed, and observers as well as images must be selected meticulously. The consequence is that subjective experiments are typically resource intensive and time demanding.

Because of the subjective nature of image quality, it may seem counterintuitive to attempt to quantify it using objective measurements. By definition, an objective measurement should be independent of human judgment, which is the very opposite to the definition of image quality. However, there is often a misconception of the terms subjective and objective in this context. By subjective, we do *not* mean an arbitrary judgment of some property by a casual observer with no real justification of the judgment given. Instead, subjective image quality assessment is the investigation of how a population responds to visual stimuli in images; varied opinions of individual persons must be combined to obtain reproducible results. Therefore, if correctly performed, a subjective experiment will produce an objective result that should be accurate and repeatable. This should make it possible, at least conceptually, to replace subjective experimentation with objective measurements.

Using objective metrics has some obvious advantages, most notably that measurements have the potential for full automation, making the assessment considerably less resource demanding. It is possible to describe two main types of objective measurement categories: those that aim to describe the image quality in a specific image, and those that aim to predict the image quality of any image produced by a particular piece of image capturing equipment (e.g., camera). The first category is further divided into *full reference, no reference,* and *reduced reference* methods (Wang and Bovik, 2006). The second type of objective metrics consists of methods for the measurement of individual attributes, such as sharpness, noise, and so on, by capturing images of special test charts and analyzing those images using particular testing protocols.

For video quality assessment, methods in the former category are extensively used. The reason for this is mainly due to the fact that the concern in this case is on the evaluation of video compression coders and encoders, where a ground truth image is available for easy comparison using full reference metrics. For the same reason, such metrics are also used for evaluating other image processing algorithms for, for example, color interpolation, noise reduction, and so on.

Since this book is concerned with camera image quality benchmarking, that is, comparing different image capture equipment and attempting to predict the image quality produced by that equipment, the emphasis will be on the second main category of objective measurements. The aim in this case is, as already mentioned, to be able to predict the image quality obtainable by a given camera. In so doing, for all the important image quality attributes, there must be a corresponding metric. In the following chapters, these attributes, together with their metrics will be described in detail.

Summary of this Chapter

- We define image quality as "the perceived quality of an image under a particular viewing condition as determined by the settings and properties of the input and output imaging systems that ultimately influences a person's value judgment of the image."
- Image quality is influenced by both camera performance, including shooting conditions such as lighting, exposure, and so on, and scene content.
- The perception of image quality is different between ordinary consumers, professional photographers, and other imaging experts.

- Image quality can be described in terms of image quality attributes, such as sharpness, colorfulness, geometrical distortions, and so on.
- There exist two main categories of image quality attributes: local and global. In contrast to local attributes, global attributes are not influenced in any significant way by changes in magnification, viewing distance, or image size.

References

Engeldrum, P.G. (2000) *Psychometric Scaling: A Toolkit for Imaging Systems Development*, Imcotek Press, Winchester, MA, USA.

Keelan, B.W. (2002) *Handbook of Image Quality – Characterization and Prediction*, Marcel Dekker, New York, USA.

Persson, M. (2014) *Subjective Image Quality Evaluation Using the Softcopy Quality Ruler Method*, Master's thesis, Lund University.

Silverstein, D.A. and Farrell, J.E. (1996) The relationship between image fidelity and image quality, in *Proc. IEEE Int. Conf. Image Proc.*, Lausanne, Switzerland, pp. 881–884.

Wang, Z. and Bovik, A.C. (2006) *Modern Image Quality Assessment*, Morgan & Claypool Publishers, USA.

3

Image Quality Attributes

As pointed out in the previous chapter, image quality can be described in terms of a limited set of *attributes*, each representing different aspects of the imaging experience. Through the use of examples, this chapter will review the impact and appearance in images of the most important attributes. The assignment of attributes as either local or global, introduced in Chapter 2, will be maintained and elaborated. In subsequent chapters, the causes and origins of the various attributes will be further described and explained. The bulk of this chapter is focused on still image attributes. However, most of the discussion can easily be carried over into video capture. In this case, additional attributes and artifacts will arise, especially in relation to motion. These additional complications will be discussed in the concluding section.

3.1 Global Attributes

A global image quality attribute, as stated in Chapter 2, is essentially independent of magnification and viewing distance. This means that differences in appearance of such attributes will appear equally visible at most viewing distances and image sizes. The most important global attributes are

- Exposure, tonal reproduction, and flare
- Color
- Optical distortion and other geometrical artifacts
- Image nonuniformities such as (color) shading

Each of these will be described in detail in the following sections.

3.1.1 Exposure, Tonal Reproduction, and Flare

The brightness of an image is certainly of great importance to the overall impression of the captured scene. Typically, the image brightness is associated with the combined effect of focal plane exposure and ISO setting, to be explained in greater detail in Chapters 4 and 6. Even though there is a great deal of artistic freedom in adjusting the exposure, it is still fairly easy to judge whether the subject of a photo is correctly exposed or not, as illustrated in Figure 3.1. What is more, the *dynamic range* of the camera, that is, the ability to capture highlights and shadows in a scene simultaneously,

Camera Image Quality Benchmarking, First Edition. Jonathan B. Phillips and Henrik Eliasson.
© 2018 John Wiley & Sons Ltd. Published 2018 by John Wiley & Sons Ltd.
Companion website: www.wiley.com/go/benchak

Figure 3.1 Example of underexposed image (left), well exposed image (middle), and overexposed image (right).

is limited. The fact that the dynamic range of the camera is frequently narrower than the range of the actual scene makes it necessary to adjust the exposure in order not to blow out highlights or introduce unnecessarily high amounts of noise, or block out details completely in the shadows. A common misconception regarding dynamic range is that a low dynamic range leads to blown out highlights in an image. However, whether highlights become too light or not depends completely on the exposure, and it is in most cases perfectly possible to decrease the exposure to avoid such a situation. At the other end of the tonal range, a low dynamic range will in that case most certainly have a negative impact on the ability to render shadow detail. Figures 3.2 and 3.3 show images of a scene captured by two cameras with very different dynamic ranges. Depending on what is important in the scene, the exposure can be adjusted to either maintain details in the highlights or the shadows. For the two example cameras, those two cases will in this example produce quite similar images. However, as shown in Figure 3.3, if the top images in Figure 3.2, where the exposure is adjusted so that no part of the scene is becoming overexposed, are enhanced by digitally multiplying the image values in order to more clearly show details in the shadows, it is obvious that the low dynamic range camera has severe difficulties in the low key areas of the image. There is clearly a large difference in the noise in the shadows, while the highlights are similar in appearance in this particular case,[1] as seen in the upper images in Figure 3.2.

For *high dynamic range* (HDR) imaging, the problem is to match the dynamic range of the scene with that of the camera and display. Even though a camera may have a dynamic range that is close to that of the scene, the display or print will usually be quite limited in this respect. In order to produce a pleasing image, the tonal range of the image must be *tone mapped* to match the range of the display or print. This may involve local amplification of low key areas. With an image produced by a low dynamic range camera, this procedure may bring out excessive noise, as well as quantization artifacts, in the tone mapped shadow areas of the image.

Also related to exposure is *tonal rendering*, which determines the relationship between lowlights, midtones, and highlights in an image. As a rule, a *tone curve* is applied to the image in order to make it more pleasing. However, if not used properly, the result can turn out quite disagreeable, as shown in Figure 3.4.

---

1 The two main properties of an image sensor that affect the dynamic range are the noise floor and charge capacity. Due to a phenomenon called photon shot noise, to be explained further in Chapter 4, a low dynamic range camera may also show excessive noise in highlights.

Figure 3.2 Examples of images produced by cameras with a low (left) and high (right) dynamic range. Top: low exposure; bottom: high exposure. As explained in the text, there is no significant difference between these images in terms of image quality.

Figure 3.3 Examples of low exposure images that have been digitally enhanced. Left: Low dynamic range camera; right: high dynamic range camera. In this case, the low dynamic range camera shows considerably less detail due to noise masking, see text.

Figure 3.4 Example of an image with a mid-tone shift, giving an unnatural appearance.

Another effect that may be considered related to exposure and tonal rendering is *flare*, which is unwanted light scattered by various parts of the optical system that reaches the image sensor. This may give rise to localized artifacts (ghosting), but can also produce a global offset in the image that decreases the contrast and color saturation, and may also increase the noise. This effect is additive, and will therefore have a greater impact for low image values, that is, in the shadows, compared to the higher values found in the highlights, as shown in Figure 3.5.

Figure 3.5 Example of flare. Note the loss of contrast in the upper left part of the image, where the flare is most noticeable.

3.1.2 Color

 The color of an object can vary over a comparatively wide range and still be considered acceptable. An example is shown in Figure 3.6. The objects in the left and right images are in both cases clearly identifiable, and even though the hue and saturation are quite different, it is not possible to judge which color is "correct" unless having visited this particular location. For some colors, the range of acceptability is narrower than other colors. These are known as *memory colors*, and among them are the colors of sky, faces, and green grass, as well as recognizable brand colors. For such colors, the color tolerance must be kept tighter when rendering images from a camera in order to maintain high image quality.

Figure 3.6 Examples of two different renditions of color, each equally realistic.

Figure 3.7 Examples of colorimetrically "correct" rendition of a scene (left), and a preferential rendition with increased color saturation and vividness (right).

The determination of color reproduction for a particular camera is usually done by comparing with colors rendered for a standard observer. Because of the preferential nature of color, the tolerance of acceptability needs to be made relatively large. However, because of memory colors, different colors have to have different tolerances, which will make the task of determining camera color reproduction a non-trivial task. This is made even more difficult by the fact that we seem to prefer images of a scene to be more vivid and saturated in color compared to what is actually colorimetrically correct, see Figure 3.7 (Fairchild, 2013; Giorgianni and Madden, 2008; Hunt, 2004).

If it is relatively difficult to determine what is good color reproduction, it is considerably easier to decide what is bad color reproduction. The image in Figure 3.8 is clearly not acceptable by any standard, at least if the goal is to obtain a reproduction of the appearance of the original scene.

A very important aspect of human color vision is *chromatic adaptation*, which is the ability to experience color in similar ways under a wide range of different illuminants. A digital camera is emulating this behavior by using automatic white balancing algorithms. If the white balancing fails, this is usually immediately visible, as illustrated in Figure 3.9 for an image that has a significant blue color shift. The white balancing will impact the overall color, but in contrast with shifts in color reproduction, a failed white balancing will always impact the appearance of white and neutral objects in the image.

3.1.3   Geometrical Artifacts

A geometrical artifact is a displacement of portions of an image, leading to a distortion of the appearance of objects. In principle, this means that the intensity information will be maintained such that the objects in the image are identifiable although their shapes are changed. As described in Chapter 1, shape is a fundamental property of objects in a scene. Therefore, geometrical distortions should be quantified because they can degrade the image quality. The various types of geometrical distortions will be described in the following paragraphs.

3.1.3.1   Perspective Distortion
If a lens with a wide field of view is used, the perspective as compared to what an observer of the scene is seeing can be changed dramatically. Furthermore, the proportions of the objects in the scene may also change. This is fundamentally due to the fact that

Figure 3.8 Example of unacceptable color, such as seen in the green sky.

Figure 3.9 Example of unacceptable white balance, here with a distinct overall blue shift.

the camera is projecting a three-dimensional world onto a two-dimensional medium. This *perspective distortion* can be manifested in two ways. The first case is seen when photographing, for instance, a group of people. If the group covers a wide enough field of view, the persons at the periphery become deformed. The reason for this is that a three-dimensional object will occupy a certain angle in space with respect to a point located some distance away from the object. The extent of this angle is constant as long as the distance between object and point is the same, and the object is rotationally symmetric, as, for example, for a sphere. If imaged by a camera, the light rays from the edges of the sphere will converge at the optical center of the lens and then diverge to form a cone with its top at the optical center of the lens and the base at the image sensor. For objects at the center of the image, the base of the cone will be perpendicular to a line from the center of the base toward its top. For off-center cones, however, the cone will be skewed, and the base area will become larger. Consequently, objects of identical size will appear different in size and form depending on position in the image. The wider the field of view, the more pronounced this effect will be.

The second kind of perspective distortion is seen when using a wide angle lens and, in order for the subject not to disappear in the background, the distance between the subject and camera must be made quite short. In a portrait situation, this makes the relative distance between nose and camera much shorter than the distance between the ears and camera. The obvious consequence is that the central region, such as the nose, becomes enlarged and the geometries of the surrounding regions become reduced, resulting in a disproportionate rendering of the face. This has become an issue in mobile phone cameras, where the demand for increasingly thinner devices has made the camera the component that restricts the total phone thickness. In order to make camera modules with lower height, it is tempting to make the focal length of the lens shorter. Consequently, a conflict arises between the ability to capture good portrait photos at full resolution and the mechanical design. This results in phones with cameras that are not well suited for portrait photographs, which is somewhat ironic considering the fact that one of the main uses of the camera in a phone is to take snapshots and videos of people.

3.1.3.2 Optical Distortion

In contrast with perspective distortion, which is solely due to the geometry of the scene, *optical distortion* is a property of the camera lens and can appear quite different between lenses and lens types. This distortion is due to variations in magnification as a function of image height. The distortion is typically negligible around the center of the image and increases toward the corners. Two examples are shown in Figure 3.10.

3.1.3.3 Other Geometrical Artifacts

Beside the effects discussed above, other factors may also affect the geometry of the image. One typical example in mobile phones is *rolling shutter distortion*. The origin of this effect is due to the fact that in a CMOS sensor, the pixel rows are exposed at different points in time, as will be discussed in more detail in Chapter 4. This affects the appearance of moving objects, sometimes with quite dramatic effects, like the one shown in Figure 3.11. Here we see an airplane propeller that is obviously reproduced in a way that does not accurately represent its actual shape: the rolling shutter distortion modifies the blades from being attached to the hub to being loose objects "floating" near

Figure 3.10 Examples of optical distortion. Left: pincushion; right: barrel.

Figure 3.11 Example of image distortion due the use of a rolling shutter. The propeller blades appear to be "floating" in mid-air instead of being attached to the hub of the propeller.

the plane. In video clips, this type of distortion may give rise to a "jello" effect where solid objects moving in different directions in front of the camera will appear to be bending in a direction dependent on the relative motion between object and camera.

3.1.4 Nonuniformities

Many of the attributes presented in this chapter may show different appearance across the field of view. This is also true for exposure and color attributes. The general term for these nonuniformities is *shading*. This is used in order to emphasize that these effects may be due to several different factors, such as lens vignetting, variations in angular sensitivity of the sensor pixels and IR cut filter, and so on.

Figure 3.12 Example of luminance shading. The sand in the corners is distinctly darker than in the center of the image.

Figure 3.13 A color shading example in which the center region is greenish and the corners are reddish.

3.1.4.1 Luminance Shading

Luminance shading is a variation in apparent brightness across the image that is not related to variations in illumination in the scene. The geometry of luminance shading can be quite variable, from strictly radial symmetry to very complex patterns with, for example, isolated bright or dark spots at arbitrary positions in the image, as well as many other manifestations. An example of radially symmetric luminance shading is given in Figure 3.12.

3.1.4.2 Color Shading

Color shading may arise from variations in angular sensitivity of the image sensor pixels as well as from the use of a reflective IR cut filter. In this type of filter, the spectral transmittance varies with the angle of the light, explained further in Chapter 4. This has the effect that the color and white balance changes as a function of image height, as shown in Figure 3.13 in which the center white balance is greenish and the corners are reddish, resulting in a significant hue shift in the sand color moving from the center to the corners. Depending on the design of the camera and image signal processing, the color shading may show a quite large variation with, for example, reddish center and greenish corners (or the opposite), blue to yellow variations, and so on.

3.2 Local Attributes

The appearance of local attributes depends on the viewing conditions with respect to viewing distance and size of the image. The main attributes in this category include

- Sharpness and resolution
- Noise
- Texture blur
- Color fringing
- Image defects
- Artifacts due to, for example, compression

3.2.1 Sharpness and Resolution

The sharpness of an image is determined by a number of factors, most notably focusing, lens aberrations, and motion. It can often be quite difficult to distinguish blur due to defocusing from blur due to lens aberrations. However, lens aberrations frequently show variations across the field of view, with more blur toward the edges of the image compared to the center, as shown in Figure 3.14. Blur due to defocusing is very often used to emphasize certain objects in an image. For example, this is routinely done for portraits, where the person of interest is kept in focus while the back- and foreground are blurry. The volume within which objects in a scene are kept in focus is known as the *depth of field*. An example is shown in Figure 3.15, where the foreground object, closest to the camera, is clearly out of focus, while the objects farther back appear sharp. By changing the f-number of the lens, the depth of field can be varied.

Figure 3.14 Example of blur due to lens aberrations. Note the deterioration in sharpness toward the edges and corners of the image.

Other attributes related to sharpness are resolution and aliasing, as well as artifacts due to excessive digital sharpening of the image. It is important to understand the distinction between these very different artifacts and how they relate to each other. Figure 3.16 illustrates the fact that there does not necessarily have to be a relation between sharpness and resolution. The left image appears considerably sharper than the right. However, looking more closely at small details in the two images, both images show a similar degree of resolution, and the right image may in fact be slightly better at resolving the very smallest details. Thus, the appearance of sharpness and resolution are not necessarily correlated.

Usually, some amount of digital sharpening is done on images in order to enhance the appearance of edges in the scene. A side effect of this treatment is the appearance of halos around sharp edges, which can sometimes become quite disturbing. Too much sharpening also has a tendency to amplify noise in the image. An example of oversharpening is shown in Figure 3.17. While the right image appears sharper than the left, it could

Figure 3.15 Illustration of depth of field where the foreground object on the right is strongly out of focus, while structures farther back appear sharp.

Figure 3.16 Images showing the distinction between sharpness and resolution. The left image is clearly sharper than the right image. However, the right image shows more detail in fine structures such as in the segmentation and edges of the reeds.

be considered to be oversharpened due to the halo artifacts, visible especially along the mountain range in the far distance.

Motion of an object in the scene or of the camera may have an impact on the sharpness of the image. For example, if the exposure time is too long, an object with sufficient

Figure 3.17 Example of sharpening artifacts. Left: original; right: oversharpened. Note especially the bright "halo" around the mountain range in the far distance, due to excessive sharpening.

Figure 3.18 Example of motion blur. Note that the person in the foreground appears blurrier compared to people in the background. This is mainly explained by the fact that objects farther away from the camera are moving at a lower angular speed, and therefore travel over a smaller portion of the field of view during a given time period, compared with objects closer to the camera.

speed will have had the chance to move a certain distance during the exposure period. In this case, only the object will appear unsharp. However, the whole image may also become unsharp if the camera is moved too fast during the exposure. Figure 3.18 shows an example which is blurry because of both object motion and, to a lesser degree, camera motion. The people walking fast are blurry because of their high speed relative to the exposure time, and the entire image is slightly blurry due to camera shake.

3.2.2 Noise

There will always be some amount of noise in any image captured by a digital camera. The cause and origin of the different noise sources will be elaborated in Chapter 4. The visibility of this noise will depend upon a variety of causes, such as the quality of the camera, the lighting conditions, noise filtering, and viewing distance. The appearance of the noise can also be different depending on the spatial content and the amount of coloration. A few examples are shown in Figure 3.19. Using the standard deviation of a uniform region in the image as a measure of noise, we can calculate the signal to noise ratio (SNR) by dividing the mean value of the region by the standard deviation. However, doing so for the three images containing luminance noise in Figure 3.19, we find that the SNR is identical. This is clearly not in correspondence with how the noise is perceived in those images—the top left image with white luminance noise appears to have the least amount of noise whereas the $1/f$ and column noise have structure that is quite apparent. Therefore, a naïve metric of noise involving only the standard deviation is evidently not adequate. In later chapters of this book, metrics with a much better correlation with perception will be presented.

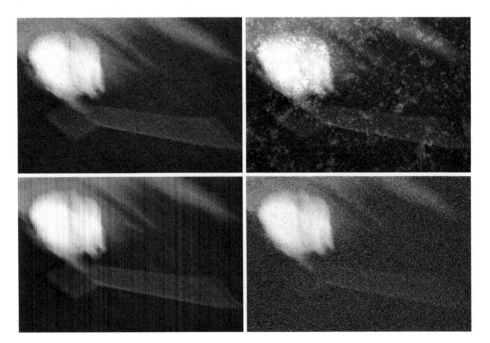

Figure 3.19 Examples of different types of noise. Top left: white luminance noise; top right: $1/f$ luminance noise; bottom left: column noise; bottom right: chrominance noise.

In the lower right image in Figure 3.19, colored noise was added to the image. This adds another dimension to the perception of noise, and this aspect will also be further explored in subsequent chapters.

Just as in the case of sharpness, noise can show variation across the field of view. This is typically due to the fact that shading compensation algorithms are used to make the image corners look brighter. Since the signal level of the corners was originally darker due to the light falloff in the optics, the signal to noise ratio was lower compared to the center. By amplifying the signal level in the corners, the noise will also be amplified by the same amount. Consequently, the signal to noise ratio will be unchanged. But since it was originally lower than in the center, the end result will be that the noise will appear stronger in the corners.

3.2.3   Texture Rendition

With modern noise reduction algorithms, it is possible to remove noise without affecting the sharpness of the image. Sharp edges will still look sharp, but the algorithms are not flawless and sometimes have problems distinguishing low contrast detail and textures from noise. As a result, such features will become blurry, but the overall appearance of sharpness is maintained. The end result is an image that bears some resemblance to, for example, an oil painting, as shown in Figure 3.20. Note that the edges of the dog's nose and geometric design in the pillow are apparent, but that the texture in the fur is extremely blurry.

3.2.4   Color Fringing

In some high contrast scenes under strong illumination, such as trees against the sky in the background, it is possible to see edges that appear colored. A simulated example is shown in Figure 3.21 in which the edges of both the tree branches and statue toward the corners have apparent colored fringing. The main reason for this phenomenon is

Figure 3.20 Example of texture blur. Even though the dog's nose as well as structures in the pillow appear sharp, the fur is blurry, giving an unnatural appearance.

Figure 3.21 Example of color fringing. The effect is most clearly seen toward the edges and corners of the image.

chromatic aberrations caused by the lens. Usually, they are more visible toward the edges of the image and are most clearly seen in overexposed regions.

3.2.5 Image Defects

The image sensor may contain pixels that are "stuck" at either completely white or black, or sometimes somewhere in between. This produces small colored spots in the image that can be quite disturbing if located in the wrong place. Also, dust or lint may have found its way to either the surface of the image sensor or on top of the cover glass or IR cut filter located a small distance away from the sensor surface. Furthermore, there might be scratches on the sensor cover glass or other optical components. In the case of dust or lint directly on the sensor surface, this will give rise to sharp structures that are considerably larger than a pixel, while defects that are located some distance above the sensor surface will appear as blurry spots. Figure 3.22 shows examples of how such defects could appear: note the blue light defect in the upper left corner, which is an example of a pixel defect, and the blurry dark circular objects in the sky, which are examples of particles in the optical path.

3.2.6 Artifacts

The attributes listed in the previous sections can all be measured using more or less accurate objective metrics. However, image processing algorithms as well as hardware

Figure 3.22 Example of image defects. Note the light blue defect in the upper left corner which is an example of a pixel defect, and the blurry dark circular objects in the sky which are examples of particles in the optical path.

properties may introduce new features into the image that were not visible in the scene. We refer to such features as *artifacts*. There is a large number of such artifact types, and the subsequent sections will only list a few of those most commonly encountered in images.

3.2.6.1 Aliasing and Demosaicing Artifacts

An artifact related to sharpness is *aliasing*, that is, low spatial frequency structures appearing in fine repetitive patterns. An example is shown in Figure 3.23. The scene contains a tatami mat that has a high frequency pattern due to the weave of the straw. In the left image, aliasing is apparent, though subtle, as the mat extends to the windows. In the right image, the aliasing is much stronger because the image was first downsampled and then upscaled, which causes a reconstruction of the scene with incorrect spatial frequency information. This type of artifact can make the interpretation of the original scene content more difficult to infer from the image.

As will be explained in Chapter 4, the image coming out of the image sensor needs extensive processing in order to become a proper image. For a color image, a very important part of this processing is *color interpolation* or *demosaicing*. To produce a color image, the image sensor pixels have color filters on top of them arranged in some pattern of typically red, green, and blue colors. This means that each pixel will only register one color; but in order to obtain a real color image, each pixel must be able to provide red as well as green and blue signals in some suitable mixture. Therefore, the missing color

Figure 3.23 Example of aliasing artifacts. The image to the right is a downsampled and upscaled version of the left image. Notice the patterns appearing in the tatami mat.

for each pixel must be interpolated using the values of neighboring pixels. Depending on the algorithm, this will produce more or less agreeable results with severe artifacts in some cases. Typical artifacts of this type are shown in Figure 3.24, particularly visible in the folds of the robe of the statue.

3.2.6.2 Still Image Compression Artifacts

In order to keep the file size of a digital image manageable, it is usually compressed. The most well-known and used compression technique is JPEG (Joint Photographic Experts Group). This is an example of *lossy compression*,[2] that is, the image quality gets compromised to some extent. Using a higher compression ratio, thereby decreasing the file size, will decrease the image quality. However, the JPEG compression algorithm was designed to take into account perceptual aspects, and therefore the compression ratio can in many cases be quite high without being noticeable. Nevertheless, in some situations, depending on the scene, viewing distance, and so on, compression artifacts may become noticeable. Examples of artifacts are shown in Figure 3.25, where "mosquito noise" can be seen around the antennas of the butterfly, and blockiness is prevalent in the entire image.

3.2.6.3 Flicker

As discussed above, a CMOS sensor typically uses a rolling shutter. Apart from distorting fast moving objects, this might also give rise to *flicker* in an image. Such an artifact

---

2 A lossless version of JPEG actually exists. However, in the majority of cases, lossy JPEG compression is used.

Figure 3.24 Example of demosaicing artifacts. Note the colored "zipper" artifacts around sharp edges.

Figure 3.25 JPEG compression artifacts, seen especially around the antennas of the butterfly and as an overall "blockiness."

can occur if the light source illuminating the scene is driven by AC power and therefore oscillates with the AC frequency. If the exposure period of the sensor is not matched to the AC period, different rows may be exposing the scene during different times within the period of the oscillating source. As a result, the exposure will vary from row to

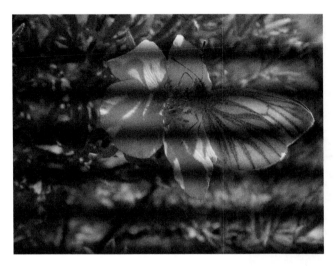

Figure 3.26 Example of flicker as seen in the parallel horizontal dark bands.

row. This will show up as parallel dark bands, which in a video can move in the vertical direction. An example of how flicker may appear is shown in Figure 3.26.

3.2.6.4 HDR Processing Artifacts

As mentioned at the beginning of this chapter, HDR imaging requires tone mapping in order to match the tonal range of the camera with the computer screen or print. There are many techniques for doing this, all of which have their own strengths and weaknesses (Reinhard *et al.*, 2006). Common to many of these methods is that they can generate specific artifacts, such as halos or false colors, as well as an overall unnatural look of the image. An example of a tone mapped image is shown in Figure 3.27. In this case, halos around edges are very visible.

Another issue related to HDR is due to the fact that in order to capture a high dynamic range scene, some cameras capture several images using different exposure times. These images are combined in software and then processed by the tone mapping algorithms. Since the final image is now made up of several images that have been captured at slightly different times, as well as with different exposure times, objects in motion will be at different positions in the images and with different amounts of motion blur, leading to more or less visible *motion artifacts*.

3.2.6.5 Lens Ghosting

As described previously in this chapter, a strong light source located inside or outside the field of view may give rise to flare (or veiling glare). The same source may show other effects in the image, called *ghosting*. These are localized features which are mostly circular, but that could also have other geometries, such as hexagonal and other shapes. By moving the camera with respect to the light source, these shapes will change and move across the field of view, or appear and disappear at certain locations. An image with lens ghosting is shown in Figure 3.28. Note the colored, translucent circular artifacts in the upper left.

Figure 3.27 An HDR processed image showing typical tone mapping artifacts, for example, the strong "halos" around the tree.

3.3 Video Quality Attributes

Contributed by Hugh Denman

Compared with still image photography, video introduces the dimension of time to the imaging experience. This has several implications regarding the quality of the footage. The final section of this chapter will raise the most important aspects of video quality, discussing frame rate requirements, exposure, focus, white balance responsiveness, and audio-visual synchronization as well as new aspects of noise that are typically not seen in still images.

3.3.1 Frame Rate

The essential temporal characteristic of video is the frame rate. As described in Chapter 1, the perception of motion requires frame rates of at least 10 frames per second (fps), and higher frame rates will improve the quality of the motion percept, especially for fast motions. Higher frame rates also reduce the likelihood of temporal aliasing, also discussed in Chapter 1, occurring in scenes containing repetitive motion, as the frame rate must be at least twice the rate of repetition to avoid temporal aliasing.

Cameras for broadcast typically capture video at a rate of 25–60 fps, equivalent to an interval of 16–40 ms between the start of each frame capture. The frame rate is fixed and consistency is paramount. Consumer cameras typically offer a broader range of frame rates, from as low as 5 fps up to 1000 fps. Lower frame rates of 5–15 fps are provided to enable improved image quality in low-light conditions, by allowing for longer exposure

Figure 3.28 Lens ghost example, seen as faint rings and circles in the upper left of the image emanating from the sun in the lower right.

times per frame. Higher frame rates, up to 1000 fps, are provided to enable a slow motion effect, when the captured footage is played back at a lower frame rate. The video frame rate may be selected by the user, but in some applications, for example videoconferencing, it is chosen automatically by the camera in response to the lighting conditions and perhaps the motion characteristics of the scene.

For the presentation of motion to be acceptable, the frame rate should be consistent through the duration of the footage (Ou *et al.*, 2010). While a smoothly varying frame rate need not cause problems, provided it is kept within an acceptable range, delays in individual frames are readily perceptible and intrusive. This sort of uneven video playback is called frame rate *jitter*, and this term is often used to encompass frame drops as well as frame delays. Even small amounts of temporal delay can have a significant impact on the perceived video quality (Huynh-Thu and Ghanbari, 2006; Claypool and Tanner, 1999). Jitter effects most commonly arise during playback, but can occur during capture, especially if the capture device is not using dedicated hardware, for example where a mobile phone camera has a signal path that involves the main CPU.

Cinema has traditionally used 24 frames per second. This is certainly high enough to induce perception of motion, but is also claimed to be low enough to encourage suspension of disbelief: footage at 24 fps has a definite "film look" which the brain can distinguish from reality and thereby accept the fictions presented. Higher frame rates become less distinguishable from reality and are associated with a "video look" that is used for news footage and television. Cinematic forays into higher frame rates have not been universally well received, partly on the basis that the footage looks "too real" and suspension of disbelief could not be sustained. A similar effect occurs when televisions artificially increase the video frame rate by upsampling, resulting in what is termed the "soap opera effect." These complaints are probably due to learned expectations for the appearance of film, and it is likely that high frame rate (HFR) cinematography will become more acceptable as it becomes more prevalent.

The camera frame rate should also be high enough to capture the motion of the scene content unambiguously. In other words, it should be high enough to avoid *temporal aliasing*. The most commonly encountered example of temporal aliasing is the "wagon wheel effect," in which a spoked wheel portrayed in a film appears to slow down or reverse its direction of motion as the vehicle accelerates, because the temporal frequency required to capture the motion of the spokes exceeds the frame rate of the camera. This effect is not limited to film, as the same phenomenon can be observed in the real world under strobe lighting. For example, care must be taken in factory installations that fluorescent lights not be supplied from the same AC power source as the machinery, otherwise a spinning lathe can appear to be stationary. Intriguingly, the effect has also been recorded in real life under continuous light, an observation which informs the debate on whether consciousness is continuous or discrete (Koch, 2004).

3.3.2   Exposure and White Balance Responsiveness and Consistency

The per-frame exposure time and white balance setting should be consistent within a video, providing that the lighting conditions remain constant, in order to avoid variations in image tone and level of motion blur. If the lighting conditions in the scene change while the video is being recorded, the exposure and white balance must adapt to the new conditions rapidly and smoothly, without overshooting or oscillating about the new optimal setting.

In general, for a well-lit scene, the best results are obtained using a short exposure time, both for still images and for video frames. However, if the scene contains motion, and the video is being captured at a frame rate of 25–40 fps, a long exposure time may be required to obtain a level of motion blur yielding the best motion effect (as discussed in Chapter 1). Thus, in some video capture contexts, the choice of exposure time is a question of compromise rather than of a single optimal value. Adjustment of the lens aperture, if available on the camera in use, may help to reconcile these factors. However, adjusting the aperture will affect the depth of field in the video, and thus have an aesthetic impact of its own.

3.3.3   Focus Adaption

As video is recorded, motion of the camera or the scene elements can result in the optimal focal distance varying over the course of the video. Thus, the focus setting needs to adapt to the scene as the video is recorded. Focus hunting while the camera is recording results in periods of noticeable blur in the resulting video.

Some scenes pose particular challenges for autofocus systems, for example where two halves of the image plane are at different depths. In this context there are two defensibly correct focus settings with regard to the scene, and the autofocus system may switch rapidly between the focus settings corresponding to the two dominant scene depths. This may be acceptable for pre-capture preview, but the effect of such rapid focus switching in recorded video is undesirable. Autofocus systems for video may be implemented with heuristics or temporal smoothing filters to avoid this effect.

3.3.4   Audio-Visual Synchronization

When audio and video are being recorded simultaneously, it is important that the synchronization between the two be maintained throughout the recording. In some cases, the audio track may be at a fixed temporal offset from the video track, for example if there is a processing delay in the signal path for video that is not compensated for in the audio. In other cases, particularly if audio is being recorded on a separate device, small differences in the sample rate or clock accuracy may result in the audio slowly drifting out of sync with the video over time.

The quality impact of poor audio-video synchronization depends on the magnitude of the discrepancy in presentation time, on whether the sound is played early or late relative to the video, and on the nature of the scene depicted. The impact on quality is particularly severe when human subjects are speaking in the video, where the problem is termed "lip sync" error. This error can negatively affect the perceived quality even when the magnitude of the error is below the threshold of conscious perception.

A number of broadcasting authorities have established acceptable limits for loss of audio-visual synchronization: these standards generally agree on a maximum discrepancy of a few tens of milliseconds, with tighter limits on advanced sound than on delayed sound. The European Broadcasting Union standard is a representative example, recommending that where sound is presented before the corresponding video frames, the offset should be no more than 40 ms, and when the sound is delayed relative to the corresponding video frames, the delay should be no more than 60 ms (European Broadcasting Union, 2007).

3.3.5 Video Compression Artifacts

Digital video compression can introduce artifacts similar to those described for image compression above, and also artifacts particular to video. An overview of methods for video compression is presented in Chapter 4, which provides some insight into how these artifacts can arise.

As will be discussed in Chapter 4, video compression most often operates on each spatial block of an image separately. Where the video bitrate is too low to effectively encode the content of the video, blurring may be introduced due to the discarding of information at high spatial frequencies within each block. Furthermore, the block structure can become apparent in the compressed video, giving rise to *blocking artifacts*. These blurring and blocking artifacts are similar to those that can occur in digital image compression, but are more common and more severe in video. The greater frequency and severity of these artifacts in video is due to the higher compression ratio required in video compression, relative to still image compression. The severity of these artifacts is also increased in video by the fact that they generally occur on a spatially fixed grid within the image space. Thus, the position of the blocking artifacts remains fixed as the elements in the video move, and so their position relative to the scene content changes continuously, increasing their visibility.

Where video content containing motion is compressed at low bitrates, moving elements in the scene may leave a motion trail, in which each frame shows the object and also a copy of the object, replicating its position in the previous frame. This effect is referred to as *ghosting*, as the copy of the object is typically desaturated. "Ghosting" is also used to describe a similar artifact that can appear in digital television, where interlaced video is incorrectly converted to progressive-scan. Analog television transmission could also exhibit ghosting, typically due to multi-path distortion of the amplitude-modulated signal. However, in the case of analog television, ghosting appeared as a spatially-offset copy of the entire image, rather than affecting only the moving elements. Note that this type of ghosting should not be confused with lens ghosting discussed previously in Section 3.2.6.5.

Another artifact associated with low bitrate is *temporal pumping*, which shows itself as a flickering effect that appears in the transition from a P-frame to an I-frame, see Chapter 4. This is due to progressively increasing errors in the predicted frames, giving rise to a sudden, very noticeable improvement in quality when an I-frame follows a group of P- and B-frames. When this effect arises, it occurs periodically, typically twice a second, resulting in a distracting fluctuation in the video quality.

3.3.6   Temporal Noise

Each frame of a video will exhibit the spatial noise described above, and much of the noise is random and thus will vary from frame to frame. This results in temporal noise—random variations in the color or intensity value in each pixel site. Such noise is most visible in darker regions of videos shot under low light levels. Temporal noise is difficult to compress, and therefore regions with high levels of temporal noise are prone to blocking artifacts.

3.3.7   Fixed Pattern Noise

Noise, in the sense of variations in pixel value that do not correspond to variations in light reflected by the scene, is chiefly random-valued, but in some camera architectures can have a significant non-random component. This will be described in detail in Chapter 4. As this non-random noise results in the same set of deviations from the desired pixel values appearing in each frame, it is termed fixed pattern noise (FPN).

FPN may be completely random across the image, but may also exhibit a row or column structure, increasing its visual impact. In a still image, it is not possible to distinguish random FPN from temporal noise. In video, on the other hand, the distinction can be very clear depending on the amount of FPN present and may give rise to a partially opaque artifact superimposed on a fixed area of the video frame. Where the fixed pattern noise is correlated across pixel sites, it can result in an even more visible effect. This gives rise to the *dirty window effect*, so-termed as the percept is similar to watching the video through a dirty window.

3.3.8   Mosquito Noise

Frequency-domain digital compression of images can result in intensity oscillations in the vicinity of sharp edges, termed *ringing*. In video, these oscillations vary from frame to frame and thus the area around an edge shimmers, with small bright spots appearing and disappearing over the course of the video. The effect is suggestive of a cloud of mosquitoes swarming around the edge, and is therefore termed *mosquito noise.*

Summary of this Chapter

- Exposure, flare, dynamic range, and tone reproduction all have an influence on the brightness of an image and may also modify other attributes such as noise.

- The range of acceptable colors for objects is, in general, relatively large. However, for certain colors, known as memory colors, the tolerance is smaller within the objects' general hues.
- Various types of distortion can appear in images. Perspective distortion is due to the geometrical arrangement of scene objects and camera focal length. Optical distortion is dependent on the lens design, while rolling shutter artifacts arise because of properties of the CMOS image sensors typically used in today's digital cameras.
- Luminance and color shading phenomena are common in images captured by mobile phone cameras and may show quite large variations in terms of geometry and intensity range as well as colors.
- Sharpness and resolution are two concepts that are related, but not necessarily correlated. Digital sharpening will increase the appearance of sharpness in images, but will also introduce sharpening artifacts and may increase the noise level.
- The appearance of noise can show great variation, and some variations may be more annoying than others, even though a simple noise metric will give the same result for all cases.
- Noise reduction algorithms will reduce the noise in images while maintaining the appearance of sharpness, but potentially with the side effect that texture and low contrast details may be blurred out.
- Color fringing is mainly caused by chromatic aberrations and is mostly visible toward the edges of the image.
- Defects in the image sensor pixels as well as dust and dirt located near or at the sensor surface will give rise to visible unwanted structures in the image.
- Artifacts are defined as structures in the image not present in the scene. These may be due to the image signal processing as well as the lens or the image sensor. Some examples are aliasing, demosaicing artifacts, compression artifacts, flicker, lens ghosting, and HDR processing artifacts.
- Video introduces an additional number of artifacts, mainly due to motion. Among these are jitter, blocking artifacts, and temporal variations in exposure and white balance as well as in focusing. In video clips containing sound, audio-video synchronization is of high importance.

References

Claypool, M. and Tanner, J. (1999) The effects of jitter on the perceptual quality of video, in *Proceedings of the Seventh ACM International Conference on Multimedia (Part 2)*, ACM, New York, NY, USA, MULTIMEDIA '99, pp. 115–118.

European Broadcasting Union (2007) The relative timing of the sound and vision components of a television signal, EBU Recommendation R37-2007.

Fairchild, M.D. (2013) *Color Appearance Models*, John Wiley & Sons Ltd, Chichester, UK, 3rd edn.

Giorgianni, E.J. and Madden, T.E. (2008) *Digital Color Management: Encoding Solutions*, John Wiley & Sons Ltd, Chichester, UK, 2nd edn.

Hunt, R.W.G. (2004) *The Reproduction of Colour*, John Wiley & Sons Ltd, Chichester, UK, 6th edn.

Huynh-Thu, Q. and Ghanbari, M. (2006) Impact of jitter and jerkiness on perceived video quality. *Second International Workshop on Video Processing and Quality Metrics for Consumer Electronics (VPQM-06), Scottsdale, Arizona, USA.*

Koch, C. (2004) *The Quest for Consciousness: A Neurobiological Approach*, Roberts and Company, Englewood, CO, USA.

Ou, Y.F., Zhou, Y., and Wang, Y. (2010) Perceptual quality of video with frame rate variation: A subjective study, in *2010 IEEE International Conference on Acoustics, Speech and Signal Processing (ICASSP)*, pp. 2446–2449.

Reinhard, E., Ward, G., Pattanaik, S., and Debevec, P. (2006) *High Dynamic Range Imaging – Acquisition, Display, and Image-based Lighting*, Morgan Kaufmann, San Francisco, CA, USA.

4

The Camera

In order to perform a meaningful image quality characterization of a camera, a firm understanding of its operation is necessary. Therefore, the goal of this chapter is to provide the necessary background for understanding the sources of the various image quality artifacts discussed in other chapters. The workings of a camera are based on principles established several hundred, if not more than a thousand (Wikipedia, 2016), years ago, and this perspective is valuable for the full understanding of digital photography.

To understand the formation of an image using a digital camera, one needs to combine several fields of science and technology. Physics, providing the underlying concepts, will be most important, and areas such as signal processing, optics and electronics, as well as color and vision science, all play an important role in describing the formation of the photographic image.

The chapter starts with a discussion of the archetypal image capture device: the pinhole camera. It then proceeds to describe how the operation of this device is refined by introducing lenses and other optical elements, photosensitive recording devices such as image sensors, and a description of image processing algorithms needed to provide a pleasing end result. The connection between the various subcomponents that make up the camera and the image quality attributes described in the previous chapter is then elucidated through examples.

As the video functionality of modern cameras has become increasingly important, the chapter ends with a description of some important aspects and technologies associated with capturing video. Of very high importance is video compression, given the fact that video is often transmitted over networks with limited bandwidth, which makes it important to decrease the data rate while maintaining a high image quality. A large part of the section is therefore devoted to describing video compression.

4.1 The Pinhole Camera

The word camera originates from the Latin name *camera obscura*, literally meaning "dark chamber." This denomination is no less valid today, since most cameras are made up of a cavity sealed away from light, inside which the light sensitive medium (film or image sensor) is contained.

The principle of the camera obscura is remarkably simple, as illustrated in Figure 4.1. At one side of the chamber, a small hole is made. Light rays from the scene will pass

Camera Image Quality Benchmarking, First Edition. Jonathan B. Phillips and Henrik Eliasson.
© 2018 John Wiley & Sons Ltd. Published 2018 by John Wiley & Sons Ltd.
Companion website: www.wiley.com/go/benchak

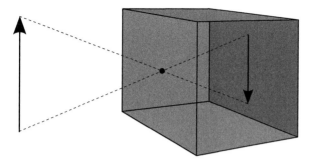

Figure 4.1 The camera obscura.

through the hole and end up illuminating the back wall inside the chamber, thus forming an image of the scene.

Three things should be immediately obvious from this figure. First, the image of the scene will be upside down. Second, the larger the hole, the brighter the image, and third, the shorter the distance between the hole and the back wall, the smaller the image.

The second point implies that there is a contradiction between image brightness and sharpness, since making the hole larger will also mean that light rays originating from one point in the scene will be spread out over a larger area on the back wall of the chamber, thus making the scene blurry. It is therefore necessary to make the hole as small as possible in order to obtain a sharp image. As will be seen later on, this is not entirely true, since optical diffraction will make the image increasingly blurry as the hole diameter is decreased below some limit.

The third point means that it is possible to shrink the camera obscura down to a more manageable size, which could even be handheld. We then have made what is commonly known as a *pinhole camera*. This could be said to constitute the archetype of the modern camera.

While it may seem to be a big leap from the camera obscura to the modern digital camera, the similarities are obvious. In order to overcome the shortcomings of the pinhole camera, a few main functional blocks have evolved. For a digital camera, these blocks can be identified as the *lens, image sensor* and *image signal processor* (ISP).

4.2 Lens

The tradeoff between image brightness and sharpness of the pinhole camera can in principle be overcome by putting a lens at the position of the pinhole. Rays diverging from a point source in the scene will be collected and focused at the image plane, as shown in Figure 4.2.

4.2.1 Aberrations

For a perfect lens, all rays emanating from one point in the scene, passing through the lens, will converge at one point in the image. However, in reality this is not the case. Figure 4.3 shows a more realistic lens. Here, parallel light rays coming from infinity are focused by the lens, but not at the same image point along the axis. Clearly, rays farther

Figure 4.2 The principle of the lens. *f* is the focal length.

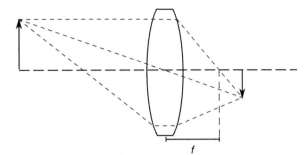

Figure 4.3 Ray trace of a thick lens focused at infinity, showing spherical aberration, see text.

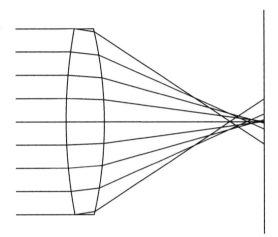

away from the optical axis (center of the lens in this case) will cross the axis increasingly distant from the position of rays closer to the optical axis. For rays closer to the optical axis, the focus position seems to converge at one point. This is known as the *paraxial* or *Gaussian* focus point. Paraxial optics rely on the fact that sines and cosines of small angles can be approximated by the angle and a constant, respectively. For this reason, it is also sometimes referred to as the *first-order theory* of optics. This simplifies calculations considerably, as, for example, demonstrated in the well-known Gaussian lens formula, relating the object position, *s*, with the image position, *s'*, and the focal length, *f* (Hecht, 1987):

$$\frac{1}{s'} + \frac{1}{s} = \frac{1}{f} \tag{4.1}$$

The deviations from paraxial behavior are called *aberrations*. These aberrations can be either *monochromatic* or *chromatic*.

4.2.1.1 Third-Order Aberrations

Usually, when discussing monochromatic aberrations, only third-order terms are considered. The sine function can be expanded in the Taylor series

$$\sin \theta = \theta - \frac{\theta^3}{3!} + \frac{\theta^5}{5!} - \frac{\theta^7}{7!} + \dots$$

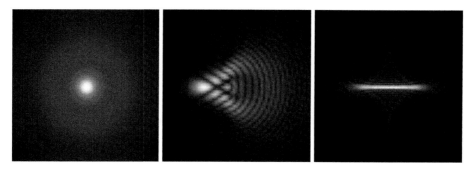

Figure 4.4 Images of a point source due to third-order aberrations. Left: spherical; middle: coma; right: astigmatism.

The paraxial description is obtained by only retaining the first term. Studying the result of incorporating the second term, of third polynomial order, yields the lowest-order aberrations: spherical, coma, field curvature, astigmatism, and distortion (Mahajan, 1998). Of these, spherical aberration is field-independent while the others will vary across the field of view of the image. Furthermore, all these aberrations, with the exception of distortion, will modify the image sharpness. Figure 4.4 shows examples of how an image of an infinitesimally small point could look due to the above mentioned aberrations. Such an image is known as a *point spread function* (PSF). Figure 4.3 shows a ray trace illustrating spherical aberration. Notice that this aberration is present also in the center of the image.

Coma gets its name from the fact that it gives rise to a PSF shaped similarly to a comet, as seen in Figure 4.4. In contrast to spherical aberration, coma is field dependent, increasing in magnitude with distance from the image center.

Field curvature is due to the phenomenon that the image plane in a simple lens is curved rather than flat. Therefore, the point of best focus will shift in the longitudinal direction as a function of field height. To counteract this in multilens systems, the focal length can be made to change with field height.

The plane of best focus can also change depending on the angle of rotation in the image plane. This is known as astigmatism. In effect, this means that the field curvature is different depending on angular position in the image plane. This produces PSFs that are oval in shape, typically in the radial direction, as shown in Figure 4.4.

In contrast with the other aberrations, distortion does not affect the sharpness of the image. Instead, the magnification is changed as a function of the image position. The effect is that objects such as straight lines will appear curved. Depending on the lens design, optical distortion can show great variety. In Figure 4.5, two basic variants are shown: pincushion and barrel distortion.

It should be mentioned that the expansion of the sine function certainly doesn't stop after the second term; higher-order aberrations will always be present in any optical system, to a greater or lesser extent depending on how well-corrected the lens design is for those aberrations.

4.2.1.2 Chromatic Aberrations

Chromatic aberrations originate from the fact that the index of refraction in the lens material (typically glass or plastic) is wavelength dependent. This will result in two main

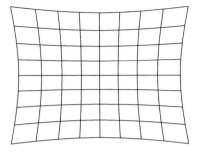 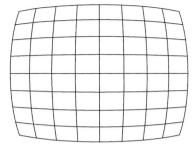

Figure 4.5 Two types of optical distortion. Left: pincushion; right: barrel distortion.

Figure 4.6 Ray trace of a triplet lens as described in a patent by Baur and Freitag (1963). *a* and *b*: principal planes; *c*: exit pupil; *d*: entrance pupil. Green rays are 550 nm and red rays 650 nm light.

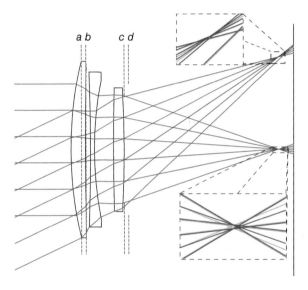

types: longitudinal and lateral. Longitudinal chromatic aberrations manifest themselves as a shift of focus depending on the wavelength, while lateral chromatic aberrations are really a change of magnification as a function of wavelength and field height. The latter can therefore be thought of as a wavelength-dependent distortion, showing increasingly visible colored fringes as one moves away from the center of the image toward the corners. Longitudinal chromatic aberrations, on the other hand, can be seen also in the image center. In Figure 4.6, a ray trace of a more complicated lens with three elements is shown. Two points at infinity are imaged through the lens, one in the center and one at the edge of the field of view. Rays for two different wavelengths are traced through the system. As seen in the insets, showing magnifications of the ray fans in image space, longitudinal chromatic aberration is clearly visible, where the green rays are brought to focus at a different position compared to the red rays. The figure also illustrates spherical aberration on-axis, as well as coma off-axis.

4.2.2 Optical Parameters

Figure 4.6 illustrates another important property of a lens system. Any arbitrarily complex lens can always be reduced to a "black box" characterized by a small set of surfaces (Hecht, 1987; Smith, 2000). In the paraxial approximation, these surfaces can be considered to be planes perpendicular to the optical axis. In the figure, four of these surfaces are shown: the front and back *principal planes* and the *entrance* and *exit pupils*. There are also two *nodal points*, but their positions on the optical axis coincide with those of the principal planes in the case when the lens is surrounded by a medium with the same index of refraction.

The distance between the back principal plane and the focal point defines the *focal length* of the lens.[1] This parameter, together with the image sensor size, effectively determines the *field of view* and magnification of the camera. It can be shown that the field of view when the object distance is large in comparison with the focal length is given by

$$\theta = 2\tan^{-1}\left(\frac{d}{2f}\right) \tag{4.2}$$

thus determined by only the image sensor size, d, and focal length, f. The image sensor size could be given as either the length of the diagonal, or the vertical or horizontal height or width. This consequently specifies three different field of view numbers: diagonal, vertical, or horizontal.

The magnification, M, of the lens is given by

$$M = \frac{f}{s-f}$$

For a fixed camera-to-object distance, s, and provided that $s \gg f$, we see that the magnification increases proportionally with the focal length. What is not obvious from this equation is how the magnification changes with image sensor size. Because of the proportionality between magnification and focal length, it has been customary to think of magnification in terms of focal length of a corresponding 35 mm camera. This makes sense in a way, since the linear magnification is dependent on the object distance. Therefore, it is more convenient to talk about magnification in terms of the equivalent 35 mm lens. The translation between the focal length, f, of some camera into the corresponding 35 mm format is then found through the expression

$$f_{35\,\mathrm{mm}} \approx 43\frac{f}{D} \tag{4.3}$$

where D is the diagonal size of the image sensor of the camera in millimeters.

The amount of light collected by the lens is determined by its f-number. This quantity is given by the ratio of the focal length to the diameter of the entrance pupil, which is the image of the aperture as seen from the front of the lens. The irradiance (illuminance) at the image plane, E, is related to the object radiance (luminance), L, according to

$$E(L,\theta) = \frac{\pi L \tau \cos^4\theta}{4N^2\left(1 + \frac{M}{m_p}\right)^2 + 1} \tag{4.4}$$

where θ is the angular field position, L the scene radiance (luminance), N the f-number, M the magnification, and m_p the pupil magnification. We have also included the transmittance of the lens as the factor τ.

---

1 More correctly, this should be referred to as the *effective focal length* to distinguish it from the back and front focal lengths, which are the distances from the back optical surface to the back focal point and from the front optical surface to the front focal point, respectively. However, when talking about the focal length for a complex lens, it is usually implied that it is the effective focal length that is referred to.

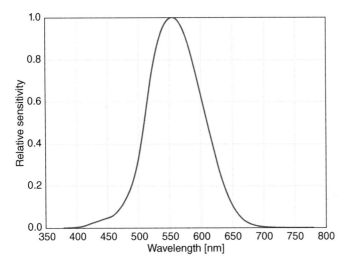

Figure 4.7 Wavelength sensitivity of the human eye.

Table 4.1 Relation between radiometric and photometric units.

Radiometric quantity	Radiometric unit	Photometric quantity	Photometric unit
Radiant intensity	Wsr^{-1}	Luminous intensity	candela (cd)
Radiance	$Wm^{-2}sr^{-1}$	Luminance	cdm^{-2}
Irradiance	Wm^{-2}	Illuminance	lux
Radiant flux	W	Luminous flux	lumen (lm)

The intensity, radiance and irradiance constitute the basic radiometric quantities. These are wavelength dependent, and therefore any reference to them must be made with respect to which wavelength interval has been considered as well as the weighting of the wavelengths within this interval. The weighting most often used is shown in Figure 4.7. This is the *CIE 1924 spectral luminous efficiency function* (CIE, 1926), representing the spectral sensitivity of a typical human observer, and using this as a spectral weighting gives us the photometric units summarized in Table 4.1. To convert from radiometric to photometric units, the following formula is used (Hecht, 1987; Smith, 2000):

$$Q_v = 683 \int_0^\infty Q(\lambda)V(\lambda)d\lambda \tag{4.5}$$

Here, Q_v is the photometric quantity corresponding to the radiometric counterpart $Q(\lambda)$, $V(\lambda)$ the human spectral sensitivity shown in Figure 4.7, and λ the wavelength. Note that the candela is the SI unit for luminous intensity. Its name originates from the fact that a common candle emits approximately 1 cd of light.

4.2.3 Relative Illumination

Equation (4.4) is often referred to as the *camera equation*. It describes the transfer of light from the scene onto the image sensor surface. Note that the irradiance is only

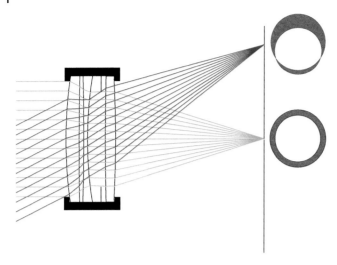

Figure 4.8 Illustration of vignetting, see text.

dependent on the object radiance, and not the distance between the lens and the object. Therefore, the only two scene properties that affect the exposure are the intensity of the light source and reflectance of the object.

In many cases, the object-to-camera distance is large, and therefore, M is much smaller than 1, leading to

$$E(L,\theta) \approx \frac{\pi L \tau \cos^4 \theta}{4N^2} \qquad (4.6)$$

The $\cos^4 \theta$ factor indicates that the image illuminance will fall off toward the edges of the image. This will become more pronounced as the field of view gets wider, and is therefore most noticeable in wide angle lenses. It should be noted that this behavior is strictly geometric in nature and therefore a fundamental property of imaging systems, not dependent on the particular design of the lens (even though design tricks, such as deliberate pupil distortion, are employed to minimize this effect for, for example, wide angle and fisheye lenses (Ray, 2002)).

The aperture stop is the element that ultimately limits the amount of light passing through the lens. Clearly, all elements in the lens have a radius outside of which light will be blocked. For off-axis light, these radii may start to overlap, thus altering the shape of the image of the aperture (called the exit pupil), as seen in Figure 4.8. In Figure 4.9, one of the stops has been made smaller. As clearly seen, there is no longer any overlap between the apertures. By stopping down the lens, the effects of vignetting may therefore be mitigated.

As a final remark, it must be noted that the $\cos^4$ "law," as expressed in Eq. (4.6) is only approximate, becoming less accurate as the distance between the aperture and image plane decreases (Smith, 2000). The combined effects of the $\cos^4$ effect and vignetting is usually referred to as *relative illumination*.

4.2.4 Depth of Field

Another effect of adjusting the size of the aperture stop is on the *depth of field* (DOF). As described by Eq. (4.1), only objects at a specified distance from the camera will be

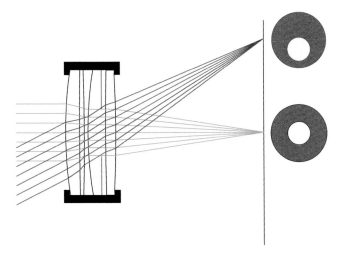

Figure 4.9 Mitigating the effect of vignetting by stopping down. In this case, the rays from both objects are equally obstructed by the aperture stop.

focused at the image plane. As objects are moved away from this position, they will gradually become more defocused, and therefore blurred. Because of properties of the human visual system as well as the camera system itself, the blurring will not become obvious until some critical level of defocus is reached.

In order for the concept of depth of field to be meaningful, we need to establish exactly how much blur we can tolerate and still regard an image to be acceptably sharp. In a geometrical optics approximation (i.e., neglecting the effects of diffraction), which is what we have been discussing so far, a point source of light will be rendered as a point in the image plane of a perfect optical system. If the system is defocused, the point will become a spot with a finite diameter. This spot is known as the *circle of confusion*.

The size of the circle of confusion is not unambiguously defined in any way. For instance, the conditions under which the image is viewed will be very important. Also the properties of the medium, for example, paper or computer screen, will play an important role. Furthermore, in reality, the circle of confusion will most certainly not be a circle with well-defined edges, but its shape will depend on the magnitude of the aberrations present as well as diffraction. The spatial metrics described in later chapters could provide a more accurate estimate of the perceived depth of field by utilizing the *through focus modulation transfer function* (MTF). The MTF concept will be explained in greater detail in Chapter 6. Figure 4.10 shows the through focus MTF as measured off-axis for a lens with some amount of astigmatism. The tangential and sagittal MTF curves are measured in two perpendicular directions, and since astigmatism can be regarded as a shift in focus between these two directions due to the asymmetric shape of the PSF (see Figure 4.4), these two curves will be shifted with respect to each other. By setting a requirement on the minimum MTF value that can be tolerated, a depth of field value that is more relevant for the actual lens under study can be obtained.

4.2.5 Diffraction

In the description of the camera lens so far, only geometrical optics have been considered. In this approximation, light is considered to consist of rays traveling in straight

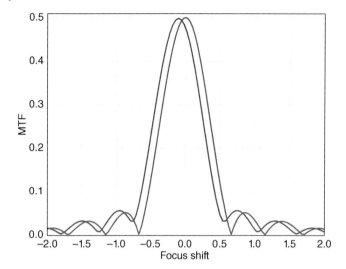

Figure 4.10 Through focus MTF. The MTF value for one specific spatial frequency is plotted as a function of focus shift. Red curve: sagittal MTF; blue curve: tangential MTF.

lines between source and destination. While this description does give a reasonably accurate result in many cases, there will be situations where it will fall short. This can happen, for instance, when the f-number of the lens is increased. One will then observe that for smaller values of this parameter, the sharpness will improve for increasing values, but beyond some critical point the sharpness will start to decrease again if the f-number is increased even further. Consequently, there seems to be an optimal f-number where the lens appears to give its sharpest images.

By increasing the f-number beyond this critical value, the lens is said to have become *diffraction-limited*. Diffraction has its origin in the wave nature of light, and the fact that waves can interfere destructively as well as constructively. This can be explained by the Huygens–Fresnel principle, which states that each point on a propagating wavefront gives rise to a new wavefront, spherical in shape, which will interfere with the waves generated from other points. In free space, this gives the result that light consisting of plane waves as, for example, originating from a source at an infinite distance, will continue as plane waves. As soon as the wave reaches an obstacle, however, the wavefront will be modified according to the shape of the obstacle, thus being diffracted. The newly formed waves will interfere with each other and form patterns at the target which might be more or less complex. For a circular opening, the diffraction pattern can be described by the following formula (Goodman, 2005):

$$h(x, y) = \left[\frac{2J_1 \left(\frac{\pi r}{\lambda N} \right)}{\frac{\pi r}{\lambda N}} \right]^2 \tag{4.7}$$

Here, $J_1(x)$ is the first-order Bessel function (Arfken, 1985), $r = \sqrt{x^2 + y^2}$ with x and y the lateral coordinates of the target, λ the wavelength of light, and N the f-number. The resulting point spread function is shown in Figure 4.11. This function describes a

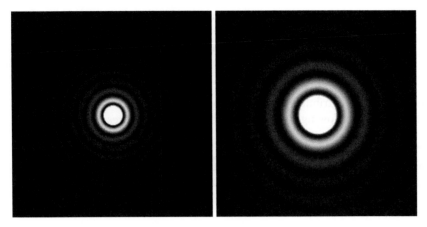

Figure 4.11 Images of the diffraction point spread function. The image shown in the right figure was generated using an f-number twice as large as in the left image.

central disk surrounded by increasingly weaker concentric circles. The central disk is usually referred to as the *Airy disk*.

In order to get an idea about the size of the diffraction PSF, the radial distance from the center out to the first minimum is often used. For the first-order Bessel function $J_1(r)$, the first zero is encountered at $r = 3.8317$. With this definition, the radius of the diffraction PSF becomes approximately equal to $R = 3.8317/\pi \times \lambda N = 1.22\lambda N$. The diameter will therefore be given by

$$D \approx 2.44\lambda N \tag{4.8}$$

Note that there is still signal remaining in the rings surrounding the center disk. Therefore, this is only an approximate assessment, but still often encountered in the literature. Being an approximate expression, we can, by observing that the product 2.44λ is very roughly equal to 1 μm for visible wavelengths, use the rule of thumb that the diameter of the Airy disk is approximately equal to the f-number in micrometers for light in the visible spectrum. Thus, the diffraction-limited PSF of an $f/2.8$ lens is in the order of 3 μm in diameter.

4.2.6 Stray Light

Stray light is unwanted light reaching the image sensor. This can originate from a number of sources. For example, strong light sources outside the field of view may enter the lens at very steep angles, and reflect off of surfaces in an unwanted way. Stray light could also originate from scattering off surface imperfections or impurities inside the lens material, as well as off the edges of apertures. Dust contamination arising from the manufacturing process, scratches on surfaces, and fingerprints could also be sources of stray light.

The infrared (IR) cut filter often encountered in the types of cameras discussed here may be of the reflective type, which means that it reflects light of wavelengths longer than approximately 650 nm while shorter wavelengths will be transmitted. Since the filter has some roll-off at longer wavelengths, some of the reflected light may be reflected back toward the sensor and then partially transmitted through the IR cut filter.

The effects of stray light will manifest themselves as ghost images or veiling glare (flare). The latter may reduce the dynamic range of the image, and also the signal to noise ratio. Its effects on the image are described in more detail in Section 3.1.1.

4.2.7   Image Quality Attributes Related to the Lens

Most of the image quality attributes listed in Chapter 3 are in some way related to properties of the lens. In the following, a summary of how those attributes relate to the lens is given.

Exposure, Tonal Reproduction, and Flare   Stray light, as already discussed previously, is caused by the lens. This will affect the tonal reproduction and dynamic range of the image. Since the f-number of the lens is variable for many cameras, this will have an influence on the image exposure.

Color   The spectral transmittance of the lens is typically wavelength dependent. Therefore, there is a possibility that the lens may affect the color and white balance of the image.

Geometrical Distortion   Optical distortion has its origin in the lens. Perspective distortion is also a purely optical/geometrical effect originating from lens properties.

Shading   Luminance shading is due to the $\cos^4$ effect as well as to lens vignetting. These effects are the predominant contributions to lens shading. Color shading may be due to the IR cut filter, explained further in Section 4.7. Also, the magnitude of the chief ray angle, which is a contributing factor to color shading, is determined by the lens design.

Sharpness and resolution   Obviously, the sharpness and resolution of an image is dependent to the largest extent on the lens.

Noise   Even though not directly related, the ability of the lens to provide light to the image sensor will affect the signal to noise ratio of the image. Furthermore, flare may also increase the photon shot noise leading to a lower SNR.

Texture Rendition   The ability to render texture is related to the image sharpness and resolution. Therefore, the lens is contributing to this attribute as well.

Color Fringing   Chromatic aberrations in the lens will give rise to color fringing.

Aliasing   Even though dependent on the relationship between the pixel count and presence of fine repetitive patterns in the scene, the sharpness and resolution of the lens will determine if aliasing artifacts will be present in the image. For instance, by defocusing the lens, aliasing artifacts can be made to disappear.

Ghosting   Ghosting is a purely optical phenomenon and therefore directly related to properties of the lens.

4.3   Image Sensor

In a digital camera, the image formed by the lens is projected onto the image sensor. For most applications, two main sensor types can be identified: Charge Coupled Device (CCD) and Complementary Metal Oxide Semiconductor (CMOS). In all cases, the role of the image sensor is to convert the optical image into a digital signal that can subsequently be processed by the image signal processor.

4.3.1   CCD Image Sensors

The CCD was invented in 1969 by Boyle and Smith (Boyle and Smith, 1970) for which they were awarded the Nobel Prize in 2009. Its main use was originally not intended to be for light detection, even though its use in such applications followed soon after.

The basic CCD was intended to transport electrical charges in a shift register-like fashion. The device is made up of an array of metal oxide semiconductor (MOS) capacitors, which are manufactured from a layer of silicon dioxide (SiO_2) on top of a p-type silicon substrate and above which metal electrodes have been evaporated (Wilson and Hawkes, 1997; Theuwissen, 1996), see Figure 4.12. By biasing the electrodes, or *gates*, with appropriate voltages and using correct timing, it is possible to move charges down the array and out of the circuit. Figure 4.13 shows one way of doing this, known as the three-phase scheme. In this type of device, every third capacitor is sharing the same voltage line. Thus, every group of three adjacent capacitors will be connected to different voltage lines. Let us first bias gate G_1 with the voltage V_b (and thus also G_4, G_7, etc.) and keep gates G_2 and G_3 at a low voltage (time t_1 in Figure 4.13). Next, if gate G_2 is biased with the same voltage (time t_2), the charge contained under gate G_1 will be shared with G_2. If in the next step we set the voltage at G_1 to zero, all charge will move to underneath G_2 (time t_3). By continuing in this fashion, we can read out the entire array. Note that in order for this to work, only the capacitors under gates G_1, G_4, and so on, should originally contain the charges we are interested in reading out. The only role of the other capacitors is to act as intermediate storage nodes.

As has been well known for a long time, semiconductors are light sensitive. When exposed to light, electron-hole pairs will be generated in the material. This is also true for the MOS capacitor. When a positive voltage is applied to it, the photogenerated electron-hole pairs separate and the electrons are trapped under the SiO_2 layer. The

Figure 4.12 Basic structure of a MOS capacitor.

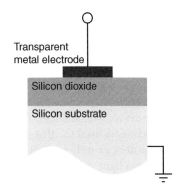

Transparent
metal electrode

Silicon dioxide

Silicon substrate

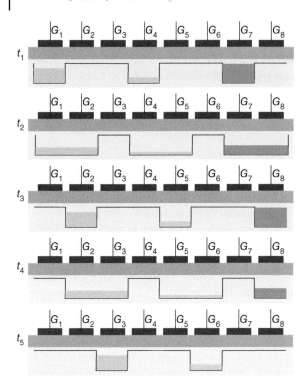

Figure 4.13 Three-phase readout scheme of a CCD, see text.

amount of charge will be proportional to the integrated light intensity. By connecting the ends of several MOS capacitor arrays to a horizontal CCD transport register and shielding gates G_2, G_3, G_5, G_6, and so on, from light, we have made a CCD area image sensor. In order to obtain an image from this device, the capacitors exposed to light, the *pixels*, are first reset by biasing them to some specified voltage. During the integration phase that follows, charge is accumulated inside the MOS capacitors until some predetermined time has passed whereafter the array is shielded from light by, for example, using a mechanical shutter. In the next phase, the charges are shifted out into the transport register, which carries them out of the sensor onto a capacitor and an amplifier, where the charge is converted to a voltage. This voltage can then be transformed into a digital number by an analog to digital converter (ADC).

Different variants of the CCD image sensor exist. The type described above, the full-frame CCD, is rarely in use since a new integration of light cannot be performed until all charge has been transferred out of the sensor. This requires a very fast readout and measurement of the charge packets, which is increasingly error-prone as the speed gets higher. A design that may be more appropriate in imaging applications is the *interline* CCD. In this device, each column consists of a light sensitive array and a transport register shielded from light, connected by a transfer gate. Typically, a *pinned photodiode* is used for this sensor, which for a long time was the predominant type for camcorders as well as digital still cameras. Since each column has its own transport register, a light shielding device such as a mechanical shutter is strictly not needed. The operation of the interline CCD is as follows. At the start of the integration, the entire array is reset. At the end of the integration, the transfer gates open and the charges in the photodiodes are transferred into the corresponding nodes of the vertical transport registers. At this point a new integration can start. The charges in the vertical transport registers are then shifted out of the sensor in the same way as for the full-frame CCD.

One of the disadvantages of the interline CCD is that a comparatively large portion of the pixel is taken up by the transport register, thereby reducing the *fill factor*, that is, the ratio of light sensitive area to total pixel area. This will reduce the light sensitivity and therefore decrease the signal to noise ratio of the sensor. A *frame transfer CCD* may be used to alleviate this problem. The transport registers are now moved outside of the light-exposed area and the image capture cycle starts with the reset of the entire array. After the integration time has passed, all the collected charge packets are shifted out into the storage area, whereafter the light sensitive area can start the next integration cycle while the charges in the storage area are shifted out of the sensor for further processing. With this scheme, no mechanical shutter is needed just as for the interline transfer CCD. The disadvantage of this solution is that the chip size grows, which may not be acceptable in some applications.

4.3.2 CMOS Image Sensors

The second main category of image sensors is the CMOS image sensor. The active pixel sensor (APS), which most modern CMOS sensors are based on, was invented in the early 1990s (Fossum, 1997; Ohta, 2008). In contrast to the CCD sensor, which requires a specialized production process, the CMOS sensor can be manufactured using standard CMOS processing used, for example, for digital memory chips. This brings with it many advantages like higher data rates, cheaper price, and the possibility to integrate logical circuits for image processing onto the same chip as the imaging array.

Most common today, as in the interline transfer CCD, is to use a photodiode for accumulation of photoelectrons. However, in contrast with the CCD, the pixel of the CMOS sensor consists of a number of components, as shown in Figure 4.14. This is a schematic of a typical four transistor active pixel sensor (4T APS) pixel, variants of which are very common in CMOS sensors on the market today. The basic operation of the pixel is as follows. At the start of integration, the floating diffusion node C_{FD} is reset by activating the Q_{RST} transistor. Immediately following this operation, the voltage of the floating diffusion is read out. Then, the charge that has accumulated in the photodiode is transferred to the floating diffusion by activating the transfer gate transistor Q_{TX}. As soon

Figure 4.14 Schematic of a 4T APS CMOS pixel.

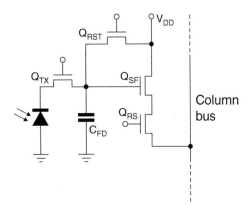

as the transfer gate is closed, the discharged photodiode can start accumulating new photoelectrons. The voltage across the floating diffusion, representing the integrated light signal, can now be read out and further processed.

The reason for reading out the floating diffusion twice is that the Q_{RST} transistor is not completely short circuited when active, and thus has a resistance. From the *fluctuation-dissipation theorem* of statistical mechanics (Reif, 1985), the dissipation of energy in a resistor is related to fluctuations in the voltage across it. The spectral density of the noise is proportional to the resistance and the bandwidth as well as the temperature. Calculating the mean square value of the charge across the floating diffusion capacitance, one finds that it is proportional to the product of the capacitance, temperature, and Boltzmann's constant. Therefore, this noise is usually referred to as *kTC noise*. Because of this, it is not possible to completely reset the floating diffusion and there will most probably be some charge remaining directly after reset. Since this charge will be different between resets, the result is that the signal read out from the pixel will be contaminated with noise. By reading out the floating diffusion directly after reset and storing that value in a memory circuit, it is possible to subtract this value from the signal transferred from the photodiode. In this way, this *reset noise* can be canceled. This scheme is known as *correlated double sampling* (CDS) and is employed not only in CMOS sensors, but also at the readout capacitor of CCD sensors.

The voltage across the floating diffusion is sensed by the source follower transistor Q_{SF}, which acts as a buffer. The connection with the readout circuits and the analog to digital converter is made through the row select transistor Q_{RS}.

In Figure 4.15, a graphical description of a CMOS image array is shown. The gates of the Q_{RS} transistors in each row will all be connected to the row select circuitry on the left side of the pixel array, allowing control over which row of pixels should be connected to the column amplifiers. At the top of the array, all the source follower outputs will be connected together through the row select transistors to the column amplifiers. In

Column amplifiers and CDS circuitry

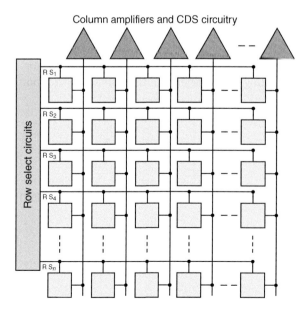

Figure 4.15 Graphical representation of a CMOS image sensor. Yellow boxes each represent one pixel.

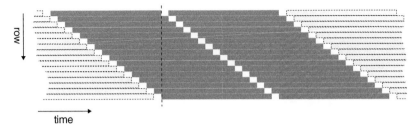

Figure 4.16 Timing diagram of the exposure of a rolling shutter CMOS sensor.

order to produce a digital number, the signals from the column amplifiers are presented to an analog to digital converter, which can be common to all column amplifiers or, to increase the conversion speed, replicated for each column.

The readout procedure of the CMOS image sensor is now performed by activating the row select signal for the row to be read, which connects the pixels of that row to the column amplifiers. The readout is a two-step procedure, where first the signal just after reset of the floating diffusion is read out and stored in a sample and hold circuit at the column level. Immediately after this, the photo signal is read out and the reset signal is subtracted, following which the result is converted to a digital number by the ADC and read out of the sensor.

Since the rows are read out in sequence, there will be a delay in time between the integration of each row that accumulates over the array. This is illustrated in Figure 4.16. This *rolling shutter* is a feature of the CMOS image sensor, which introduces distortion in images containing moving objects. The amount of distortion will be dependent on factors such as the frame rate (number of frames captured per second), and the speed and direction of the moving object. The "jello" impression seen in videos captured by mobile phones when the phone or object is moving is due to this phenomenon. An image example of the rolling shutter effect was shown in Figure 3.11.

It is possible to realize a *global electronic shutter*, where all pixels are exposed at the same time, in a CMOS sensor (Nakamura, 2005). However, at present this involves putting an analog memory inside each pixel. This will reduce the fill factor and thereby the sensitivity. Other effects may also occur, which causes a global shutter CMOS sensor to often have a considerably worse performance compared to the ordinary rolling shutter type. Since for most applications, the image quality loss due to a lower signal to noise ratio is judged to be worse than the effect of the rolling shutter, global shutter CMOS sensors have a limited market today. It is, however, possible to obtain a global shutter for CMOS sensors by using an external mechanical shutter. In this case, a *global reset* is employed. In practice this means that the mechanical shutter is first opened fully and all photodiodes are reset at the same time, whereafter the integration starts (see Figure 4.17). At the end of the integration period, the mechanical shutter is closed. The pixel rows can then be read out in sequence, just as in the rolling shutter case. Since no light will reach the pixels during readout, a correctly exposed image is obtained. One drawback of this scheme is that if the dark current (see below) of the pixels is sufficiently large, the image will contain a dark level offset that is changing across the image. This will also lead to a variation in dark noise across the image. In today's CMOS sensors, the dark current is usually so low that this is most likely not a problem.

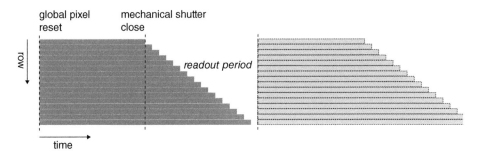

Figure 4.17 Timing diagram of the exposure of a global reset CMOS sensor with a mechanical shutter.

The CCD sensor has the advantage of having a global shutter "for free." For the interline CCD, all pixels are shifted into the transport register simultaneously. The frame transfer CCD will similarly allow for a global electronic shutter, since the charge packets can be shifted down into the light shielded area so fast that the accumulated extra charge for the last rows will be negligible compared to the first rows. However, a phenomenon known as *smear* may occur in this type of sensor when a strong light source is present in part of the image. When the charge packets are shifted down through the area illuminated by the light source, they may accumulate enough signal to produce a bright stripe extending vertically from top to bottom through the image of the light source. Interline CCDs can also suffer from smear, but in this case the origin is due to photons scattering into the vertical transport register.

Apart from eliminating the rolling shutter effect, another important application of a global shutter is in flash photography. In the case of a xenon flash, described in more detail below, the flash pulse is so short that the top and bottom rows will most likely not be exposed at the same time by the flash, leading to bright and dark bands in the image. The only situation when all rows are exposed at the same time occurs when the exposure time equals the inverse frame rate. Then, if the highest frame rate is, for example, 30 frames per second, the shortest achievable exposure time when using a flash is 1/30 second. For a mobile phone camera, with an f-number around 2.0–2.8 that cannot be changed, this means that it is only possible to use the flash when the scene is so dark that the background, not illuminated by the flash, would not risk being overexposed by the ambient light. In order to extend the usefulness of the flash, one would have to allow shorter exposure times. Therefore, a global mechanical or electronic shutter must be used.

As with the interline transfer CCD, the extra components inside the CMOS pixel will limit the fill factor. This can be remedied by adding *microlenses* on top of the pixel array. This technique is used both for CCD and CMOS sensors, and becomes increasingly important as the pixel size continues to shrink.

One important development within the image sensor field is *backside illumination* (BSI) for CMOS sensors. Even though it has been used for large sized CCD sensors for a long time, it has only very recently become available for small sized CMOS sensors. In order to counteract the inevitable performance degradation as the size of the pixels continues to shrink, the sensor chip is turned upside down and by means of etching, the rear face of the photodiode is exposed. In *frontside illuminated* (FSI) sensors, the photodiode is located at the bottom of a "well" made up of the layers of metal needed to connect the pixels with the column amplifiers, for the row select signals, and so on. As the pixel shrinks in size, this well gets narrower. The effect of this is that light from large angles from the normal will not reach the photodiode, or may leak into adjacent pixels. This will reduce the sensitivity considerably. By flipping the sensor upside down, the photodiode will be directly exposed to the light without any metal layers in between. The end result is a higher sensitivity and much improved angular acceptance. In principle, all mobile phones today use BSI image sensors.

4.3.3 Color Imaging

A typical CCD or CMOS pixel has a spectral sensitivity resembling the dashed line in Figure 4.18. If all pixels in an imager have the same spectral sensitivity like the one shown in this figure, color imaging is not possible. In order to accomplish this, at least three different types of pixels with spectral sensitivities resembling that of the human eye have to be implemented. Ideally, each photosite of the sensor would have pixels sensitive to all three or more types of spectral profiles. In reality, this is difficult to realize, even though the company Sigma (Sigma, 2015) has a line of cameras using the Foveon image sensor (Foveon, 2015) with stacked red, green, and blue pixels. Most other camera manufacturers are using other types of color filter arrays (CFAs), where each pixel has its own spectral sensitivity, resembling the colored graphs shown in Figure 4.18.

Differently colored pixels are spread out across the array in some kind of pattern. Dominating today is the *Bayer pattern* (Bayer, 1976), shown in Figure 4.19. Since different colors are sampled at different locations, interpolation has to be done in the raw image coming from the sensor in order to obtain three colors from each pixel. This *color interpolation* will introduce more or less severe artifacts, as discussed below and also in Chapter 3. As can be seen from the figure, the green color is sampled more densely

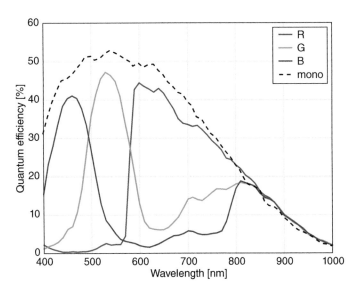

Figure 4.18 Spectral sensitivity of the ON Semiconductor KAC-12040 image sensor. Dashed lines: no color filters; full lines: with red, green, and blue color filters. *Source*: Reproduced with permission of SCILLC dba ON Semiconductor.

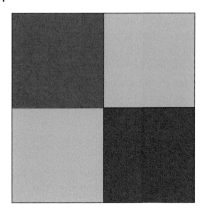

Figure 4.19 The Bayer pattern.

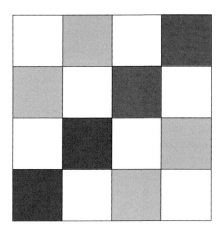

Figure 4.20 Alternative color filter array pattern including clear pixels.

(2 samples per group of 4 pixels, compared to one sample each of the red and blue colors). The reasoning behind this is that the green color channel mimics the human luminance channel, in which a large part of the detail information is carried in the medium visible wavelengths centered around 555 nm.

There have been many attempts to improve the Bayer pattern. A drawback of using colored filters in front of the pixels is that they attenuate the light reaching the pixel, thereby effectively decreasing the signal to noise ratio. In order to improve this, there have been suggestions to incorporate unfiltered, clear, pixels into the color filter array. As an example, a pattern introduced by Compton and Hamilton (2012) is shown in Figure 4.20.

4.3.4 Image Sensor Performance

Independent of the type of image sensor, CCD or CMOS, it is possible to describe the imaging performance of the sensor using the same set of parameters. Roughly, one can divide the parameters into four main groups: noise, artifacts, blurring, and light sensitivity. In the following, each category will be described separately.

Noise

Noise in image sensors can be classified in two main groups: *temporal* and *fixed pattern* noise (FPN). Temporal noise changes as a function of time, while fixed pattern noise does not change between image frames. While it is impossible to distinguish the two types in a still picture, they will show distinctly different behavior in a video sequence. In particular, excessive FPN will make the video appear as if viewed through a veil or a dirty window.

Noise may also be classified as being *structured* or *randomly distributed*. Both types can be either temporal or fixed. The structured noise usually appears as column or row noise.

A final classification can be made in terms of *signal dependent* or *signal independent* noise. Signal dependent noise will increase as the signal is increased, while signal independent noise will remain constant for all signal levels.

The noise types typically encountered in a CCD or CMOS sensor are summarized in Table 4.2. *Photon shot noise* is due to the fact that the electrons being generated in the pixel by the incident light are discrete events, having statistics following the Poisson distribution (Chatfield, 1983). Here, the probability $P(r)$ of encountering r events when the mean number of events is μ is given by

$$P(r) = \frac{e^{-\mu}\mu^r}{r!} \tag{4.9}$$

The variance of the Poisson distribution is equal to the average value, μ. Therefore, the standard deviation is equal to the square root of the mean value, that is, $\sigma = \sqrt{\mu}$. Thus, for an average amount of electrons of, say, 100 per second, individual pixels will receive anything between approximately 70 and 130 electrons per second. Consequently, light with some average intensity uniformly distributed across the image sensor area will not produce a uniform image, but instead the pixel values will fluctuate about some mean value, where the magnitude of the fluctuation is determined by the standard deviation. Thus, the standard deviation is a direct measure of the noise amplitude.

Table 4.2 Noise types typically encountered in CMOS and CCD sensors.

Noise type	Temporal/ Fixed	Structured/ Random	Signal Dependent/ Independent
Photon shot noise	T	R	D
Dark current shot noise	T	R	I
Dark signal nonuniformity (DSNU)	F	S/R	I
Column noise	T/F	S	I
Row noise	T	S	I
Photo response nonuniformity (PRNU)	F	R	D
Circuit noise (read noise)	T	R	I
Quantization noise	T	R	I

The light signal itself will therefore give rise to noise in the sensor image, which will increase as the square root of the mean pixel signal. This may seem counterintuitive, since noise is generally perceived to decrease as the signal is increased. However, the perceived noise in an image is related to the SNR, that is, the ratio of average signal value to the noise. In the case of photon shot noise, this gives

$$\text{SNR}_{\text{photon}} = \frac{\mu}{\sigma} = \frac{\mu}{\sqrt{\mu}} = \sqrt{\mu} \qquad (4.10)$$

Accordingly, the signal to noise ratio will increase as the light intensity, that is, average number of photons, is increased, and the increase is equal to the square root of the average photon count.

The *dark current shot noise* originates from fluctuations in the dark current of the pixel. Dark current arises because of a thermally induced flow of electrons (thermionic emission) in the pixel. Since the amount of accumulated charge is proportional to time, a dark signal will build up in the pixel. Because of the discrete nature of the electrons, the dark signal will not be the same in every pixel, but will follow a Poisson distribution, just as the photon shot noise. Thus, the dark current shot noise is equal to the square root of the dark current and will therefore grow with the integration time of the pixel. Since the dark current is a thermal phenomenon, it is strongly temperature dependent. The dark current is generally said to be doubled every 6 degrees centigrade, but this figure can differ substantially between different sensors, so it should be regarded more as a rule of thumb than an exact number.

The dark current generation may be different between pixels and this will show up as a fixed pattern noise usually referred to as *dark signal nonuniformity* (DSNU). For a CMOS sensor, differences in other components, whether at pixel or column level, may also give rise to nonuniformities. These may be random or structured, depending on whether their origins are in the pixel itself or in the column circuitry of the sensor. If originating from the column circuitry, the result will manifest as vertical stripes in the image.

Row noise could arise from fluctuations in, for example, power lines. If the column amplifiers or even the ADC of a CMOS sensor are disturbed by these fluctuations during readout, the signal in a whole row could be affected simultaneously, thereby generating noise showing up as horizontal stripes.

Variations in the light sensitivity between pixels will give rise to a fixed pattern noise usually called *photo response nonuniformity* (PRNU). This type of noise can be regarded as random changes in "gain" between pixels, and will therefore be proportional in magnitude to the signal. PRNU may occur in CCD as well as CMOS sensors.

Apart from the noise sources listed above, the electronic circuits in the sensor will generate other types of noise as well. We refer to these noise types as circuit noise or *read noise*. Examples of noise types in this category are $1/f$ (flicker) noise, phase noise, Johnson noise, and so on. For this discussion, it is quite sufficient to put all these different noise sources into one broad category.

The final source of noise is due to the analog to digital conversion. The digitization of a continuous signal will introduce errors, since the value being sampled is being approximated by an integer value. The error introduced is known as *quantization noise* and will depend on the number of bits in the ADC. It can be shown that the quantization noise voltage is given by (Reinhard *et al.*, 2008)

$$v_q = \frac{V_{\text{LSB}}}{\sqrt{12}} \qquad (4.11)$$

where V_{LSB} is the voltage step corresponding to a change in one digital code value, or one *least significant bit* (LSB). This also means that in the digital domain, the root mean square value of the quantization noise is equal to $1/\sqrt{12}$ LSB.

A very powerful tool to characterize the noise of an image sensor is to make a *photon transfer curve* (PTC) (Janesick, 2001, 2007). As can be noted from the discussion above, it should be possible to separate those noise sources that are different in terms of signal dependence by plotting the measured noise as a function of signal in a graph. We can find three different behaviors: either the noise is signal independent, or it is proportional to the square root of the signal (photon shot noise), or proportional to the signal itself (PRNU). By noting that these three types of behavior become linear functions with slope 0, 1/2, and 1, respectively, if their logarithm is calculated, it should be easy to separate them in a log-log plot, such as the one shown in Figure 4.21. In practice, a PTC is made by capturing raw images directly from the sensor, before any processing has been performed on the images. The sensor is illuminated by a uniform light source and image pairs are captured at different light levels, spanning from complete darkness up to saturation of the pixels. From the images, the average value from one frame and the standard deviation of the difference of the image pairs are calculated. Frequently only the temporal noise is analyzed in a photon transfer analysis, since the presence of PRNU can make it difficult to obtain a good estimate of the conversion factor. Since the temporal noise of two subsequent frames can be considered independent, the covariance of the temporal noise is zero, and consequently the standard deviation of the noise in the difference image has to be divided by $\sqrt{2}$.

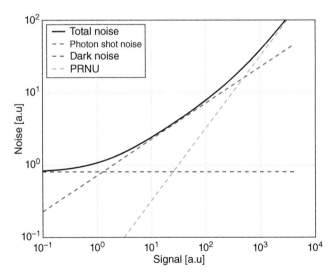

Figure 4.21 Example photon transfer curve. Three distinct regions can be distinguished: at low signal levels, signal independent dark noise dominates; at intermediate levels, the photon shot noise has the largest influence; at the highest levels, the photo response nonuniformity is the main noise source.

Once a PTC has been obtained, a few important performance figures can be extracted from the data. The values representing the raw image are proportional to the number of electrons in the pixel. To find the proportionality factor, or *conversion factor*, we observe that both the noise and the signal must have the same conversion factor. Furthermore, in the photon shot noise limited part of the PTC, the noise, N, in electron units is equal to the square root of the signal, S, in electrons. If we denote the conversion factor k (in units of digital numbers per electron), we then have

$$N' = kN = k\sqrt{S} = k\sqrt{\frac{S'}{k}} = \sqrt{kS'} \tag{4.12}$$

where N' and S' are the noise and signal, respectively, measured directly from the raw images. We can easily determine k by curve fitting. The signal value where the maximum noise value is obtained gives us the *full well capacity* of the pixel. This indicates the maximum number of electrons a pixel can hold before saturation and is an important factor when determining the dynamic range of the sensor.[2]

From difference images captured in complete darkness we can measure the temporal dark noise (e.g., dark current shot noise plus read noise). By dividing the full well capacity by the dark noise, we obtain a value for the *dynamic range* of the sensor. In effect, this metric tells us the ratio of the maximum possible signal to the minimum possible signal that the sensor can produce. This can therefore be regarded as a metric of the "span" of the sensor, that is, its ability to render highlights and lowlights at the same time. Note that dynamic range is often interpreted as the ability to capture highlights with full detail. However, it is important to point out that it is possible in most cases to adjust the exposure in such a way as to avoid blown out highlights. Depending on the dynamic range of the sensor, the shadows will then be more or less noisy. If the image is exposed correctly, the interpretation of dynamic range is then the ability to keep the noise so low that it is possible to distinguish details even in the deepest shadows (see Chapter 3 for image examples).

As discussed above, the signal to noise ratio is a metric that is better correlated to the image quality in terms of noise. Figure 4.22 shows a plot of the SNR as a function of signal level. The three regions identified in the PTC above will now show a different appearance, with SNR proportional to the signal in the lowest region, followed by a region where it is proportional to the square root of the signal (photon shot noise limited), and finally reaching a plateau of signal independent SNR (PRNU limited).

Artifacts

Apart from noise, the image captured by the sensor may be corrupted by other artifacts, such as smear, blooming, and defects. Also variations in the dark signal across the image can occur; this is usually known as *dark current shading*.

Smear mostly occurs in CCD sensors and appears as vertical stripes in the image. An explanation of this effect for an interline CCD sensor was given in Section 4.3.2.

Blooming may occur when the pixel saturates and charge spills over into adjacent pixels. This effect is also most frequently encountered in CCD sensors.

---

2 To be strict, the signal value at maximum noise level will tell us either the maximum number of electrons the pixel can handle or the number of electrons corresponding to ADC saturation. If the analog gain is increased, the full well capacity will decrease in units of electrons. However, the reason it decreases is due to the fact that the ADC starts saturating for a lower number of electrons.

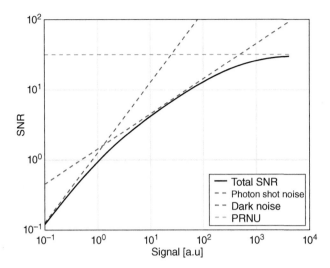

Figure 4.22 Signal to noise ratio of an image sensor as a function of signal value. Just as for the noise versus signal graph in Figure 4.21, three distinct regions can be distinguished in the graph.

The dark current in an image sensor is highly temperature dependent. As mentioned previously, a rule of thumb is that the dark current doubles with every increase of temperature of 6 degrees centigrade. Consequently, the dark current shot noise then doubles every 12 degrees. Since CMOS sensors offer the possibility of also including other circuitry on the sensor chip, problems with dark current shading are most notably found on these devices, since that circuitry may dissipate heat into the pixel array close to where it is located, thereby increasing the temperature and consequently the dark current locally. It should be noted, however, that any device that for some reason shows a temperature gradient across the imaging array may have this problem.

Another type of artifact is *flicker*, described in Section 3.2.6.3. This is associated with the sensor technology rather than the performance of a particular sensor. In CCD sensors using a global shutter, flicker will be difficult to identify in a single frame. In a video sequence, it might however be seen as variations in exposure between frames. In CMOS sensors using a rolling shutter, the flicker will typically show up as dark horizontal bands moving across the image. This will be noticeable also in still pictures. The cause of flicker is the fact that most light sources are intensity modulated by the power source. Some sources will be more afflicted than others. For instance, fluorescent tubes are more likely than incandescent bulbs to give this effect. If the frame rate of the sensor is not synchronized with the blinking of the light source, the result will be variations of intensity in the image. In the case of the CCD, the pixel integration is synchronized across the whole array, and consequently the whole image will fluctuate in intensity over time. For the CMOS sensor, pixel rows are exposed at different times and therefore the image values will show variations in both time and image array position. The remedy for flicker has traditionally been to make sure that the integration is performed in pace with the flickering light sources. This has worked comparatively well in the past with 50/60 Hz AC current powering the light sources. The LED sources that are now starting to appear on the market could however be modulated with just about any frequency, and this might prove to be a challenge moving forward, even though the higher frequencies typically used for these sources may make the problem less severe.

Some pixels in the sensor array may for different reasons not function properly. For instance, excessive dark current may cause the pixel to have a signal that is always large, independent of light level. It will therefore appear as a constant bright pixel in the image if not corrected. Very high dark currents in individual pixels may be due to process variations, but could also appear due to accumulated exposure to cosmic radiation resulting from, for example, airplane travel. Since a histogram of the dark current over the pixel array usually shows a long tail toward high numbers, it is not always easy to distinguish defective pixels and DSNU. Furthermore, some defective pixels might even turn on and off at irregular intervals—these are known as "blinkers." There might also be black or dark pixels in the array, that have partial or no response to light. These are usually referred to as "dead" pixels.

Blurring

Not immediately obvious is the fact that not only the lens may give rise to blurry images, but also the image sensor itself. The reason for this is twofold. First, the size of the pixel itself means that it cannot resolve arbitrarily small objects, but is limited to the actual pixel size. To connect with the discussion about the lens in the above sections, one can say that the point spread function of the pixel is given by the shape of the pixel. We will discuss this in more detail in Chapter 6.

The other effect leading to image blurring is *crosstalk* between pixels. This may be due to purely optical effects, where the light intended for one pixel is instead leaking into adjacent pixels. It may also be due to leakage of charge within one pixel into its neighbors. These two crosstalk contributions are often referred to as optical and electrical crosstalk, respectively. The result in both cases is that the image of a point source will be spread out over more than one pixel, thus producing a larger point spread function and as a result blurring of the image. As will be mentioned below, crosstalk can also have the unwanted effect of increased noise in color imagers. This is due to the nature of the color filter array, where each pixel has its own color. If there is crosstalk, there will be mixing of colors over adjacent pixels which will lead to more desaturated colors in the image. This has to be corrected by a more aggressive *color matrix*, which will amplify the noise.

Light Sensitivity

The sensitivity to light, that is, the ability to accumulate electrons for a certain light level, is dependent on the pixel size, the fill factor of the pixel, and the *quantum efficiency* (QE). The last is a measure of the amount of electrons generated out of each photon hitting the pixel, typically expressed as a percentage value. This is a wavelength-dependent quantity, since the photon energy is dependent on the wavelength. The graphs in Figure 4.18 show the QE for both a monochromatic and a color sensor. Obviously, a higher sensitivity means that a larger quantity of electrons is accumulated in a given amount of time.

In most digital still cameras, it is possible to adjust the ISO sensitivity. Changing the ISO sensitivity can be considered to be equivalent to changing the gain of the sensor, that is, the electrical amplification of the pixel signal. This will certainly not affect the amount of electrons produced by the pixel. Therefore, an increase in gain will not change the SNR, since it will amplify the signal and the noise by equal amounts. However, this is not strictly true, since the dark noise may actually increase more slowly than the signal as a function of gain. Why this is the case can be understood by considering this noise as composed of two noise sources, one located before and one after the analog gain stage of the sensor. The first noise source, n_1, will then be amplified by the gain stage with a factor g, while the second noise source, n_2, will be unaffected by the amplification of the gain stage. Assuming the two noise sources are uncorrelated, the total dark noise may be written:

$$n_{tot} = \sqrt{g^2 n_1^2 + n_2^2} \tag{4.13}$$

By dividing the total dark noise by the gain, we can more clearly see what happens:

$$\frac{n_{tot}}{g} = \sqrt{n_1^2 + \frac{n_2^2}{g^2}} \tag{4.14}$$

Therefore, as the gain is increased, the influence of the noise source after the gain stage will become increasingly less. This can be understood intuitively, since by increasing the gain, both the signal from the pixel as well as n_1 will increase, but n_2 will remain constant. Thus, at a sufficiently high gain, the dark noise will be completely dominated by the n_1 term. Therefore, for very low light levels (or short exposure times or large f-numbers), it may be advantageous to increase the analog gain by some appropriate amount. To determine the appropriate gain level, one should observe that an increase of gain will lead to a lower apparent full well capacity, since the ADC will saturate for a smaller amount of electrons. This will hamper the dynamic range, so excessive gain levels should be avoided. At the same time, it is usually found that the dark noise drops off and reaches values close to n_1 rather quickly as the gain increases, which means that even moderate increases of gain can give substantial SNR benefits.

A final remark concerning sensitivity can be made regarding the pixel size and the dynamic range. One often encounters the statement that small pixels have a low dynamic range due to faster saturation since the full well capacity is lower. However, a smaller pixel also means that the light sensitivity decreases. All other parameters being constant, both the light sensitivity and the full well capacity should scale in the same way with pixel area. Therefore, the pixel will fill up at the same rate independent of the size of the pixel. In reality, this may of course not be the case exactly, but is still true within reasonable bounds. The accurate way of describing the impact on dynamic range is instead in terms of SNR: a smaller capacity pixel will have a smaller number of electrons at its disposal compared to a pixel with higher full well capacity, which will lead to a lower SNR across the whole range, especially at low light levels. Consequently, the range over which objects can be distinguished will inevitably be less in such a situation, and thus the dynamic range will be lower.

4.3.5 CCD versus CMOS

For many years, CCD sensors were considered superior in terms of image quality compared to CMOS sensors. Some artifacts that are present in CMOS sensors are simply not found in CCD sensors. The fact that charges are read out from the CCD array in a, for all practical purposes, lossless fashion and then converted to a voltage and amplified off-chip means that nonuniformities are generally much lower in CCD sensors. On the other hand, the special fabrication process needed for CCD sensors combined with the fact that they need peripheral circuits to operate and also have a higher power consumption due to the readout mechanism, means that the cost of a comparable CMOS device is considerably less. Also, it is substantially more difficult to accomplish a high readout speed in a CCD sensor compared to CMOS. The issues with nonuniformities can in principle be remedied by calibration, and the possibility of including image processing on-chip in a CMOS sensor means that it is possible to perform this calibration even before the image leaves the sensor.

Putting all these factors together, the CMOS image sensor is today capable of replacing the CCD sensor in the majority of applications, and has more or less already displaced CCD sensors completely in consumer imaging, even at the high end.

4.3.6 Image Quality Attributes Related to the Image Sensor

This section provides a summary of how the image quality attributes discussed in Chapter 3 relate to the image sensor.

Exposure, Tonal Reproduction, and Flare   The light sensitivity of the image sensor will most certainly influence the amount of light needed to obtain a well exposed image. Also, the ability to adjust the exposure time by means of an electronic shutter will affect the image exposure.

Color   In order to obtain a color image, the image sensor must be able to discriminate between at least three colors. This is readily achieved by putting color filters on top of the pixels. The quality of these filters will have a great impact on the color rendering.

Shading   As discussed above, some sensors may suffer from dark current shading. Therefore, shading seen in images could potentially have its origin in the image sensor.

Sharpness and resolution   Crosstalk and size of the pixel will influence the sharpness and resolution in the final image.

Noise   The major noise sources all arise in the image sensor, making this component the dominant source of image noise.

Texture Rendition   On this level, the rendering of texture is dependent on the sharpness, and therefore the image sensor may have an impact on this attribute.

Color Fringing   This attribute may be partly due to the image sensor, especially if a CCD sensor with blooming issues is used.

Aliasing   The number of pixels in the sensor will certainly have a large impact on aliasing artifacts in the image.

Image Defects   Defects in terms of dead and hot pixels arise solely from the image sensor. Also, dust falling on the sensor surface or cover glass may be considered to be contributions from the sensor to this attribute.

Flicker   In case of a CMOS sensor, flickering has its sole origin in the image sensor.

4.4   Image Signal Processor

The raw image data provided by the image sensor is far from presentable to a user as a pleasing still image or video. Therefore, it has to be processed before further use. Furthermore, in order to guarantee an image with high quality, some type of control logic must take care of adjusting parameters such as exposure and white balance. In addition, the processed image or video usually takes up so much space that it must be compressed. The *image signal processor* (ISP) is responsible for all these tasks.

4.4.1   Image Processing

In order to obtain a usable image from the raw data provided by the image sensor, at least the following steps must be performed:

- Black level correction
- White balance
- Color interpolation
- Color correction
- Gamma curve application

Black level correction is necessary to remove any offset present in the image data. As explained above, the dark current in the image sensor pixels will produce a signal having a level dependent on exposure time and temperature. This black level must be subtracted from the data. However, since image data is represented by digital numbers from zero up to the maximum allowable value set by the ADC, a signal close to zero will risk clipping the noise, which can lead to incorrect estimations of average signals as well as noise. Therefore, a fixed black level is usually added to the image values that are output from the sensor. If this black level is not subtracted from the data, signal level dependent color errors will be introduced. This is shown in Figure 4.23, comparing images with and without black level subtraction before the color correction matrix has been applied. Apart from this, having an offset present in the image data will make the whole image look "hazy." In order to find a reliable estimate of the dark signal, pixels shielded from light are added at the periphery of the sensor array. The dark signal can be obtained by reading out the signal from those pixels together with the signal from the active array (the part of the sensor that is exposed to light), thereby making sure that both exposure time and temperature are equal or close to that of the imaging pixels.

The human visual system has a remarkable ability to distinguish colors under a very wide range of illuminations. This is known as *color constancy* (Ebner, 2007). Since the camera is trying to emulate human vision, it also has to be able to account for this behavior by performing *white balancing*. The actual operation of white balancing is straightforward: it merely amounts to multiplying each individual color channel by a fixed value, or *white balance gain*. However, determining the gain values for each color channel is by no means trivial and many algorithms exist to perform this task. The simplest one, which still works surprisingly well, is the *gray-world* algorithm. This basically assumes that the average image values in a scene are gray. Therefore, the individual white balance gains are found by calculating the average red, green, and blue values of the image, and then for each color channel calculating the ratio of the average green value to the average red and blue values. This leads to a white balance gain value of the green

Figure 4.23 Illustration of color error introduced by having an incorrect black level correction. Left: black level subtracted before color correction; right: no black level subtraction before color correction.

channel of 1. The reason for this choice is that the majority of the luminance information is considered to be carried by the green channel. Therefore, this definition will lead to the least modification of the overall image level. Figure 4.25 shows an image where white balancing has been switched on and off.

The next step in the processing of the raw image is to perform *color interpolation*. Since each pixel only contains information about one color (red, green, or blue in the case of the Bayer pattern), the missing colors have to be calculated for each pixel. This is done through interpolation, where information from adjacent pixels is used to find the missing values. One of the simplest types of interpolation is *bilinear* interpolation, where the average of the surrounding pixel color values are used to find the missing information for a particular pixel. Aided by Figure 4.24, the algorithms for calculating the missing values at position (m, n) for red at green pixels in a blue row, R^{G1}, red at green pixels in a red row, R^{G2}, red at blue pixels, R^B, green at red pixels, G^R, etc., become:

$$R^{G1}_{m,n} = \frac{R_{m-1,n} + R_{m+1,n}}{2}$$

$$R^{G2}_{m,n} = \frac{R_{m,n-1} + R_{m,n+1}}{2}$$

$$R^{B}_{m,n} = \frac{R_{m-1,n-1} + R_{m+1,n-1} + R_{m-1,n+1} + R_{m+1,n+1}}{4}$$

$$B^{G1}_{m,n} = \frac{B_{m,n-1} + B_{m,n+1}}{2}$$

$$B_{m,n}^{G2} = \frac{B_{m-1,n} + B_{m+1,n}}{2}$$

$$B_{m,n}^{R} = \frac{B_{m-1,n-1} + B_{m+1,n-1} + B_{m-1,n+1} + B_{m+1,n+1}}{4}$$

$$G_{m,n}^{R} = \frac{G_{m-1,n} + G_{m+1,n} + G_{m,n-1} + G_{m,n+1}}{4}$$

$$G_{m,n}^{B} = \frac{G_{m-1,n} + G_{m+1,n} + G_{m,n-1} + G_{m,n+1}}{4} \qquad (4.15)$$

As discussed in Chapter 3, the color interpolation algorithm may produce several kinds of artifacts. For instance, the simple bilinear algorithm described here will give images that are blurrier than necessary and with "zipper" artifacts across edges, which are quite noticeable. In order to cope with such artifacts, adaptive algorithms are used where, for instance, the direction of interpolation is determined based on the local contrast of the image. For an overview of the state of the art in this field, the reader is referred to review papers on the subject (Li *et al.*, 2008; Menon and Calvagno, 2011).

The spectral sensitivities of the image sensor deviate from human vision. Therefore, images coming straight out of a sensor will not show correct colors and some variety of *color correction* is needed. A simple way to perform this correction is by way of multiplying the sensor RGB (red/green/blue) values by a 3×3 *color correction matrix* (CCM). This implies that there is a linear relationship between the image sensor spectral sensitivities and the corresponding human sensitivity. If this is the case, the *Luther–Ives condition* (Reinhard *et al.*, 2008) is satisfied. In reality, this condition is never completely met, which means that it will not be possible to obtain a perfect match between the color vision of the human visual system and the camera. Other color correction methods exist, which can take into account nonlinearities. Such methods may use

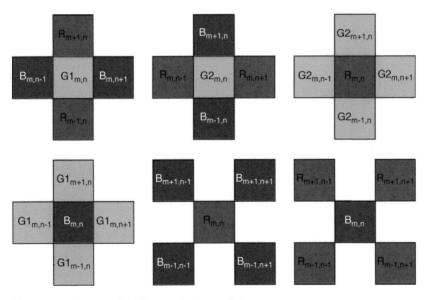

Figure 4.24 Geometry for bilinear color interpolation.

Figure 4.25 Example of white balancing and color correction. Left: no white balance; middle: white balanced; right: white balanced and color corrected.

either nonlinear matrices or three-dimensional lookup tables (3DLUTs). An example of a corrected and an uncorrected image from a camera using a linear CCM is shown in Figure 4.25.

As a last step to obtaining an acceptable image, a *gamma correction* has to be performed. The cathode ray tube (CRT), commonly used as a display during the last part of the 20th century, had a nonlinear relationship between the voltage that had to be applied to generate an image and the light intensity of the phosphor pixels. This nonlinearity is approximately given by a power law with an exponent γ, typically found to be in the range 2.0–2.4. In order to correct for this, an inverse gamma correction has to be applied to the image data. Even though more modern screens, using, for example, liquid crystals, do not show the same behavior, the gamma correction is still applied when encoding images for display. As a matter of fact, the gamma encoding has been standardized and is used with most RGB color spaces, but in slightly different shapes. The most common encoding is the one used by the sRGB standard (Poynton, 2003; IEC, 1999), where the input, X_{in}, is transformed to the output, X_{out} according to

$$X_{out} = \begin{cases} 1.055X_{in}^{1/2.4} - 0.055 & X_{in} > 0.0031308 \\ 12.92X_{in} & X_{in} \leq 0.0031308 \end{cases} \qquad (4.16)$$

One advantage of applying a gamma curve to the image data is that since usually the data is quantized down to 8 bits in this step from 10 or maybe even 12 or 14 bits, the perceived quantization errors will be less because of the compression of the data in the highlights and expansion in the lowlights.

Additionally, modern ISPs typically include the following processing steps:

- Defective pixel correction
- Shading correction
- Noise filtering
- Sharpening

- Global and/or local tone mapping
- Resizing
- Compression

As discussed earlier, the pixels in an image sensor will not all be fully functioning. Some may have excessively large dark currents, which in practice will mean that they get "stuck" at large values. Others may instead be "dead," producing no signal at all. Since such pixels will be very noticeable in an image, they must be removed. This can be done either by dynamic methods, where, for example, variants of a median filter can be used, or statically by detecting the defect pixels in production and storing their locations in a lookup table. The pixel values at the stored locations will then be replaced by the average of their surrounding pixels before being presented to the user.

As discussed previously in this chapter, as well as in Chapter 3, the lens in a camera will typically show some light falloff toward the corners of the image. In order to compensate for this, many ISPs will employ some kind of shading correction. For this to work, the camera must be calibrated. It is usually not enough to compensate only for intensity variations over the image. As also mentioned previously, color shading may occur in the image as well. In order to remove this effect, separate correction needs to be applied for the individual color planes.

The signal to noise ratio in an image is affected by several parameters, such as the noise level of the image sensor, the f-number of the lens, and also the integration time. In order to improve the image quality, *noise reduction* is usually employed. A simple approach to filter out noise in an image is simply to use a spatial lowpass filter. An example of such a filter is a moving average, where the pixel to be modified gets its original value replaced by the average of its surrounding pixels. The larger the surround, the more filtering. This is an example of a *linear filter* using shift-invariant convolution, which in its discrete form can be expressed mathematically as:

$$I_{i,j}^{out} = \frac{\sum\limits_{m=-M}^{M} \sum\limits_{n=-N}^{N} F_{m,n} I_{i-m,j-n}^{in}}{\sum\limits_{m=-M}^{M} \sum\limits_{n=-N}^{N} F_{m,n}} \quad (4.17)$$

Here, $I_{i,j}^{in}$ and $I_{i,j}^{out}$ are the input and output image values at position (i,j), and $F_{m,n}$ the filter function or convolution kernel. In our simple example with the moving average, all its values are the same. More sophisticated kernels could certainly be used, with the most common example being the Gaussian function

$$F_{m,n} = e^{-\frac{m^2+n^2}{w}} \quad (4.18)$$

where w is a width parameter. The side effect of this filter is that the image in most cases will become unacceptably blurry. To improve this, an adaptive filter that tries to avoid blurring edges is used. An example of such a filter is the *sigma filter* (Lee, 1983). Here, the pixel to be modified is replaced by the average of those surrounding pixels that have a value that deviates from the central pixel by no more than some value σ. This gives rise to a considerably sharper image than the simple averaging filter. Figure 4.26 shows examples of the averaging filter and the sigma filter. Even though applying the sigma filter does result in a sharper image, there is a price to be paid. In Figure 4.26 the

Figure 4.26 Example of noise filtering. Left: original image; middle: linear filter; right: sigma filter. Note how sharp edges are retained for the sigma filter, while low contrast texture is smeared out.

sharpness benefit of the sigma filter is clearly visible. However, one can also see local blurring occurring in low contrast textured areas. This gives an overall "oil painting" impression.

The sigma filter is related to a type of filter known as the *bilateral filter* (Reinhard *et al.*, 2006; Tomasi and Manduchi, 1998), named in this way because it takes into account both spatial and intensity information in the calculation. This type of filter has become quite popular in image processing and is found in many different applications. Mathematically, this filter can be expressed in the following way:

$$
I_{i,j}^{\text{out}} = \frac{\sum_{m=-M}^{M} \sum_{n=-N}^{N} F_{m,n} G\left(I_{i-m,j-n}^{\text{in}} - I_{i,j}^{\text{in}}\right) I_{i-m,j-n}^{\text{in}}}{\sum_{m=-M}^{M} \sum_{n=-N}^{N} F_{m,n} G\left(I_{i-m,j-n}^{\text{in}} - I_{i,j}^{\text{in}}\right)}
\tag{4.19}
$$

where, in comparison with Eq. (4.17), a kernel dependent on the intensity information, G, has been added.

Many cameras, usually not camera phones, however, use an *optical lowpass filter* in order to avoid aliasing effects, which could give rise to moiré patterns in the image. This will make the image appear blurrier. Also, both the pixels themselves and the lens will affect the sharpness negatively. In order to compensate for this, a sharpening filter is used. This is in essence a high pass filter that will increase the contrast for intermediate and high spatial frequencies. A popular variant is the *unsharp mask* filter, which basically works by subtracting a lowpass filtered version of the image from the original image (Sonka *et al.*, 1999). The result of such an operation is shown in Figure 4.27. Here one can also see some typical artifacts coming from the sharpening. The sharpening process enhances edges, which gives rise to the typical "halos" that are clearly seen around the window frames in Figure 4.27.

The gamma correction that was discussed above is a simple implementation of a more general operation known as *tone mapping*. This is a technique to change the contrast

Figure 4.27 Example of unsharp masking. Left: no sharpening; right: unsharp masking applied. The two bottom images show crops of the images above. Note how the right image appears significantly sharper, but also introduces "halos" around edges.

in the image according to some preferential criteria. A *global tone mapping* applies the same change to the entire image, while a *local tone mapping* uses different contrast modifications depending on the image content (Reinhard *et al.*, 2006). In high dynamic range (HDR) imaging, tone mapping is a very important step in the processing of images in order to compress the usually very wide tonal range of the input signal to a range that is suitable for display.

The image may also be resized in order to fit the application at hand. This could be done either by cropping or scaling the image. The first case involves just selecting pixels within some region of interest, thus producing a smaller image, while the latter includes an interpolation step to either enlarge or reduce the image. Depending on the type of interpolation used, the image may show more or less artifacts in terms of aliasing or blur.

4.4.2   Image Compression

Contributed by Hugh Denman

As a final step in the image processing, the output image is usually compressed in order not to take up an unnecessarily large amount of disk space, or to consume too much bandwidth when distributed over a network. The size of the image file for a given image produced by a camera is directly proportional to the number of pixels in the image. For an 8 megapixel camera with one byte per pixel and color channel, this amounts to a file size of 24 megabytes. Clearly, this can quickly fill up storage space, and a compression scheme is therefore usually needed to bring file sizes down to manageable proportions.

As mentioned in Chapter 3, the most commonly used image compression algorithm currently in use is JPEG (Joint Photographic Experts Group). This is a *lossy* compression, in that it will destroy information in the original file, which cannot be retrieved upon decompression. However, the image quality degradation due to compression is in most cases negligible, and the possibility to obtain a high compression ratio without too noticeable image quality loss has made this scheme highly successful. The steps involved in JPEG compression are *chroma subsampling*, *transform coding*, and *entropy coding*, as explained in the sections below.

4.4.2.1   Chroma Subsampling
As shown in Figure 7.2, the chrominance information of an image can be substantially blurred without noticeable quality degradation. This can be exploited in order to reduce the image file size. In JPEG compression, this is accomplished by transforming the RGB image to the YCbCr[3] color space and then subsampling the Cb and Cr channels by a factor 2 in the horizontal direction (referred to as 4:2:2), or in both directions (4:2:0). Consequently, this reduces the data size by a factor $1/3$ or $1/2$, respectively. The conversion between RGB and YCbCr is typically done using the transformation

$$\begin{pmatrix} Y \\ Cb \\ Cr \end{pmatrix} = \begin{pmatrix} 0.299 & 0.587 & 0.114 & 0 \\ -0.168736 & -0.331264 & 0.5 & 128 \\ 0.5 & -0.418688 & -0.081312 & 128 \end{pmatrix} \begin{pmatrix} R \\ G \\ B \\ 1 \end{pmatrix} \qquad (4.20)$$

4.4.2.2   Transform Coding
The natural basis for digital images is a spatial grid of intensity samples. This is intuitive and well-suited for display, but for analysis and compression, a spatial frequency basis is often more efficient. This is because in real world images, the spatial distribution of intensity values does not, in general, have particular statistical properties, but real world images do have exploitable spatial frequency characteristics. The frequency spectrum

---

3 YCbCr is a color space frequently used in digital image processing. The elements are luma (Y), blue difference (Cb) and red difference (Cr).

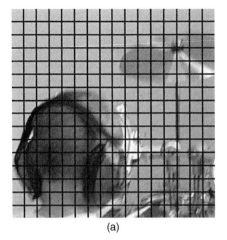
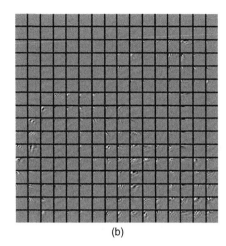

(a)   (b)

Figure 4.28 An example of the blockwise Discrete Cosine Transform. (a) Image blocks in the spatial domain. (b) Image blocks in the frequency domain (via DCT).

of natural images follows a power-law distribution: most of the energy is at low spatial frequencies.

The best-known frequency basis for signal analysis is the *Discrete Fourier Transform* (DFT). The DFT decomposes a signal into a fixed set of sinusoidal components, where each component is parameterized by its amplitude and phase. This basis is often used for analysis and filtering, but for compression the *Discrete Cosine Transform* (DCT) is preferred. The DCT is cheaper to compute than the DFT, as it does not explicitly represent the phase of the sinusoidal components, but this is not a limitation for encoding. Figure 4.28a shows a sample image divided into blocks of 16 pixels on a side, and figure 4.28b shows the corresponding Discrete Cosine Transform of each block. Some correspondences are immediately apparent. A flat block in the spatial domain corresponds to a flat DCT block; in the absence of noise, such a block can be coded using a single DCT coefficient (the only DCT component present is the 0-cycles-per-pixel, or "DC," component). Where blocks contain texture or edges, the energy in the DCT is concentrated at lower spatial frequencies (located toward the top-left corner of the DCT block). This is in accordance with the power-law tendency described above.

4.4.2.3 Coefficient Quantization

The perception of visual detail depends on the frequency domain characteristics of the visual stimulus. The great advantage of transform coding is that knowledge of how spatial frequencies are perceived enables a scheme to be adopted for discarding some of the signal with predictable, progressive impact on visual quality.

In particular, the human visual system is not highly sensitive to the exact amplitudes of the frequency components in a signal, and so the DCT coefficients can be *quantized*, that is, some precision can be discarded. This is described as the *quantization parameter*; for instance, a quantization parameter of 10 would indicate rounding of each coefficient to the nearest multiple of 10. Moderate levels of quantization can be applied without introducing any visual degradation of the signal.

The human visual system is more sensitive to amplitude changes at lower spatial frequencies. This means that the DCT coefficients for higher spatial frequencies can be quantized more coarsely than those for lower spatial frequencies. Thus, rather than using a fixed quantization parameter, image coding schemes use a *quantization matrix*—effectively a separate quantization parameter for each frequency component.

Furthermore, the human visual system is susceptible to *spatial masking*: a frequency component that would be perceptible on its own can be rendered imperceptible, or masked, by a stronger frequency component in the same region. Thus, the quantization matrix can adapt to take into account the range of frequency component amplitudes, or, equivalently, the variance of the intensity values in the spatial domain. Low-amplitude components can be quantized more coarsely if high-amplitude components are present in the block.

This quantization of DCT coefficients is the only point at which data is lost in still image and video compression schemes, and it greatly improves the compressibility of the coefficients. For example, in many blocks the high frequency coefficients represent only low-amplitude noise, and after quantization these will be set to zero.

For the interested reader who wants to learn more, the classic reference for quantization-based signal compression remains Gersho and Gray (1991).

4.4.2.4 Coefficient Compression

Once quantized, the transform coefficients are losslessly compressed; it is at this point that the reduction in data rate is attained by the compression.

The coefficients are usually represented as a two-dimensional matrix, with the lowest spatial frequencies in the top-left corner and the highest frequencies in the bottom-right corner. The elements of this matrix are scanned for compression along the diagonals, in a zig-zag pattern from top-left to bottom right. This scan order maximizes the likelihood of runs of zeros, as the coefficients for higher spatial frequencies are often zero-valued. Longer runs improve the compressibility of the data.

Coefficient compression, then, involves transmitting symbols identifying the sequence of [run-length, value] pairs describing each block, in diagonal-scan order. The earliest coefficient compression scheme defined a fixed set of symbols, for a fixed set of pairs of run-length and coefficient value. Modern codecs use more complex, and more efficient, *entropy coding* compression methods.

Entropy coding algorithms maintain a model for expected symbol frequencies, or symbol probabilities, and try to ensure that more frequent symbols are encoded using fewer bits than less frequent symbols. The compression becomes less efficient if these symbol frequency statistics evolve away from the model over the data stream. This can be countered by dividing the stream of data into blocks and encoding each separately (where each block has its own statistical model), or by using *adaptive entropy coding*, in which the probability model is continuously updated by both the encoder and the decoder as each symbol is processed.

Assuming that the symbols are statistically independent, the appropriate model is just the probability of each symbol, p_i for symbol i in our alphabet, with $\sum_{i=1}^{N} p_i = 1$, where there are N symbols in the alphabet overall. The *entropy* according to this model is then

$$H = -\sum_{i=1}^{N} p_i \log_2(p_i)$$

measured in bits per symbol. This is the best attainable compressed rate for data following this model, that is, this rate is attained if symbol i is encoded using $-\log_2(p_i)$ bits.

The best-known entropy coding technique is *Huffman coding*. Here, each symbol is encoded using an integer number of bits. In consequence, this scheme can attain the optimal rate only if the symbol probabilities happen to be negative powers of two. *Arithmetic coding* is a more general scheme in which symbols can, in effect, be encoded with a non-integer number of bits. This can achieve close-to-optimal results, but is considerably more complex than Huffman coding.

Because coefficients are losslessy compressed, this step does not directly affect the image quality; therefore, no particular quality deficiency or coding artifact can be attributed to the coefficient compression algorithm. However, more sophisticated coefficient compression algorithms do contribute to better image quality at a given compression ratio or data rate.

4.4.3 Control Algorithms

In order to obtain a well-exposed image with correct white balance and focus, an ISP also needs to incorporate control algorithms for:

- Auto exposure
- Auto white balance
- Autofocus

These algorithms are at times referred to as the *3A algorithms*. The auto exposure algorithm tracks the light level in the scene and controls the exposure time, gain, and aperture so that an appropriate exposure level is maintained. The auto white balance algorithm gathers statistics of the colors of the scene, and this data is used to estimate a proper *white point*, which should ideally correspond to the color of the light source illuminating the scene. If the camera is equipped with an autofocus lens, this is controlled by algorithms that analyze the sharpness in the image and gather statistics during successive frames and are used to make decisions about which direction the lens should move and when optimum focus is found so that it can be stopped in the correct position.

In some cameras, automatic image stabilization is also used, where either the image sensor or the lens (or individual lens elements) is moved in response to camera movements. Control logic for this may also be implemented in the ISP. In Section 4.6.1, video stabilization is described. Since image stabilization for still and video capture is in principle very similar, the concepts described there can be applied equally to the still capture case.

4.4.4 Image Quality Attributes Related to the ISP

As may have become evident by reading this chapter, the ISP is responsible for so much processing of the original raw image that all attributes discussed up to this point will be affected in some way or the other. Some examples for each of these attributes are given below.

Exposure, Tonal Reproduction, and Flare   The exposure is controlled by the 3A algorithms in the ISP. Therefore, the ISP most certainly has a great impact on the tonal reproduction. Furthermore, the ISP may incorporate algorithms for flare compensation. Depending on how successful such algorithms are, flare may or may not be present in images and may also be overcompensated, leading to other artifacts.

Color Since the ISP is responsible for the color correction, the end result is critically dependent on the performance of the ISP.

Geometrical Distortion In some ISPs, there may be algorithms for correcting geometrical distortion. Once again, the end result will be depending on the effectiveness of these algorithms.

Shading The ISP will in most cases include shading correction. If not performed properly, the end result may even turn out worse than if no shading correction had been used at all.

Sharpness and resolution The sharpening and noise reduction filters in the ISP, as well as image compression, will affect the sharpness of the image.

Noise The noise reduction filters together with the image compression in the ISP will typically reduce the image noise considerably.

Texture Rendition A side effect of the edge preserving noise reduction filtering that is generally employed is the introduction of texture blur. Also the image compression may contribute to texture loss.

Color Fringing ISPs may include algorithms for reducing color fringing.

Aliasing The color interpolation in the ISP may introduce aliasing artifacts.

Image Defects The correction of image defects is to some part usually performed in the ISP. Therefore, the visibility of those defects is directly dependent on the performance of the ISP.

Ghosting Ghosting is a purely optical phenomenon, and as such therefore not directly related to the ISP. However, algorithms for, for example, tone mapping may make these artifacts more visible.

Flicker The standard way of avoiding flicker is to set the exposure time to a multiple of the frequency of the flickering light. Since the ISP is responsible for the exposure, it will indirectly affect this artifact as well.

4.5 Illumination

In low light situations, the camera may not be able to deliver an acceptable image. In such cases, some kind of illumination may be used to provide the scene with more light during the moment of exposure. For still images, a flash has been used since the early days of photography. Today, there are mainly two different types of flash encountered in a camera, *LED flash* and *xenon flash*. In mobile phones, the most commonly used flash is the LED based variant, while digital still cameras as well as DSLRs most commonly use a xenon flash.

4.5.1 LED Flash

Light emitting diodes (LEDs) have developed substantially during recent years and are now so powerful and efficient that they can be used in more demanding applications. Advantages of using these devices are that they are small, run at low voltages, and do not require much control circuitry. This is especially important in mobile phones where space is of the utmost concern. Disadvantages are that they do not provide as much energy in as short time as a xenon flash, which limits their range and increases the risk of motion blur in images.

4.5.2 Xenon Flash

A xenon flash is based on a discharge tube connected to high-voltage capacitors (Allen and Triantaphillidou, 2009). It is able to deliver a very short, intense light pulse capable of illuminating scene objects at comparatively large distances. Since the light pulse is so short (on the order 10 to 100 microseconds), it also becomes possible to freeze motion for fast-moving objects.

The range of a xenon flash is expressed by its *guide number*. This is given by the following formula:

$$\text{Guide number} = \text{range} \times \text{f-number} \tag{4.21}$$

As we have seen previously, the intensity of light falls off as a function of inverse squared distance from the light source. Since the guide number is proportional to the range of the flash, this must mean that it is proportional to the square root of the intensity of the flash. Disadvantages of using xenon flash are that they take up more space and that it is necessary to handle high voltages (several hundreds of volts). It is also more difficult to implement in a camera using a rolling shutter (in practice, most CMOS sensor based cameras today). Since the flash has a very short duration, all pixels of the sensor must be integrating simultaneously in order to register the scene when it is illuminated by the flash. Therefore, a CMOS camera most likely needs a mechanical shutter, which further increases the cost of the product.

4.6 Video Processing

Contributed by Hugh Denman

Video capture by digital cameras involves significant additional complexity in the ISP. In particular, the very high raw data rate of video means that the video must be compressed as it is captured. In consumer cameras, this compression will be highly lossy, however, the compression algorithms are sufficiently sophisticated that this lost data will be imperceptible in most circumstances. Video capture also requires that image quality be sustained over time as well as spatially. These video-specific image processing aspects are described here.

4.6.1 Video Stabilization

Video captured using hand-held cameras often exhibits *shake*, where small movements of the camera result in the scene displacing rapidly, vertically and horizontally, from frame to frame. This phenomenon is especially pronounced when the camera is operated with a telephoto lens, or when a digital zoom is applied.

Shake is aesthetically undesirable in most contexts, and also makes the video more difficult to compress. Shake is avoided in professional filming using stabilizing camera mounts; for consumer devices *optical image stabilization* (OIS) can reduce shake by detecting camera motion and applying compensatory shifts to the lens or sensor. Severe shake can be removed after capture, using algorithms implemented on the ISP: this is referred to as *digital image stabilization* (DIS), *electronic image stabilization* (EIS), or video stabilization.

Digital image stabilization relies on estimating the motion between successive frames, and then applying a shift to remove this motion. The motion arising due to camera movement is termed *global motion*, as it affects the entire scene; this is in contrast to the estimation of *local motion*, which describes the individual movements of the elements of the scene. Further details on this distinction are available in Tekalp (2015).

In some camera systems, information describing the camera motion at the time of capture may be relayed to the ISP to inform the motion estimation—see, for example, Karpenko *et al.* (2011). In general, however, the estimation is performed on the basis of the captured digital frames.

4.6.1.1 Global Motion Models

Motion estimation in general involves an image equation of the form

$$I_n(\mathbf{x}) = I_{n-1}(\mathbf{F}(\mathbf{x}, \mathbf{\Theta})) + \epsilon(\mathbf{x}) \tag{4.22}$$

$I_n(\mathbf{x})$ denotes the intensity (or tristimulus color) value of the pixel at site $\mathbf{x}$ (a 2-vector containing the row and column) in frame n, that is, I_n and I_{n-1} are successive frames within the video. The vector function $\mathbf{F}(\mathbf{x})$ represents a coordinate transformation according to the motion model, parameterized by $\mathbf{\Theta}$. The $\epsilon(\mathbf{x})$ term accounts for errors arising due to occlusion or revealing of scene content, model inadequacy, or noise in the image sequence. So for example, if the image has shifted four pixels to the left, and the pixel site [10, 10] depicts the scene background, the image equation should specify that

$$I_n([10, 10]) = I_{n-1}([10, 14]) + \epsilon' \tag{4.23}$$

where ϵ' is a small contribution from noise.

Global motion estimation methods attempt to fit a single, low-order motion model to the video content as a whole. An in-depth discussion of global motion models has been presented in Mann and Picard (1997). Properly speaking each depth plane in the scene undergoes a different global motion when the camera is moved, but in general for video stabilization the assumption is made that the scene content is effectively planar (the zero-parallax assumption).

The "correct" motion model for a plane imaged by a moving camera (disregarding lens distortion) is the eight-parameter projective transform:

$$\mathbf{F}(\mathbf{x}, \mathbf{\Theta}) = \frac{\mathbf{A}\mathbf{x} + \mathbf{d}}{\mathbf{c}\mathbf{x} + 1} \tag{4.24}$$

This is a nonlinear coordinate transform and is not generally used in on-camera stabilization as the complexity required to solve for this model is not justified, relative to simpler approximations that can be solved more robustly.

The simplest choice for the form of $\mathbf{F}$ is the *translational model*, $\mathbf{F}(\mathbf{x}, \boldsymbol{\Theta}) = \mathbf{x} + \boldsymbol{\Theta}$, with $\boldsymbol{\Theta} = [\Delta_x, \Delta_y]^T$, where Δ_x and Δ_y are the horizontal and vertical displacements. Using this model for global motion compensation can dramatically improve the appearance of shaky footage, but it is not adequate to completely remove the shake.

The *affine model* is given by $\boldsymbol{\Theta} = (\mathbf{A}, \mathbf{d})$, $\mathbf{F}(\mathbf{x}, \boldsymbol{\Theta}) = \mathbf{A}\mathbf{x} + \mathbf{d}$. Here, $\mathbf{A}$ is a 2×2 transformation matrix accounting for scaling, rotation, and shear, and $\mathbf{d} = [\Delta_x, \Delta_y]^T$ as before. The affine model is commonly used in on-camera stabilization as translational and rotational shake can easily be introduced by handheld camera operation, and shear results when a rolling shutter camera is panned horizontally.

Rolling shutter cameras can introduce *wobble* in video, also known as the *jello effect*, introduced in Chapter 3. If the frequency of the camera is much higher than the frame rate, individual slices of the frame may be stretched or sheared independently. The affine model cannot account for this distortion when applied globally. However, it is adequate if the frame is divided into a grid and a separate affine model is fit to each patch within the grid, as within each patch the distortion is approximately affine (Baker *et al.*, 2010; Grundmann *et al.*, 2012).

4.6.1.2 Global Motion Estimation

Whatever the model for global motion in use in the system, it is necessary to estimate the parameters that best account for the observed image data.

The error associated with some parameter values $\boldsymbol{\Theta}'$ is assessed via the *displaced frame difference*, or DFD:

$$\text{DFD}_{\boldsymbol{\Theta}}(\mathbf{x}) = I_n(\mathbf{x}) - I_{n-1}(\mathbf{F}(\mathbf{x}, \boldsymbol{\Theta})) \tag{4.25}$$

In general parameter estimation terminology, the DFD measures the residuals for parameters $\boldsymbol{\Theta}$. The aim then is to find the optimal parameters $\hat{\boldsymbol{\Theta}}$ minimizing some function of the DFD. A least squares solution, for example, entails solving

$$\hat{\boldsymbol{\Theta}} = \arg \min_{\boldsymbol{\Theta}} \sum_{\mathbf{x}} \text{DFD}_{\boldsymbol{\Theta}}^2(\mathbf{x}) \tag{4.26}$$

At pixel sites $\mathbf{x}$ undergoing global motion, $\text{DFD}_{\boldsymbol{\Theta}}(\mathbf{x})$ will be small if the parameters $\hat{\boldsymbol{\Theta}}$ describe the motion at that site, and large otherwise. Thus, the parameters $\hat{\boldsymbol{\Theta}}$ minimizing the DFD values overall are likely to describe the global motion accurately, within the limits of the model. At pixel sites not undergoing global motion, however (for example, where an element within the scene is moving independently of the camera, or where scene content is being occluded or revealed), the value of DFD($\mathbf{x}$) will be unrelated to the accuracy of the parameter estimate, and is likely to take a high value for all parameter values. Therefore, the minimization technique adopted must be robust against outliers.

Global motion estimation as described here is more properly termed *dominant motion estimation*: the motion parameters minimizing the DFD values are those that describe the motion affecting most of the scene. This will most commonly correspond to the background motion, but if there is a competing motion, for example a large object filling the foreground, the motion estimation step may "lock on" to the motion of that large object, rather than to that of the background. A classic example of this problem is a video stabilization algorithm choosing to minimize the motion of a closeup of a hand waving in the video clip, mistaking the hand motion as camera motion. The problem can be partly avoided by heuristics such as tuning the global motion estimator to be biased toward motion at the edge of the frame.

The computational cost of motion estimation can be reduced by estimating against *image projections*, rather than by using full frames of image data. In the simplest form of this idea, a fit for the translational motion model can be found by computing the horizontal and vertical projections (the sums of rows and sums of columns) for each of the two frames, and then estimating a displacement value between the horizontal projections to find the vertical motion, and a displacement value between the vertical projections to find the horizontal motion. This approach can be extended to higher-order motion models, albeit with diminished computational savings. Further details can be found in Milanfar (1999) and Crawford *et al.* (2004).

4.6.1.3 Global Motion Compensation

Once the global motion parameters relating frame n to frame $n-1$ have been estimated, they can be used to *compensate* frame n to align it with frame $n-1$, thus minimizing the appearance of any shake. This compensation will involve some digital interpolation of pixel values, and therefore will affect the sharpness of the compensated image. Therefore, sharpness assessment of cameras featuring digital image stabilization should be performed with some realistic level of shake applied to the camera while the video is recorded. This aspect of video quality assessment is covered in detail in Chapters 6 and 7.

The global motion cannot be compensated away entirely, as this would negate intentional camera motion, such as pans, initiated by the operator. The stabilization system will typically apply a highpass filter to the estimated motion parameters before compensation, so that shake is removed but low frequency motions, such as pans, are preserved. Depending on the implementation, this filtering may result in a delay between the physical start of the pan and the pan motion passing through the filter and appearing in the video (Grundmann *et al.*, 2011).

When shake is removed from a frame via global motion compensation, there will be some missing data at the border of the compensated frame. For example, if the frame is shifted to the right by four pixels, then the values of the four pixels at the left edge of each row must be obtained. In some camera systems the sensor is larger than the frame capture size, and the missing data can be recovered from this larger area. A complement to this approach is to apply a small digital zoom, which may be fixed or adaptive with a slow rate of change, so that no values are missing from the zoomed, compensated frame. This zoom will also affect the sharpness of the stabilized video. More sophisticated stabilization systems may attempt to reconstruct the missing pixels using data from previous frames or image extrapolation (Matsushita *et al.*, 2006; Litvin *et al.*, 2003), but this approach is more common in off-line video editing software than in on-camera systems. Whichever approach is used to handle missing data, there will be a limit on the magnitude of shake that can be compensated; shake exceeding this limit can only be attenuated.

Video stabilization, as described, does not encompass compensating for motion blur in the video frames, which can arise if the camera shake is fast relative to the exposure time of the frames. Where motion blur has occurred due to camera shake, stabilizing video will render this motion blur more visible, and in fact the subjective quality of the video may be reduced by the stabilization rather than enhanced. The video capture system may have some means of dealing with motion blur, such as discarding excessively blurry frames or applying a rudimentary deblurring filter. Independently of these possibilities, the stabilization algorithm could opt to retain some shake in regions of high motion blur to reduce its effect.

4.6.2 Video Compression

A great deal of engineering effort has been spent on devising video compression schemes, but all currently popular methods follow the same essential structure. The most salient aspect of video data is the vast temporal redundancy present in the data, since successive video frames typically have most of their image content in common. To exploit this, most of the frames will be encoded differentially, that is, as a prediction derived from a previously encoded frame, along with an error compensation signal, the *residuals*, accounting for deficiencies in the prediction.

Each frame, whether encoded as spatial data or residuals, is then *transform encoded*, as described in Section 4.4.2. The transformed data is quantized to improve compressibility; this is the step at which data fidelity is lost, but the quantization is tuned so as to minimize the perceptibility of this loss. The quantized coefficients can then be losslessly encoded for storage or transmission.

4.6.2.1 Computation of Residuals

Figure 4.29a shows a single frame from a video; we will refer to this as the reference frame (frame N). Figures 4.29b–d show the three following frames. Figures 4.29e–g show the corresponding residual images: the result of subtracting each frame from the reference frame. In these images, a difference of zero is presented as mid-gray, while positive difference values are brighter and negative difference values are darker.

The residual images are in a sense "flatter" than the video frames, in that the signal values are close to zero over most of the image. This idea of flatness can be formalized as the *signal variance*, denoted σ^2, and computed via

$$\sigma^2 = \frac{1}{N} \sum_{n=0}^{N-1} (x_n - \bar{x})^2$$

where the values x_n are the intensity levels of each pixel in the image, and $\bar{x}$ is the mean intensity level of the image (for the purposes of illustration, we are concerned only with the luminance channel). Broadly speaking, lower signal variance corresponds to easier signal compression: at the limit, a zero-variance signal can be summarized simply by its mean value.

Because the following frames are similar to the reference frame, the residual images have lower signal variance, and thus are more amenable to compression, than the frames themselves. The decoder can reconstruct these frames by adding the residuals to the reference frame.

It is clear from figures 4.29e–g that the residual signal is nonzero in regions where the scene content is moving. Taking the motion of scene content into account when computing the residuals results in a lower-variance residual signal, and thus allows for more effective differential coding. This refinement is known as *motion compensation*. It should be noted that this is distinct from the global motion estimation described for video stabilization, in that global motion estimation is concerned with motion due to movement of the camera, and seeks to disregard motion of the elements of the scene. In the video compression context, the aim is to compute a correspondence map between blocks of

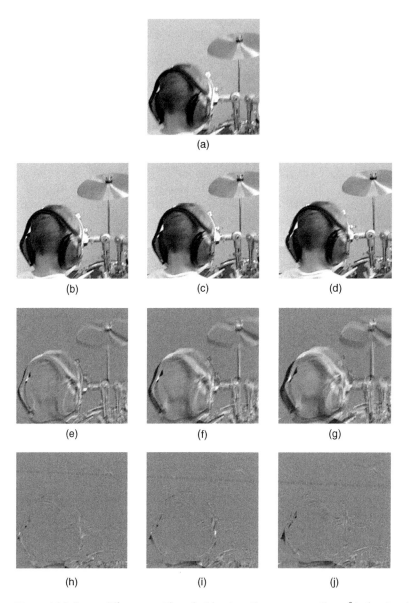

Figure 4.29 Frame differences with and without motion compensation. σ^2 is the signal variance, a measure of the difference between images, see text. (a) Frame N (reference). $\sigma^2 = 3478$. (b) Frame $N + 1$. $\sigma^2 = 3650$. (c) Frame $N + 2$. $\sigma^2 = 3688$. (d) Frame $N + 3$. $\sigma^2 = 3745$. (e) Difference of frames N and $N + 1$. $\sigma^2 = 1426$. (f) Difference of frames N and $N + 2$. $\sigma^2 = 2265$. (g) Difference of frames N and $N + 3$. $\sigma^2 = 3020$. (h) Motion-compensated differences of frames N and $N + 1$. $\sigma^2 = 205$. (i) Motion-compensated differences of frames N and $N + 2$. $\sigma^2 = 299$. (j) Motion-compensated differences of frames N and $N + 3$. $\sigma^2 = 363$.

image data in the reference and target frames, regardless of whether this is due to camera movement, moving elements in the scene, or a false correspondence due to repetitive structures. This correspondence data is termed the *motion vectors*. Additional details on motion estimation can be found in Tekalp (2015) and Stiller and Konrad (1999).

The reference frame is then deformed by displacing each block according to the motion vector; the residuals for the target frame are then computed by subtraction from this deformed (that is, motion compensated) reference frame, rather than from the original. The motion vectors are losslessly compressed and transmitted along with the residuals for the target frame. The decoder can then reconstruct the target frame by using the same motion compensation step using the supplied vectors.

Figures 4.29h–j show the residuals after motion compensation has been applied. The residual images are clearly "flatter" and the signal variance has been considerably reduced.

Motion compensation is a key part of nearly all video codecs, and motion estimation accounts for much of the computational cost of video encoding. There are many approaches in use, but most are variants of *block matching*, in which the block to be matched (in the reference frame) is subtracted from each candidate block in the target frame, and the candidate yielding the smallest difference is chosen as the match. In the video compression context, there is no requirement that the motion map found should correspond to the "true motion" of the scene, but merely that it should provide a good basis for differential encoding. The estimated motion vectors resulting can be incoherent and spatially varying and thus difficult to compress. Codec implementations can use heuristics to ensure that differential encoding is only used when the resulting residuals and motion vectors are more compressible than the image data itself.

4.6.2.2 Video Compression Standards and Codecs

The techniques described above represent the sorts of tools used for video compression. For commercial use, specific combinations of these tools have been standardized as video compression standards, such as MPEG-4 and H.264. An implementation of a video compression standard is a *video codec* (for video coder/decoder). For most standards there are numerous codecs available, including free and open source software implementations, commercial software, and hardware implementations. A consumer camera may use a commercially available video encoding chipset, or may license a hardware codec design for inclusion within a custom system.

The encoding tools described by a video coding standard result in a performance curve, where more computational effort can be used to attain higher video quality at a fixed bandwidth (or, equivalently, lower bandwidth for a fixed level of video quality). The specific performance curve attained depends on the particular codec implementation. As video coding standards have evolved, the toolset provided has grown enormously in sophistication and computational requirements. This has resulted in much more efficient compression, in terms of the data rate per megapixel for a given level of quality. At the same time, the demands made of video coding systems have grown to include higher resolutions, higher frame rates, and delivery over non-dedicated networks (i.e., the Internet), and so video coding remains a very active area of research and standardization.

4.6.2.3 Some Significant Video Compression Standards

The first digital video standard, established in 1984 by the International Telecommunication Union–Telecommunication Standardization Sector (ITU-T), was H.120 (ITU-T, 1984). It used differential encoding of intensity values in the spatial domain, but this technique did not give usable performance with the networking technology of the day. The designers realized that practical video codecs would need to reach coding rates of less than one bit per pixel, and this informed the move to systems that encoded blocks of pixels at a time.

The successor standard, H.261, was standardized in 1988 (ITU-T, 1988). This was developed to enable videoconferencing over ISDN, at spatial resolutions of 176×144 and 352×288 pixels. The H.261 standard introduced the structure and principal features used by all major codecs since: 4:1 spatial downsampling of chroma channels, motion compensation, transform coding, and quantization of coefficients with diagonal scan. It also introduced the terminology of an *intra-coded frame* (now more commonly denoted I-frame), for frames encoded without reference to other frames, and *inter-coded frame*, for differentially encoded frames.

Block-based transform coding can introduce *blocking artifacts*, which are the most commonly encountered degradation of digital video. Some H.261 implementations included a *deblocking filter* after the decoder, to smooth out the blocking artifacts. More recent coding standards have mandated a deblocking filter, incorporating it not only in the decoder but also in the encoder, so that the action of the deblocking filter can be taken into account when compressing the signal.

The MPEG-1 standard, first published by the ISO/IEC Moving Picture Experts Group in 1990 (ISO, 1993), introduced *bidirectional prediction*. With this refinement, a differentially-encoded frame may use data from a succeeding frame, as well as from a preceding frame, as its base for prediction. Frames using this encoding are termed *B-frames*, for "bidirectionally predicted," and forward-predicted frames are called *P-frames*. Use of B-frames implies that frames cannot be encoded in display order, as both the forward and backward reference frames must have been received in order to decode the B-frame. This results in increased memory requirements for both the encoder and decoder, as both reference images must be retained in memory while processing the B-frame.

The first video compression standards to bring digital video to a very broad consumer base were MPEG-2 (ISO, 1996), the standard used for DVD video, and H.263 (ITU-T, 1996), which was the first major standard used for Internet video.

The H.264 standard (ITU-T, 2003), first published in 2003, contained significant technical advances that enabled much better quality at a given bandwidth than its predecessors. It is used in Blu-ray and became dominant in Internet video to the point of ubiquity. Among the new features, the ability to use varying block sizes for motion estimation improves the motion compensation performance significantly, and the ability to use multiple reference frames (where previous codecs had a maximum of one forward and one backward reference) improves handling of large, repetitive motion and cuts between alternating scenes. A good overview of the technologies found in H.264 is available in Richardson (2010).

The advent of ultra-high definition (UHD) displays and 4k cinema production pushed consumer resolution requirements to above 8 megapixels per frame, or 240–720 megapixels per second, depending on the frame rate. The two compression standards deployed to deal with this data rate are H.265 (also known as High Efficiency Video Coding, or HEVC) (ITU-T, 2013) and VP9 (Google, Inc., 2013). H.265 is the successor to H.264, developed as part of the work of MPEG, and contains some patent-protected mechanisms. VP9 is an open standard developed by Google. Each of these standards introduced advanced techniques such as hierarchical subdivision of coding blocks, the use of transforms other than the DCT (in particular, the asymmetric discrete sine transform, or ADST), and more sophisticated entropy coding methods. Relative to previous standards, H.265 and VP9 incorporate finer-grained motion compensation and have been structured to be amenable to simultaneous encoding and decoding of multiple regions in each frame, in order to exploit parallel processing platforms.

At the time of writing, development is in progress on the VP10 standard, intended to succeed VP9. There is also a new open-source entrant under development, the Daala format, sponsored by the Mozilla Foundation. Daala departs from the block-wise transform-coding approach common to all mainstream codecs since H.261, using instead an overlapping transform so the blocks within each frame are not encoded independently. This promises to eliminate blocking artifacts and improve coding efficiency.

4.6.2.4 A Note On Video Stream Structure

All coding standards since MPEG-1 have incorporated the notion of intra-coded frames (I-frames), forward-predicted frames (P-frames), and bidirectionally-predicted frames (B-frames), with some having some additional frame types broadly analogous to these. In most encoding scenarios, the arrangement of I, P, and B frames will be fixed to a repeating sequence; this is the *group of pictures* structure, or GOP structure. For example, "IBBPBBPBBP" is a reasonable 10-frame GOP structure, where every 10th frame is intra-coded, and of the remaining nine, three are forward-predicted and six are bidirectionally predicted. A codec may be configured to use an adaptive GOP structure, or to be able to use an I-frame at some point within a GOP if there is a coding benefit, for example at a scene change.

The choice of a GOP structure is a tradeoff between quality, bandwidth, and the granularity of random access ("seekability") within the stream. Differentially encoded frames typically consume far less bandwidth than intra-coded frames. However, over-use of these encodings can reduce video quality in challenging material, for example video containing fast motion, explosions, or rapid scene cuts. The potential quality impact depends on the motion analysis capabilities of the codec and the encoding parameters chosen, in particular, how much CPU time to devote to motion estimation. Methodologies to quantify the impact on image quality and benchmarking of cameras are discussed in detail in later chapters.

Where bandwidth is the principal concern, a GOP structure may be chosen with an I-frame appearing only once every 60 seconds, for example, with all other frames differentially encoded. Apart from the possible quality degradation, this results in a stream that can only be joined (seeked into) at 60-second intervals, as decoding cannot be initiated at a P- or B-frame, but only from an I-frame.

4.7   System Considerations

So far, the performance of the individual components of the camera has been studied. When putting everything together into a working unit, other considerations that are not immediately obvious have to be made. As discussed above, shading in a camera is to a large extent dependent on the camera lens. However, the pixel angular sensitivity will also play a role. The mobile phone lens is faced with severe restrictions on size, and this will lead to an optical design with a very short total track length. The effect is that the *chief ray angle* (CRA) will be quite large toward the corners of the image. Values around 30° are not uncommon. As a result, the off-center pixels will receive light at an angle that will increase toward the edges of the image. If the sensor has poor angular response, the signal will get increasingly lower as the CRA grows. In most mobile phone sensors, this is counteracted by a shift of the microlenses determined by the position in the image. In order for the camera to have color vision resembling the human eye, an infrared cut filter needs to be used. In many cases, especially in mobile phone cameras, this filter is of the interference type (also called reflective type) built up of thin layers of materials with different refractive indices. Even though this filter type has many advantages, such as a very sharp cutoff at the wavelength band limits and a high overall transmittance, it also suffers from substantial angular dependence. When the CRA increases, the long wavelength limit especially will shift markedly. The end result will be *color shading*, that is, the color reproduction will change depending on the position in the image. Remedies to this can either be to use an absorptive filter or a combination *hybrid* filter. The ISP also has algorithms to compensate for the shading. In cameras with less severe height constraints, the lens design may be done in such a way as to avoid excessive CRAs. Such a *telecentric* design will improve the shading performance considerably.

Another important aspect of the pixel angular sensitivity is the fact that poor angular sensitivity often correlates with large amounts of crosstalk. This will not only be a problem for large CRA lenses, but also in lenses with a small f-number. In this case, the cone of light with its base at the exit pupil will be quite large, and consequently the light at the edges of the light cone will have large angles with respect to the pixel. This might result in pixel crosstalk, which, in a color sensor, will lead to mixing of color signals and desaturation of the color signal. In order to obtain correct colors, a more aggressive color matrix has to be employed, resulting in a higher noise level. Therefore, it is possible that the sensitivity of a camera will not increase by the same amount as the lens is opened.

These considerations are just examples of the fact that in order to obtain the best possible camera, it is necessary to consider not only the individual parts, but also how they interact and adjust their performance accordingly.

Summary of this Chapter

- In principle, a digital camera consists of three main functional blocks: lens, sensor, and image processing.
- The lens projects and focuses the image onto the photosensitive surface of the image sensor. On a general level, it can be characterized by the parameters focal length and f-number. The focal length determines the scale of the imaged scene while the f-number determines the depth of field and limits the amount of light that reaches the sensor from the scene. The smaller the f-number, the more light enters and the shorter the depth of field becomes.
- The image sensor converts the optical signal into an electronic signal which is subsequently transformed into digital numbers. Two main types of image sensors exist:

CMOS and CCD. At the time of writing of this book, CMOS is dominant in the consumer segment. A few performance parameters that are important to understand for the image sensor are:
- full well capacity
- noise: shot noise, dark current shot noise, read noise, fixed pattern noise
- light sensitivity.

- Image signal processing (ISP) transforms the raw digital image data from the sensor into a visible and pleasing image. To produce a color image, at least the following steps must be performed:
  - black level correction
  - white balance
  - color interpolation
  - color correction
  - gamma curve.

- Additionally, modern ISPs typically include the following processing steps:
  - defect pixel correction
  - shading correction
  - noise filtering
  - sharpening
  - global and/or local tone mapping
  - resizing
  - compression.

- In order to obtain a well-exposed image with correct white balance and focus, an ISP also needs to incorporate the 3A algorithms:
  - auto exposure
  - auto white balance
  - autofocus.

- The loss in image quality caused by JPEG compression is due to the quantization of the components of the discrete cosine transform.

- Video compression has many similarities with JPEG still image compression, for instance DCT encoding. However, video compression also employs similarities between frames in order to obtain a more efficient compression and includes sophisticated motion estimation algorithms. This led to the notion of I-frames and P/B-frames. I-frames are compressed standalone, while P- and B-frames are estimated from surrounding frames.

- In order to obtain the best possible image out of a camera, it is not sufficient to analyze and optimize only its individual components. The different components must be matched to each other and the algorithms of the ISP must be tuned for the particular combination of hardware at hand.

References

Allen, E. and Triantaphillidou, S. (eds) (2009) *Manual of Photography*, Focal Press, Burlington, MA, USA.

Arfken, G. (1985) *Mathematical Methods for Physicists*, Academic Press, San Diego, CA, USA, 3rd edn.

Baker, S., Bennett, E., Kang, S.B., and Szeliski, R. (2010) Removing rolling shutter wobble, in *IEEE Computer Society Conference on Computer Vision and Pattern Recognition (CVPR 2010)*, San Francisco, CA, USA.

Baur, C. and Freitag, F. (1963) Triplet wide-angle objective lens. https://www.google.com/patents/US3087384, (accessed 29 May 2017), US Patent 3,087,384.

Bayer, B.E. (1976) Color imaging array. https://www.google.com/patents/US3971065, (accessed 29 May 2017), US Patent 3,971,065.

Boyle, W.S. and Smith, G.E. (1970) Charge coupled semiconductor devices. *Bell Syst. Tech. J.*, **49**, 587–593.

Chatfield, C. (1983) *Statistics for Technology*, Chapman & Hall, London, UK, 3rd edn.

CIE (1926) *Commission Internationale de l'Éclairage Proceedings, 1924*, Cambridge University Press, UK.

Compton, J.T. and Hamilton, J.F. (2012) Image sensor with improved light sensitivity. https://www.google.se/patents/US8139130, (accessed 29 May 2017), US Patent 8,139,130.

Crawford, A.J., Denman, H., Kelly, F., Pitié, F., and Kokaram, A.C. (2004) Gradient based dominant motion estimation with integral projections for real time video stabilisation, in *Proceedings of IEEE International Conference on Image Processing, ICIP'04*, vol. **5**, vol. 5, pp. 3371–3374.

Ebner, M. (2007) *Color Constancy*, John Wiley & Sons Ltd, Chichester, UK.

Fossum, E.R. (1997) CMOS image sensors: Electronic camera on a chip. *IEEE Trans. Electron Devices*, **44**, 1689.

Foveon (2015) Foveon website. http://www.foveon.com, (accessed 29 May 2017).

Gersho, A. and Gray, R.M. (1991) *Vector Quantization and Signal Compression*, Kluwer Academic Publishers, Norwell, MA, USA.

Goodman, J.W. (2005) *Introduction to Fourier Optics*, Roberts & Company, Eaglewood, CO, USA, 3rd edn.

Google, Inc. (2013) VP9 bitstream & decoding process specification, *Draft*, Google, Inc.

Grundmann, M., Kwatra, V., Castro, D., and Essa, I. (2012) Calibration-free rolling shutter removal, in *IEEE International Conference on Computational Photography*.

Grundmann, M., Kwatra, V., and Essa, I. (2011) Auto-directed video stabilization with robust l1 optimal camera paths, in *IEEE Conference on Computer Vision and Pattern Recognition (CVPR 2011)*.

Hecht, E. (1987) *Optics*, Addison-Wesley, Reading, MA, USA, 2nd edn.

IEC (1999) IEC 61966-2-1:1999 Colour Measurement and Management in Multimedia Systems and Equipment – Part 2-1: Default RGB Colour Space – sRGB.

ISO (1993) ISO/IEC 11172-3:1993 Information Technology – Coding of moving pictures and associated audio for digital storage media at up to about 1,5 Mbit/s. ISO.

ISO (1996) ISO/IEC 13818 Information Technology – Generic coding of moving pictures and associated audio information. ISO.

ITU-T (1984) Recommendation H.120: Codecs for videoconferencing using primary digital group transmission. ITU-T.

ITU-T (1988) Recommendation H.261: Video codec for audiovisual services at p x 64 kbit/s. ITU-T.

ITU-T (1996) Recommendation H.263: Video coding for low bit rate communication. ITU-T.

ITU-T (2003) Recommendation H.264: Advanced video coding for generic audiovisual services. ITU-T.

ITU-T (2013) Recommendation H.265: High efficiency video coding. ITU-T.

Janesick, J.R. (2001) *Scientific Charge Coupled Devices*, SPIE Press, Bellingham, WA, USA.

Janesick, J.R. (2007) *Photon Transfer*, SPIE Press, Bellingham, WA, USA.

Karpenko, A., Jacobs, D., Baek, J., and Levoy, M. (2011) Digital video stabilization and rolling shutter correction using gyroscopes. *Stanford CS Tech Report*.

Lee, J.S. (1983) Digital image smoothing and the sigma filter. *Computer Vision, Graphics, and Image Processing*, **24**, 255–269.

Li, X., Gunturk, B., and Zhang, L. (2008) Image demosaicing: A systematic survey. *Proc. SPIE*, **6822**, 68221J.

Litvin, A., Konrad, J., and Karl, W. (2003) Probabilistic video stabilization using Kalman filtering and mosaicking. *Proc. SPIE*, **5022**, 663–674.

Mahajan, V.N. (1998) *Optical Imaging and Aberrations – Part I: Ray Geometrical Optics*, SPIE Press, Bellingham, WA, USA.

Mann, S. and Picard, R.W. (1997) Video orbits of the projective group: a simple approach to featureless estimation of parameters. *IEEE Trans. Image Process.*, **6**, 1281–1295.

Matsushita, Y., Ofek, E., Ge, W., Tang, X., and Shum, H.Y. (2006) Full-frame video frame video stabilization with motion inpainting. *IEEE Trans. Pattern Anal. Mach. Intell.*, **28**, 1150–1163.

Menon, D. and Calvagno, G. (2011) Color image demosaicking: An overview. *Signal Process. Image Commun.*, **26**, 518–533.

Milanfar, P. (1999) A model of the effect of image motion in the radon transform domain. *IEEE Trans. Image Process.*, **8**, 1276–1281.

Nakamura, J. (ed.) (2005) *Image Sensors and Signal Processing for Digital Still Cameras*, Taylor & Francis, Boca Raton, FL, USA.

Ohta, J. (2008) *Smart CMOS Image Sensors and Applications*, Taylor & Francis, Boca Raton, FL, USA.

Poynton, C. (2003) *Digital Video and HDTV: Algorithms and Interfaces*, Morgan Kaufmann Publishers, San Francisco, CA, USA.

Ray, S.F. (2002) *Applied Photographic Optics*, Focal Press, Oxford, UK, 3rd edn.

Reif, F. (1985) *Fundamentals of Statistical and Thermal Physics*, McGraw-Hill, Singapore, intl. edn.

Reinhard, E., Khan, E.A., Akyuz, A.O., and Johnson, G.M. (2008) *Color Imaging – Fundamentals and Applications*, A. K. Peters, Wellesley, MA, USA.

Reinhard, E., Ward, G., Pattanaik, S., and Debevec, P. (2006) *High Dynamic Range Imaging – Acquisition, Display, and Image-based Lighting*, Morgan Kaufmann, San Francisco, CA, USA.

Richardson, I.E. (2010) *The H.264 Advanced Video Compression Standard*, John Wiley & Sons Ltd, Chichester, UK, 2nd edn.

Sigma (2015) Sigma corporation website. http://www.sigmaphoto.com/cameras/dslr, (accessed 27 May 2017).

Smith, W.J. (2000) *Modern Optical Engineering*, McGraw-Hill, USA.

Sonka, M., Hlavac, V., and Boyle, R. (1999) *Image Processing, Analysis, and Machine Vision*, Brooks/Cole Publishing Company, Pacific Grove, CA, USA.

Stiller, C. and Konrad, J. (1999) Estimating motion in image sequences. *IEEE Signal Process. Mag.*, **16**, 70–91.

Tekalp, A.M. (2015) *Digital Video Processing*, Pearson Education, Westford, MA, USA, 2nd edn.

Theuwissen, A.J.P. (1996) *Solid-State Imaging with Charge-Coupled Devices*, Kluwer, Dordrecht, The Netherlands.

Tomasi, C. and Manduchi, R. (1998) Bilateral filtering for gray and color images, in *IEEE Sixth International Conference on Computer Vision*, pp. 839–846.

Wikipedia (2016) Camera obscura on Wikipedia. https://en.wikipedia.org/wiki/Camera_obscura, (accessed 29 May 2017).

Wilson, J. and Hawkes, J.F.B. (1997) *Optoelectronics: An Introduction*, Prentice-Hall, Upper Saddle River, New Jersey, USA.

5

Subjective Image Quality Assessment—Theory and Practice

Now that we have described the essence of photography and camera benchmarking concepts in Chapter 1, image quality and its attributes in Chapters 2 and 3, and the camera itself in Chapter 4, the book continues with chapters on the fundamentals of subjective and objective image quality assessment. These fundamentals are necessary for performing comprehensive camera image quality benchmarking. The topic of this chapter is subjective assessment.

Human observation is an important and fundamental aspect of defining image quality. We shouldn't forget that photography is a technology that is typically assessed by the human visual system (HVS). As such, subjective evaluation should be incorporated into the image quality benchmarking process to ensure that the camera output is meeting this aspect of photography. Though the term "subjective" can have negative connotations of "unpredictable," "not quantifiable numerically," and "soft science," subjective studies can be conducted in ways that provide repeatable and conclusive results. Designing experiments to generate reliable and repeatable conclusions about perceived image quality is an achievable practice when using established methodologies. While this type of experimentation is not necessarily understood by, or familiar to, those in the field of camera component design, subjective evaluation is not a new practice. In fact, the scientific field of psychophysics, that of the study of the relationship between the perceptual and physical realms, has existed for nearly two hundred years. Thus, an understanding of how this established field can be applied to image quality evaluation should benefit the practitioner and strengthen the premise of why image quality benchmarking should include both objective and subjective elements. Many psychophysical methods exist for quantifying subjective image quality with human observers. When such methods utilizing human observers to measure and quantify sensory percepts are used, the resulting metrics are termed psychometrics.

The goal of this chapter is to define the subjective methodologies that relate directly to obtaining meaningful subjective evaluation for still and video image quality. A review of key psychophysical techniques such as category scaling, forced-choice comparisons, acceptability ratings, and mean opinion score (MOS) will emphasize the strengths and weaknesses of each methodology. The review will also explore the similarities and differences between still and video subjective evaluation techniques, including how these are able to quantify important perceptual aspects of the HVS. Particular focus will be on the anchor scale method and how that can be used to quantify overall image quality in just noticeable differences (JNDs) for still images. These methodologies

Camera Image Quality Benchmarking, First Edition. Jonathan B. Phillips and Henrik Eliasson.
© 2018 John Wiley & Sons Ltd. Published 2018 by John Wiley & Sons Ltd.
Companion website: www.wiley.com/go/benchak

provide means for performing image quality assessment that feed directly into the task of camera benchmarking.

5.1   Psychophysics

Perceptual psychology originates from research of Ernst Heinrich Weber and Gustav Theodor Fechner in the early 19th century, which forms the psychophysical foundation for subjective evaluation relevant to image quality. Fundamental to psychophysics is the relationship between the *perceptual response*, Ψ, in the psychological realm to the *stimulus*, ϕ, in the physical realm. For example, one can have a sensation of coldness when grabbing a glass jar from a refrigerator and an even stronger sensation of coldness when grabbing an ice cube from a freezer. Weber established the relationship between the smallest stimulus generating a perceptual response and the starting intensity of the stimulus as noted in the following form of Weber's Law (Trotter, 1878):

$$\Delta\phi = K_w \phi \tag{5.1}$$

Here $\Delta\phi$ is the smallest stimulus change generating a perceptual response and K_w is the constant fraction of the starting intensity of the stimulus, ϕ. For a given type of stimulus, for example, weight, brightness, or loudness, the smallest physical amount of the stimulus that produces a perceptual response is the *absolute threshold*. As shown in the relationship above, this smallest stimulus change, or *difference threshold*, varies with the starting intensity. Weber also discovered that the value of K_w varies depending on the type of stimulus. For example, the Weber constant, K_w, for weight perception, is 0.05. Thus, if the starting intensity of a stimulus is the weight of holding a 20 kg dumbbell, then the smallest stimulus generating a perceptual change in weight when added to the said dumbbell is 1 kg. To put it explicitly, a K_w value of 0.05 will generate a $\Delta\phi$ value of 0.05 × 20 kg = 1 kg. This means that 1 kg is considered the *just noticeable difference* (JND) for this dumbbell. In contrast, 1 kg would be substantially greater than a JND for the example of a perceptual change in weight of holding a seemingly weightless feather; a person holding a feather would know without a doubt when 1 kg was added to the weight of a feather compared to the 1 kg being added to the dumbbell. We will consider the JND for the visual perception of image quality later in the chapter.

Note that Weber's Law does not explicitly include a variable of the psychological realm. However, the fact that the stimulus change is proportional to the stimulus itself implies a logarithmic response. Assuming the perceptual response change, $\Delta\Psi$, is 1, that is, 1 JND, we may write

$$\frac{\Delta\phi}{\phi} = K_w \Delta\Psi \tag{5.2}$$

In the limit of infinitesimal changes of ϕ and Ψ, this becomes a differential equation that is easily solved and gives the following explicit relationship between stimulus intensity and perceived response:

$$\Psi = K_f \log \phi + M \tag{5.3}$$

where $K_f = 1/K_w$ and M is a constant due to integration, which could be set to zero to yield the usual form of what is known as Fechner's Law (Trotter, 1878). Notice that this mathematical form identifies the compressive nature of the human nervous system regarding the intensity range of the physical stimuli. This means that as the intensity of the stimulus increases, the difference in perception decreases. For example, if the stimulus is loudness, the perceived magnitude will compress as the stimulus grows louder. This is why the unit of decibels used to quantify loudness is necessarily a logarithmic function.

A resurgence of the works of Weber and Fechner occurred nearly 100 years later when Stevens established that the perceived magnitude of a stimulus, Ψ, is a constant, K_s, times the intensity of the stimulus, ϕ, raised to the power α, which is specific to the type of stimulus (Stevens, 1957). Stevens' contribution identifies that sensory perception is more complex than the Weber and Fechner model implies. Stevens' Power Law can be stated as:

$$\Psi = K_s \phi^\alpha \tag{5.4}$$

In addition to proposing his Power Law, Stevens provided modality-specific exponent values for α. Stevens noted that the perceived response of the human nervous system could be compressive as with Fechner's Law, but also had expansive or even linear behavior, depending on modality. Thus, Fechner's Law needed to be expanded to include a more complete representation of sensations. For example, $\alpha = 1$ for estimation of visual length, $\alpha = 1.4$ for assessment of saltiness, and $\alpha = 0.67$ for loudness. This means that the capability of an observer to accurately estimate small distance differences, such as 1 cm, or larger distance differences, such as 15 meters, remains constant. However, small changes in the amount of salt have significant changes in the perception of saltiness, as indicated by the expansive α term. Thus, when adjusting the saltiness of a food dish to taste, a small amount of salt "goes a long way" as more salt is added. Conversely, loudness perception is compressive in nature when compared to the increase in volume stimulus: the noticeable increase in loudness diminishes as the sound intensity increases. Table 5.1 contains a more comprehensive list of α values for given physical stimuli. Note the range in α from highly compressive (brightness, $\alpha = 0.33$) to highly expansive (electric shock, $\alpha = 3.5$), indicating the broad range of perceptual responses to physical stimulation.

Table 5.1 Stevens' Law α values for various modalities. Note the range from 0.33 to 1.0 to 3.5, which includes both compressive, linear, and expansive perceptual response, respectively. *Source*: Adapted from Stevens 1975. Reproduced with permission of Wiley. Copyright (c) 1975 by John Wiley & Sons, Inc.

Modality	Measured α	Stimulus condition
Brightness	0.33	5° target in dark
Viscosity	0.42	Stirring silicone fluids
Brightness	0.5	Point source
Brightness	0.5	Brief flash
Vibration	0.6	Amplitude of 250 Hz on finger
Smell	0.6	Heptane
Loudness	0.67	Sound pressure of 3000 Hz tone
Visual area	0.7	Projected square

(continued)

Table 5.1 (Continued)

Modality	Measured α	Stimulus condition
Warmth	0.7	Irradiation of skin, large area
Discomfort, warm	0.7	Whole body irradiation
Taste	0.8	Saccharine
Tactual hardness	0.8	Squeezing rubber
Vibration	0.95	Amplitude of 60 Hz on finger
Brightness	1.0	Point source briefly flashed
Visual length	1.0	Projected line
Cold	1.0	Metal contact on arm
Thermal pain	1.0	Radiant heat on skin
Pressure on palm	1.1	Static force on skin
Vocal effort	1.1	Vocal sound pressure
Duration	1.1	White noise stimuli
Lightness	1.2	Reflectance of gray papers
Taste	1.3	Sucrose
Warmth	1.3	Irradiation of skin, small area
Finger span	1.3	Thickness of blocks
Taste	1.4	Salt
Angular acceleration	1.4	5-sec rotation
Heaviness	1.45	Lifted weights
Tactual roughness	1.5	Rubbing emery cloths
Warmth	1.6	Metal contact on arm
Redness (saturation)	1.7	Red-gray mixture
Discomfort, cold	1.7	Whole body irradiation
Muscle force	1.7	Static contractions
Electric shock	3.5	Current through fingers

5.2 Measurement Scales

Quantification of perceptual response can be obtained using various methodologies. As such, the degree of quantification can vary, depending on what and how data is collected. Four basic categories of *measurement scales* are most frequently used in perceptual psychology: nominal, ordinal, interval, and ratio (Stevens, 1946). These scales are listed in the order of increasing degree of relationship between the values of the scale and the stimulus of interest.

For the *nominal scale*, values are simply used as designations and do not have any quantitative relationship between themselves. While a nominal scale may contain numbers, the association between the numbers and the stimuli of interest is arbitrary. An example of this scale in a psychophysical study would be when stimuli are arbitrarily labeled with numbers to provide designations used in administering the study. An observer could be presented with five fragrance samples, labeled 1 through 5, as part of a study. The instructions could state, "Pick which sample number represents your favorite fragrance." If the observer chooses sample 4, for example, the numeric value does not imply anything about the fragrance; the number is only a label for fragrance sample 4.

An *ordinal scale* indicates the relationship between stimuli as being greater or less than each other, though the exact spacing between the perceptual responses is ambiguous. The relationship of the numbers on the scale is monotonic. When an observer ranks favorite flavors, for example, the relationship between the rankings indicates the order or preference. The observer could rank the flavors of cherry, grape, sour apple, and mint in the order of 1 through 4. Cherry is most favored, followed by grape, then sour apple, and finally mint. But, the rank order does not provide specificity as to how favored the cherry is above the others (even though ranked 1 through 4), nor the specificity of the differences between any pair of flavors.

The measurement scale that provides both the order and the distance between individual perceptual responses is the *interval scale*. Here, the numbers on the scale have more specificity because the scale is set up to be necessarily linear. As such, a linear transform on the scale does not impact the relative differences between numbers. An example of this scale is related to how an observer rates the pitch of a series of musical notes. For this particular example, the numerical scale is typically designated by an alphabetical scale from A to G# and half steps in between. The pitch perception between the half steps is perceptually equal throughout the scale. For observers versed in pitch perception, the relative pitch change can be identified between the auditory stimuli with specific indications of perceptually linear difference in pitch.

The fourth basic category of measurement scales is the *ratio scale*. In addition to providing both the order and the distance between numbers on the scale, the ratio scale contains an absolute zero point. This scale is beneficial for applying to perceptual response because numbers on the scale are linear for a given stimulus series and the absence of a stimulus results in a meaningful zero on the scale. The L^* component of CIELAB (see Chapter 6), that is, the perceived lightness level of a visual stimulus, is an instance of the ratio scale because L^* is formulated in such a way that the numbers on this scale of the perception of lightness are both perceptually linear and have a zero point. For example, the difference of 20 L^* units between 10 and 30 should be approximately the same perceptual lightness difference as between 80 and 100. The zero point on the L^* scale is the black point—no lightness is perceptible to the HVS because the luminance level of the black is zero compared to the luminance level of the reference white point.

A summary of measurement scales appears in Table 5.2.

Table 5.2 Measurement scales related to psychophysical testing.

Scale	Definition	Example
Nominal	Scale with arbitrary labels that organize individual elements	Numbers used on basketball team uniforms, e.g., 16, 1, 5, 37, etc.
Ordinal	Scale with labels that represent the order of individual elements	Rank order of favorite flavor of candy, e.g., 1 = cherry, 2 = grape, 3 = sour apple, 4 = mint, etc.
Interval	Scale with labels that indicate both the order and the distance of individual elements	The Fahrenheit temperature scale
Ratio	Scale with labels that indicate both the distance and an absolute zero point within the order of individual elements	The Kelvin scale with absolute zero

5.3 Psychophysical Methodologies

Once laws of perception and measurement scales are understood, then a further level of understanding relates to how to measure perceived magnitude of the stimulus, Ψ. The psychophysical methodology employed has a significant impact on observer assessment—some methods are useful for obtaining general perceptual information, while others can be used to determine precise JNDs. The types of stimuli, types of questions being asked of the observer, and viewing conditions all have a direct impact on the understanding gained by the experimentation. Thus, selection of the method has both analytical and pragmatic implications. Fundamental psychophysical methodologies pertaining to evaluation of image quality include rank order, categorical scaling, acceptability scaling, anchored scaling, forced-choice comparison, and magnitude estimation (which will be described further in the subsections). For image quality evaluation, methodologies are best utilized when judging a single image quality attribute (such as sharpness, distortion, or color) or overall quality. Judging images for multiple individual quality attributes is possible, but the attributes must be specified and assessed individually.

Several key book references exist on the topic of psychophysical methodologies and methods pertinent to image quality assessment. A fundamental work by Gescheider (1997) discusses general psychophysical theory and methods, including examples from the literature and sample problems. Experts with ties to the Kodak Research Labs and Xerox have had significant contribution to the application of psychophysics to photographic assessment. Bartleson and Grum (1984) include a six-chapter section related to psychophysical methods in their book volume on visual measurements, including discussion on which method is to be used when. The majority of the content and cited examples were derived from their work in the Kodak Research Labs. Keelan's Handbook of Image Quality (Keelan, 2002a) discusses many methods with pragmatic perspective from Kodak such as how long a particular method might take when running a study. Comments regarding the conversion of the subjective data to ratio scales, ideal intervallic scales, or JNDs are stated for each method. Keelan also discusses the quality ruler method and how it can be utilized to obtain overall quality or individual attributes of quality. Engeldrum's Psychometric Scaling (Engeldrum, 2000) was written specifically from a practical standpoint and thus includes items such as selection of stimuli, stimuli presentation, and observer, all within the context of psychometric scaling methods.

Photographic standard committees such as CPIQ and ISO have also published materials on the topic of subjective evaluation based on experience of participating experts who worked in research and development for companies such as Fujifilm, Hewlett-Packard, Kodak, Micron, Polaroid, and Sony Ericsson. Examples include ISO 20462, of which Part 1 is an overview of practical psychophysical components, including a useful glossary and necessary elements to obtain meaningful psychophysical reporting such as minimum number of observers and scenes, contents of experimental instructions, and viewing condition requirements (ISO, 2005a). Parts 2 and 3 of the ISO 20462 document spell out the experimental details for the methodologies of triplet comparison and quality ruler (ISO, 2005b, 2012). The CPIQ Phase 1 document contains a section on subjective evaluation, detailing techniques to use for obtaining observer data, selection of images, data collection, and reporting of subjective results (I3A, 2007). Example psychometric results are given for an experiment using preference scaling for paired comparisons to analyze color rendering options. CPIQ published a document entirely on the topic of subjective evaluation methodology as part of the CPIQ Phase 2 documents (I3A, 2009). The content describes the methods of rank order, paired comparison, and various types of scaling that can be used to obtain a reliable prediction of the response of a particular population to the image quality of a set of images. Based on the learnings by the companies involved, the Phase 2 document expands on the specifics of designing, administering, and evaluating psychometric studies for two of the methodologies: acceptability scaling and the use of a softcopy quality ruler for anchored scaling.

The following subsections describe in more detail the elements of key subjective methodologies mentioned in the previous discussion. For more information, the reader is directed to the reference books and documents referred to therein.

5.3.1 Rank Order

Rank order is part of everyday comparisons, such as ranking favorite sports teams, styles of music, or grocery stores. This *rank order* method can also be part of decision making, such as ranking the specifications and aspects of different car models when figuring out which vehicle to purchase. While this method is familiar to most, the process is fundamentally limited to an ordinal measurement scale. As such, the order is determined but the spacings between the rankings are unknown. Thus, there are fundamental limitations in describing the relationship between rank orders, particularly the "how much" difference between the ranks.

5.3.2 Category Scaling

Many types of scaling methodologies can be utilized in subjective evaluations. A *categorical scaling* approach provides a simple assessment for understanding how an observer perceives the quality level of a test image. The experimenter selects a group of categories to describe a range of quality and the observer determines the category that best describes the quality of the test image. Category sets often have an odd number of levels intended to be symmetrically and equally spaced in quality around the center point. Engeldrum points out that odd numbers with a midpoint help the observer to keep the categories intervallic when making an assessment, a desirable characteristic for quantifying quality (Engeldrum, 2000). Seminal experimental work by Zwick (1984) on the topic of categorical names for photographic quality used five qualitative terms of "excellent," "good," "fair," "poor," and "bad". However, the categorical scaling results from the experiment indicated that "excellent" and "good" were separated by a quality difference twice as large as those between other categories on the scale, so a sixth category of "very good" was necessary.[1] As such, a commonly used set of intervallic categories for photographic quality includes the even number of qualifiers below. Note "bad" is replaced with "not worth keeping" (Rice and Faulkner, 1983).

- Excellent
- Very Good
- Good
- Fair

---

1 Note that these fundamental experiments were carried out in English with presumably a significant majority of native speakers. Caution should be applied when assuming the intervallic relationship with observers whose native language is not English or in literal translation of the categories into other languages.

- Poor
- Not Worth Keeping

Other observer data has supported the finding that the six categories above are intervallic, specifically with spacings of approximately 6 JNDs (Keelan, 2002a). This intervallic nature increases the usefulness of the categorical scaling because quality placement and movement along the scale is more easily calculated and interpreted. Instructions do play a role in communicating the intervallic intention. For example, the sample observer instructions in Bartleson and Grum (1984) explicitly state and explain that differences between categories represent equal intervals. However, Engeldrum does caution that many category scales are not intervallic, at least for portions of the scale. He strongly notes that calculations should not assume equal widths, specifically with arbitrary categories (Engeldrum, 2000). Ordinal or even nominal scales should be assumed unless known to be otherwise.

Typically, the data is evaluated by calculating the percentage per category or identifying the highest average category obtained from the pooled data submitted by multiple observers for a given stimulus. Calculating the standard deviation or standard error provides an understanding of the degree of unanimity among the observers.

5.3.3 Acceptability Scaling

When a category assignment is determined, the acceptability of that category may not be implicit. Therefore, an *acceptability scaling* provides additional information to the experimenter. Note that in order for an observer to assign an acceptability level, there must be a defined scenario to make that assessment. An observer might perceive the quality of a stimulus to be fair, but until the scenario is stated for the given assessment, the decision to assess the acceptability is ambiguous. For example, if told by the experimenter that the image quality acceptability is to be assigned for a scenario in which the image is to be temporarily on display on the kitchen refrigerator for passersby to see, then the acceptability would generally be higher than the scenario in which the image is to be used as an entry for an international photography contest. Thus, acceptability levels are strongly dependent on the use case. Similar to categorical scaling, percentage per acceptability category or mean acceptability levels can be calculated for the observer population data.

A minimum acceptability scale contains two ratings: acceptable and unacceptable. However, more levels on either side of the threshold of acceptability can provide more insight. An example scale that includes more levels of acceptability is listed below (I3A, 2009).[2]

- Completely acceptable, no improvement needed
- Acceptable, slight improvement needed
- Borderline, moderate improvement needed
- Unacceptable, significant improvement needed
- Completely unacceptable, of little or no value

Note that, from a pragmatic point of view, acceptability scaling results are useful for determining how well the particular attribute quality or overall quality level meets the observer's expectations of quality. Suppose an observer describes the category of a video

---

2 Adapted and reprinted with permission from IEEE. Copyright IEEE 2012. All rights reserved.

as being "Fair." While this level is just above "Poor" using the categories described in the previous section, the image quality for the observer may be completely acceptable if the purpose of the video is to count the number of people walking by an alleyway. However, if the purpose of the video is to provide face recognition, the "Fair" category may be unacceptable to the observer. This aspect of acceptability scaling provides valuable insight into subjective evaluation.

5.3.4 Anchored Scaling

Anchored scaling provides the observer with reference stimuli along the scale continuum. A goal of this methodology is to stabilize the internal scale for the observer when making an assessment of a test stimulus. By providing discrete reference stimuli along the scale, considered to be anchors, the decision making becomes similar to matching for the observer. When the quality of the anchor stimuli is calibrated, the scale becomes strongly intervallic. Responses from observers are collected and averaged to determine the position of quality on the scale.

5.3.5 Forced-Choice Comparison

A common practice of *forced-choice comparison* in imaging is to present pairs of test stimuli that are compared for a given attribute. In order to be a full experimental study, each test sample is paired with every other test sample, thus resulting in a methodology with many comparisons per observer. The observer is instructed to select one of the pair as the preferred sample for a given assessment. Because the observer is not allowed to indicate that the samples have no difference or that there is no preference over either sample, this methodology is "forced-choice" comparison, or 2AFC (two-alternative forced choice). Analysis is based on Thurstonian scaling (Thurstone, 1927), explained in more detail below. As such, the analysis is not suited for experiments where stimuli differences in the pair are large enough to generate unanimous judgments for all participating observers. Understandably, the difference in the physical stimuli is necessarily small for this to be the case. This methodology is therefore the most sensitive technique regularly used for visual psychometric studies. However, the large number of comparisons needed (i.e., $n(n-1)/2$ comparisons for n images) makes this a very time consuming method that puts high demands on the observers. The number of comparisons can be decreased by using a forced-choice approach with more choices such as triplet comparisons (3AFC, three-alternative forced choice). This triplet comparison method will be described in further detail later in the chapter.

5.3.6 Magnitude Estimation

The typical approach for the methodology of *magnitude estimation* is to assign a standard response value, or modulus, for a given attribute level. Once this modulus is defined for the observer, responses are collected for subsequent images. The observer assigns a response based on how many times greater or lesser the sensation is compared to the modulus of the first image. For example, if the modulus is 50 for the roughness of a reference surface and the test stimulus feels twice as rough as the modulus for a given observer, then that observer would assign the test stimulus a value of 100. Responses from multiple observers are collected and combined, usually by calculating the median or geometric mean of the responses for each experimental stimulus.

Table 5.3 Advantages and disadvantages of fundamental psychophysical methods. Based on Table 2 in CPIQ Phase 2 Subjective Evaluation Methodology (I3A, 2009). Adapted and reprinted with permission from IEEE. Copyright IEEE 2012. All rights reserved.

Method	Advantages	Disadvantages
Rank Order	• Simple to conduct and analyze • Meaningful to non-technical audience	• Typically provides only ordinal scale • Requires large numbers of observers to convert to interval scale
Category Scaling	• Simple to conduct and analyze • Meaningful to non-technical audience	• Requires large numbers of observers for reliability • May not transfer to other languages
Acceptability	• Simple to conduct and analyze • Meaningful to non-technical audience	• Requires large numbers of observers for reliability • Acceptability drifts over time
Anchored Scaling	• Can provide absolute scores • Good for large numbers of test samples	• Anchors difficult to produce • Difficult if test and anchor scenes are not similar in content
Forced-Choice Comparison	• Good for detecting very small differences—test samples are compared directly	• Limited to relatively small range of stimuli • Analysis is complex
Magnitude Estimation	• Simple to conduct and analyze • Good for large numbers of samples	• Requires large numbers of observers for reliability

5.3.7 Methodology Comparison

Table 5.3 provides a summary of how the presented psychophysical methodologies compare. While an important aspect of deciding which methodology to use is in defining what questions need to be answered, there are also fundamental aspects regarding the corresponding perceptual statistics that can drive the selection process.

5.4 Cross-Modal Psychophysics

Cross-modal psychophysics is the field of studying how the various human senses, that is, sight, taste, hearing, smell, and touch, can interact. For perspective on psychophysics specific to image quality evaluation and benchmarking, cross-modal psychophysical experiments provide examples of how the HVS is able to interact with other sensory perceptions. Knowledge of successes with psychometrics between multiple senses should enable more confidence by those questioning whether complex visual-only psychophysical studies for camera benchmarking applications can be performed successfully.

Many studies have been performed on cross-modal psychophysics, firmly establishing the relationship between such different sensations as smell and color, lightness and sound pitch, and so on. Example cross-modal psychophysical studies in the following subsections describe, firstly, two research comparisons and, secondly, a comparison that can be used as a live demonstration in a classroom setting. The first research to be described is a study of the relationship between colors assessed with visual perception and fragrance assessed by olfaction. The second research study determined how the lightness of a gray patch assessed with visual perception related to the pitch of a note assessed by auditory perception. The live demonstration example is a variant of the second experiment.

5.4.1 Example Research

Gilbert *et al.* studied the cross-modal relationship between vision and olfaction using Munsell color samples and odorants used in commercial perfumery, including odors familiar and unfamiliar to the general public (Gilbert *et al.*, 1996). Observers were asked to select a color chip from the Munsell Book of Color or Supplementary 80-Hue Colors[3] that matched their perception of the odor of 20 unique samples.

Calculations were made with the Munsell hue, value, and chroma results from the observers' color chip selections. The hue data was determined to be the most relevant component of the Munsell colors with respect to odor. Results were sorted into order of significance based on the concentration of observations around the hue center for each odor. For the four most significant results, one can infer that observers used natural associations of colors related to those four odors. The most significant, caramel lactone, has an odor of caramel and was associated with a medium brown hue similar to typical caramel sauce. Cinnamon aldehyde, with a cinnamon odor, was associated with a red hue similar to American cinnamon candies. The odor of aldehyde C-16 is commonly used for artificial strawberry flavoring. Thus, an associated pink hue should not be surprising. Finally, bergamot oil (from the rind of the citrus, bergamot) was associated with a yellow color similar to citrus. Thus, the associations of the significant concentration of observations around the visual hue with that of odor in this experiment provide evidence of successful cross-modal subjective evaluation.

In another cross-modal experimentation, Marks studied the cross-modal relationship between the visual perception of lightness and the aural perception of pitch and loudness (Marks, 1974). Due to the nature of the stimuli, the physical aspects of both modalities are less complex compared to the Gilbert *et al.* research. In one of Marks' experiments, observers were instructed to select a pitch that matched the lightness of a uniform shade of gray on a white background. For the experiment, the eight visual stimuli were gray patches of equally spaced Munsell values 2 through 9 (i.e., dark to light gray patches), each on a white background. The pitch selection ranged from 20 to 20 000 Hz (i.e., the full range of pitch perceivable to the human auditory system) at constant 80 dB, though the starting pitch for all adjustments was fixed at 1000 Hz (approximately two octaves above middle C). The geometric average reveals that the observers selected a 10-fold range of pitch for the range of lightness in the eight gray patches.

A plot of the compiled observer results reveals two main observations, including some fundamentals of human perception. First, the observers intuitively chose higher pitches for lighter patches and lower pitches for darker patches. Second, the observers chose pitches in such a way that the logarithm of the frequency of the pitches is nearly linear

---

3 The Munsell Book of Color and Supplementary 80-Hue Colors are color index books based on Albert H. Munsell's theory of color, which organizes colors into three dimensions: hue, (e.g., red, green, blue, and yellow), value (i.e., the level of lightness), and chroma (i.e., the degree to which a color departs from a gray of similar lightness). The content of each book is organized into pages of color chips with matching hue but varying in value and chroma in continuous degrees. In total, there are over 1500 unique chip colors included in these books.

with the Munsell values of the patches. Note that pitch perception is compressive with respect to the physical stimulus in Hz and that Munsell values are designed to be perceptually linear. Thus, the slope of a semi-log plot of the geometric averages of the observers is theoretically linear. The empirical results corresponded well with the theoretical expectations.

5.4.2 Image Quality-Related Demonstration

A demonstration that can convey the concept of cross-modal psychophysics to a live audience familiar with image engineering uses a representation of an OECF (opto-electronic conversion function) chart (see Chapter 6) and sound at various single pitches. Note that the range of lightness of gray patches in the OECF chart is similar to the range of lightness in patches used by Marks above. Observers are tasked with judging which patch of the OECF has the best corresponding lightness of the presented pitch from 146.83 Hz (a D below Middle C) to 1760 Hz (an A two octaves above A 440 Hz, the typical orchestral tuning note). Prior to judging the set of actual test stimuli, the observers hear the range of pitches and have three practice examples as training.

This demonstration has been used by the authors for many years with class participants taking the technical short course that is the foundation of this book. The room conditions have been uncontrolled illumination levels under 500 lux. Each observer received a printed representation of the OECF from which the measured L^* (D50/F2, 2°) values for each patch had been determined. None of the participants were screened for visual or auditory perception. However, a replicate stimulus at 587 Hz in the middle of the pitch range (a D above A 440) was included in order to determine the consistency of the observer's selection. Observers with standard deviation beyond 4 for the patch number selection of the replicate pair were eliminated from the pooled data because their individual data proved to be of variable and questionable quality. A sample relationship of lightness (L^*) versus pitch (Hz) for one class's data appears in Figure 5.1. Each point

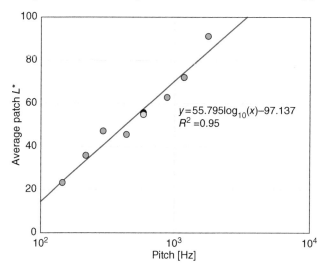

$$y = 55.795\log_{10}(x) - 97.137$$
$$R^2 = 0.95$$

Figure 5.1 The relationship of lightness (L^*) versus pitch (Hz) for 16 observers comparing lightness of OECF patches to a series of single pitches. Note the predictable and linear relationship for the range tested with scales plotted in perceptually linear units of L^* and log(Hz).

is an average L^* of the sixteen observers' responses for the given pitch stimuli. Note the nearly identical L^* results, 55.7 and 54.8, for the replicate 587 Hz stimuli plotted in black and cyan, respectively. This indicates that the observers were quite consistent in their translation of pitch of those stimuli into interpretation of the lightness of the printed OECF chart patches from which they chose. When plotting L^* versus the logarithm of pitch, the coefficient of determination for this set of observers is $R^2 = 0.95$, indicating that the pooled data strongly supports a natural relationship between the visual and auditory cross-modal perceptions for this type of study.

These cross-modal experiments and demonstration example provide insight regarding the plausibility of performing human sensory psychophysical studies when stimuli vary in sensation modality. Putting this in perspective, image quality subjective evaluation of still photography is typically limited to visual sensation, and video image quality subjective evaluation is typically focused on visual sensation (though auditory sensation is often included in the video quality evaluation and can impact perception of the visual sensation). Thus, subjective image quality evaluations are fundamentally less complex for an observer than those cross-modal experiments successfully quantified above involving multiple sensation modalities. As such, those exploring subjective evaluation of camera output should remember that this type of psychophysics is simpler compared to that employed in cross-modal psychophysics. Even for tasks of comparing various visual attributes of image quality to each other, the subjective evaluation skeptic should be reminded that the task is limited to a single modality of vision and further reminded that more complex cross-modal tasks are commonplace in the field of psychophysics. This realization is relevant and important to the applications of the quality ruler psychophysical method described later in the chapter.

5.5   Thurstonian Scaling

As noted in the Psychophysical Methodologies section above, analysis of the forced-choice comparison method involves the use of Thurstonian scaling. Louis Leon Thurstone established the Law of Comparative Judgment in which he formalized a model for the process of stimuli judgment, defined different cases of the model, and identified assumptions needed for obtaining an interval scale from the data (Thurstone, 1927; Engeldrum, 2000). In Thurstonian scaling, the variance in the paired comparison judgments is used to calculate response distributions for each stimulus. Overlaps in the distributions are then used as a measure of perceived distances between the stimuli. As such, indirect scaling is used to derive an interval scale.

Thurstone defines the perceived dimension of an attribute, for example, colorfulness of an image, as the *psychological continuum* of that particular attribute. He describes the process of comparing two or more stimuli, asking, for example, "Which stimulus is more colorful?", as the *discriminal process*. The *discriminal deviation* refers to the separation on the perceptual scale between the discriminal process for a given attribute, for example, colorfulness, of a particular stimulus and the modal discriminal process for that attribute. The *discriminal dispersion* is the standard deviation of the distribution of discriminal processes on the scale for a particular stimulus, for example, the standard deviation of repeated judgments of the colorfulness for a particular image. The scale difference between the discriminal processes of two stimuli being rated is the *discriminal difference*, for example, the quantified difference in colorfulness between two images. Thurstone assumed that repeated judgment of the stimulus could produce a response described by a normal distribution on the psychological continuum. The variation over a large number of trials for discriminal difference is also assumed to follow a normal distribution.

Therefore, Thurstone generated the Law of Comparative Judgment to be

$$S_1 - S_2 = z_{12} \sqrt{\sigma_1^2 + \sigma_2^2 - 2r_{12}\sigma_1\sigma_2} \tag{5.5}$$

where S_1 and S_2 are the psychological scale values of the two compared stimuli, z_{12} is the z-score[4] with stimulus 1 exceeding stimulus 2 (i.e., the proportion of times that the attribute of stimulus 1 is judged greater than stimulus 2), and the remaining square root is the standard deviation of the probability density function describing the difference between the two compared stimuli in which σ_1 and σ_2 are the discriminal dispersions of stimuli 1 and 2, respectively, and r_{12} is the correlation between the discriminal deviations of stimuli 1 and 2 in the same judgment. The variables in the square root term cannot be measured experimentally, so they must be determined by assumption or estimated from experimental data (Gescheider, 1997). Note that the law is specific to paired comparisons.

Thurstone established five cases to describe different sets of assumptions. Case I follows the general law for replication over trials for an individual observer, while Case II follows the same law for a group of observers in which each judge makes each comparison only once. Thus, both follow Eq. (5.5). Case III represents the situation in which the evaluation of one stimulus has no influence on evaluation of the second stimulus. Thus, the r_{12} term is zero and the equation becomes

$$S_1 - S_2 = z_{12} \sqrt{\sigma_1^2 + \sigma_2^2} \tag{5.6}$$

In addition to the Case III assumption, Case IV assumes that the discriminal dispersions σ_1 and σ_2 approach equality. Thus, the following Case IV relationship can be derived:

$$S_1 - S_2 = \frac{z_{12}}{\sqrt{2}}(\sigma_1 + \sigma_2) \tag{5.7}$$

Case V is Thurstone's simplest case, in which the assumption additional to Case III is that the discriminal dispersions are identical. Thus, the general equation becomes

$$S_1 - S_2 = z_{12}\sigma\sqrt{2} \tag{5.8}$$

Case V is utilized most often in subjective testing, providing a common means of establishing scales of perceived attributes, for example, image quality.

An application of Thurstonian scaling is contained within the second part of the ISO 20462 standard, which describes a triplet comparison method (3AFC) (ISO, 2005b). As the method points out, two shortcomings of the paired comparison method are the large number of assessments to be evaluated (e.g., $n(n-1)/2$ comparisons for n levels) and the lack of estimability as the stimulus difference approaches unanimity, that is, approximately 1.5 JNDs or larger. An advantage of simultaneous triplet comparison is reduction

---

4 A z-score is the obtained value subtracted from the mean and divided by the standard deviation. It therefore constitutes a measure of the number of standard deviations by which a value differs from the mean.

Figure 5.2 A diagram showing an example triplet comparison on the left of stimuli 1, 2, and 3 versus the equivalent of three paired comparisons of the same three stimuli on the right. For the left case, the observer compares all three stimuli at one viewing time, whereas on the right case, three separate comparisons are necessary. Even though observer judgment time is longer for the left triplet comparison, an experiment with triplet comparisons can be judged more quickly than the experiment with the same stimuli using separate paired comparisons because less presentations are necessary.

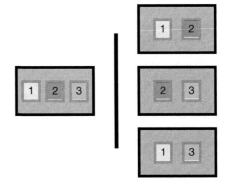

of the number of sets of images presented to the observer because a triplet comparison is equivalent to three separate paired comparisons. Note that if the number of levels to be evaluated is limited to $n' = 6k + 1$ or $6k + 3$ ($k = 1, 2, 3, 4, 5$, etc.), then the number of triplet comparisons is fundamentally $n'(n' - 1)/6$, that is, three times less stimuli presentations than required for a paired comparison study.

Figure 5.2 has a representation of the triplet comparison on the left and its equivalent in three sets of paired comparisons on the right. Fundamentally, the triplet comparison experiment can be judged more quickly than the three separate paired comparisons because there are less comparisons, though the judging of an individual presentation of a triplet takes longer than an individual paired comparison because more comparisons are made with one presentation of samples. Annex D in the ISO 20462-2 standard states that the triplet experiment time was "almost 50%" faster for the comparisons of 21 stimuli by 15 observers (ISO, 2005b).

Once the comparisons have been made, Thurstone Case V assumptions are made to obtain a cumulative frequency distribution of each experimental comparison. These individual values are then converted into probabilities, p, for each stimulus, which are ultimately used in the following formula to obtain JND values:

$$\text{JNDs} = \frac{12}{\pi}\sin^{-1}\left(\sqrt{p}\right) - 3 \tag{5.9}$$

5.6 Quality Ruler

The psychophysical method using a quality ruler, that is, a form of evaluation using anchor images, has been shown to reduce experimental variability within an observer session, improve repeatability between sessions, and increase evaluation speed from methods such as single stimulus (sequential judgment of individual images without any reference stimulus) (Redi *et al.*, 2010). As mentioned previously, because the quality of the anchor stimuli for a quality ruler is calibrated, the scale becomes strongly intervallic. In addition, if the anchors remain constant, results from testing over time can be compared and combined. These attributes of a quality ruler have resulted in a specific implementation being standardized by ISO (ISO, 2012), which will be described in more detail next.

The quality ruler described in the ISO 20462-3 standard (ISO, 2012) consists of a set of images having known stimulus differences varying in sharpness. The attribute of sharpness was chosen for the ISO quality ruler because this meets the ideal quality ruler criteria presented in Keelan (2002a). The criteria and their respective fulfillments are:

- Low variability of observer sensitivity and scene susceptibility
- Capable of strongly influencing quality so that a wide quality range can be spanned
- Produced at widely varying levels so as not to appear contrived
- Easily and robustly characterized by objective measurements
- Readily simulated through digital image processing

The ruler image sets are calibrated in overall quality JNDs. Overall quality JNDs quantify the perceptual judgment of stimulus differences with respect to their quality, regardless of the specific univariate attributes which lead to the differences. These are distinct from attribute JNDs, which quantify the detectability of appearance changes due to a given univariate attribute, for example, noise, texture blur, or geometric distortion. Keelan (2002a) has found that paired comparisons generate univariate attribute JND increments approximately half as large as overall quality JND increments, which follows the properties that univariate attribute JND increments measure detectability thresholds and multivariate quality JND increments measure suprathresholds. That is, it takes suprathreshold amounts of an attribute JND to achieve one quality JND.

The overall quality JND value associated with each ruler image was derived from psychometric paired comparison experiments with hundreds of real world images. The images used to generate the overall quality JNDs consisted of many combinations of attributes spanning variations in image quality associated with consumer and professional photography (Keelan, 2002a). Fundamentally, the image quality assessment of these types of stimuli generates multivariate JND increments because many individual attributes comprised the content as established. While the univariate attributes assessed in this quality judgment such as noise and color balance may have surpassed the threshold detection, the observers were asked to judge their value and impact on overall quality. For example, observers may have detected many attributes in the images, but the task was to determine how those univariate attributes influenced the multivariate overall quality.

The specific type of JNDs considered here are 50% JNDs because they were determined when 50% of the observers were able to correctly identify the stimulus difference in forced-choice paired comparisons used to establish the JNDs while the remaining 50% of the observers were guessing. In the ISO 20462-3 standard (ISO, 2012), the 50% JND is defined as the "stimulus difference that leads to a 75:25 proportion of responses in a paired comparison task" because, for the 50% of observers who are guessing, half of them (25% of the total observers) are still correctly identifying the stimulus. Thus, the total "correct" observers compared to "incorrect" observers fits the 75:25 proportion.

Note that the ruler sets comprise 31 anchor images, which have fixed values on the *standard quality scale* (SQS) in units of JNDs. This scale has a zero point, which corresponds to a quality level for which the image content is difficult to identify due to the strong level of blur. Test images are compared to the ruler images and the observer chooses where on the SQS the quality of the test image falls by selecting a ruler image that best matches the quality of the test image. Relative to the zero point, the SQS increases in overall quality and numeric designation of JNDs. Currently, the maximum SQS is 32 for the quality achievable and the span is up to 30 JNDs in range in the stimulus sets, depending on specified viewing distance of the ruler and test

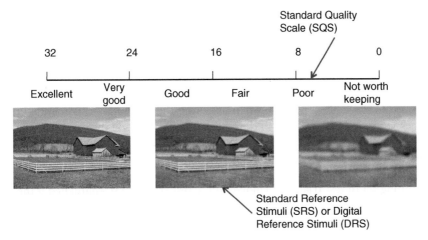

Figure 5.3 The diagram demonstrates key aspects of the ISO 20462 Part 3 quality ruler components, including the calibrated scale from 32 to 0, the associated quality categories, and representations of ruler quality levels for a given scene (ISO, 2012).

images when judging. See Figure 5.3 for a visual representation of the SQS and selected sample images to represent the range on the scale. Note that there are two forms of the ruler images: the hardcopy *standard reference stimuli* (SRS) and the softcopy *digital reference stimuli* (DRS). When using the former, the SQS is referred to as the *primary SQS* (SQS_1), while the latter is referred to as the *secondary SQS* (SQS_2). This is because the SQS_1 calibrated values were generated directly from subjectively assessed hardcopy photographs, whereas the SQS_2 calibrated values were derived from an empirical model of the subjective findings used for the primary calibrated values and then associated with the softcopy DRS. Recall that a calibrated anchor scale becomes highly intervallic. As such, the category scales of "excellent," "very good," "good," "fair," "poor," and "not worth keeping" (also highly intervallic) can be applied to the SQS as indicated in Figure 5.3, noting that 6 JNDs represents an interval for such image quality categories (Keelan, 2002a).

An important aspect regarding the use of the ISO quality ruler approach is that the scale is quantification of overall quality of a photograph even though the ruler images vary only in the univariate attribute of sharpness. That is, the associated quality JNDs are JNDs of overall quality and not JNDs of the attribute of sharpness. This means that while other quality rulers can be generated for a univariate attribute such as noise, the ISO quality ruler is calibrated in quality units, that is, Standard Quality Scale Just Noticeable Differences (SQS JNDs). In practice, this means that the observer judges a test image by comparing it to a set of quality ruler scenes and determining which overall image quality level (not sharpness level) matches the quality (not sharpness level) of the test images; in this manner, the judgment is a value judgment. This paired comparison task is with multivariate stimulus pairs. Note how the differentiation of the definitions of an attribute JND versus a quality JND in the ISO document relate to this distinction (ISO, 2012). For an attribute JND, the comparison task is for assessment of univariate stimulus pairs in terms of a specified single attribute.

In order to convey this important aspect of the quality JNDs, the experimental instructions play a key role. Annex B in the ISO 20462-3 standard (ISO, 2012) contains sample instructions, which include details for both the tester and the observer. The instructions highlight that the study is for assessing overall quality (not sharpness) and that the paired comparison is between the test stimuli and ruler reference (digital reference stimuli). A phrase in the standard, "ask yourself which image you would keep if this were a treasured image and you were allowed only one copy," describes the key essence of the task—choosing which image to keep as a better representation of an important photographic moment when assessing the tradeoffs of blurriness in the ruler series versus other attributes in the test scene that degrade the overall image quality. These sample instructions clearly state what attributes are to be considered (here, geometric distortion and unsharpness) when comparing overall quality of the test stimuli with the ruler reference. In order to assist the observer in carrying out the task, a training set of test trials is included. These test trials are strengthened if the tester is present with the observer to make certain that the observer is assessing the overall quality rather than a comparison of sharpness of the test stimuli.

As should be understood, the observers need to be screened for their vision acuity and color normalcy prior to running a study so that the tester knows the visual capability of the observer. Just as tasks such as calibration and accuracy tests are utilized with measurement instrumentation such as colorimeters and spectrophotometers, an analogous process should be applied with observers involved in psychometric testing. Details on screening will be discussed in Chapter 8.

5.6.1 Ruler Generation

A number of softcopy quality ruler sets of various scene content are made available as part of the ISO 20462 standard (ISO, 2012). The scenes depicted span a broad range of environments and photographic situations, with an emphasis on portraiture and natural landscapes, and some exemplars of architecture. Of course, the standard set cannot be expected to be entirely comprehensive, and there will be circumstances in which no supplied ruler scene corresponds to the photographic situation being assessed. Examples of contexts not covered include scenes depicting vehicles, astronomical bodies, industrial design models, street scenes, and low light conditions.

The quality ruler can be used where the images being assessed depict a wholly different scene to the ruler images. However, in general, the assessment task is easier and more intuitive when the reference and assessed scenes have some similarity: a portrait, for example, is ideally compared with another portrait, and the subject of the portraits should be similar. Where the scenes are not similar, the results are still valid, but the assessment exercise takes longer to run, and the observers report increased fatigue. These factors increase as the similarity between the reference and assessed scene decreases.

In such cases it may be expedient to generate a new quality ruler, or multiple rulers, which closely or exactly match the scenes being assessed. Both the ISO 20462-3 standard (ISO, 2012) and Keelan's Handbook of Image Quality (Keelan, 2002a) describe how to generate a univariate quality ruler varying in sharpness. In outline, the procedure to generate a softcopy quality ruler is as follows:

- Measure the modulation transfer function (MTF) of the camera used to image the quality ruler scene (this is the reference image), making use of the techniques described in Chapter 6.

- If needed, compute the MTF of the downsampling filter used to resize the reference image suitable for the softcopy quality ruler display.
- Measure the MTF of the display used to present the quality ruler.
- Combine the camera, downsampling, and display MTFs to compute the *system MTF*.
- Compute the *aim MTF*: the MTF required to generate the amount of blur appropriate for the quality ruler level at hand.
- Design and apply a filter to compensate for the difference between the system MTF and the aim MTF.

This process is repeated for each quality ruler level to form a complete set.

5.6.2 Quality Ruler Insights

Use of quality ruler methodologies does have accompanying challenges and idiosyncrasies. Extensive studies carried out at various corporations were included in the work by CPIQ and ISO to expand the original quality ruler approach with printed rulers to that of rulers displayed on color monitors (Jin *et al.*, 2009; Jin and Keelan, 2010; I3A, 2009). From these studies, several key findings are gleaned:

- With proper care, calibrated images displayed on a color monitor can be used to obtain reproducible and predictive image quality assessment.
- Similar, but not identical, lab conditions can provide statistically similar psychophysical results.
- As image quality degrades, variability in the subjective data increases: observers diverge on what constitutes poor image quality. Conversely, variability decreases as image quality increases: observers agree on what is excellent image quality.

5.6.2.1 Lab Cross-Comparisons

In the process of developing subjectively correlated metrics in CPIQ, several companies collaborated by sharing raw data from their psychophysical studies, which provided insight into the use of quality rulers. For example, four companies were involved in initial psychophysical research to study the correlation between results from quality ruler anchored scaling and preliminary versions of objective texture metrics (Phillips *et al.*, 2009). Three of the companies set up their labs in a similar manner to present observers with anchored pairs using the softcopy quality ruler method described in Jin *et al.* (2009) with an LCD monitor having a native resolution of 2560×1600 pixels with a pixel pitch of 0.250 mm. Ruler and test images were judged at a viewing distance of 43 inches using affiliated SQS_2 JNDs for that controlled distance. A fourth company used the same anchored pairs, but used an alternative method with a monitor having a native resolution of 1920×1200 pixels with a pixel pitch of 0.258 mm. In addition, the viewing distance was not fixed and was approximately 16 inches. Ten photographic scenes, each with eight levels of increasing texture blur (see Chapter 3), were judged by all observers at their respective companies with their respective protocols. A total of 17 observers used the softcopy quality ruler method and 30 observers used the alternative method.

An important aspect of the softcopy quality ruler method to understand is that the specific monitor and images are used as a calibrated and controlled visual tool, with defined SQS_2 JNDs for each ruler image specifically calculated for three viewing distances: 25, 34, and 43 inches. These viewing distances should not be confused with

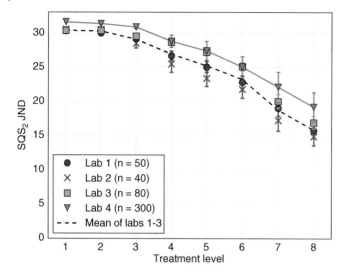

Figure 5.4 A comparison of subjective evaluation with anchored pairs performed in four different labs with four different sets of equipment. Increased treatment resulted in increasing amount of texture blur. This was corroborated by the psychometric results. Error bars are standard error values. *Source:* Data for graph from Phillips *et al.* (2009).

everyday usage of monitors to view images from any distance, most often at distances closer than the closest distances for the calibrated quality rulers. The SQS_2 JNDs for each associated ruler is determined by the perceived sharpness of the system as a function of the spatial response of the HVS. Thus, the alternative method having observers judge at a much closer distance to a monitor having higher pixel pitch than the formalized method results in a protocol that alters the relationship between the ruler and test images.

Figure 5.4 contains the SQS_2 JND values for each of the three labs using the formalized softcopy quality ruler method and the ruler results for the alternative method. Error bars are standard error values, which are dependent on the number of evaluations per data point. Note that the standard error increases as the image quality of the stimulus degrades, that is, when the SQS_2 JND values decrease. Lab 1 had 5 observers, Lab 2 had 4 observers, Lab 3 had 8 observers, and, as stated earlier, Lab 4 had 30 observers for their alternative method. Notice that as the texture blur level increases for the 10 scenes with increasing treatment level, the SQS_2 JND values decrease, which indicates that observers saw significant degradation in overall quality from the original images to the images with greatest texture blur. An average of the three labs using the softcopy quality ruler method shows that the alternative method is highly correlated, though with an offset as expected given the closer viewing distance and different monitor specifications.

5.6.2.2 SQS₂ JND Validation

The softcopy quality ruler methodology underwent additional scrutiny and was later incorporated into an updated version of Part 3 of the ISO 20462 method (ISO, 2012). As mentioned earlier, the primary SQS (SQS_1) calibration values associated with the hardcopy SRS were obtained directly from subjective evaluation, whereas the secondary SQS (SQS_2) calibration values associated with the softcopy digital reference stimuli (DRS) were obtained from an empirical model derived from the subjective findings used for the primary calibrated values. So, fundamentally, the SQS$_2$ are modeled rather than psychometrically obtained. There were also accommodations made to the MTF requirements of the hardcopy rulers when generating the softcopy version of images displayed on monitors, which have limitations in higher frequency capability at shorter viewing distances (Jin *et al.*, 2009). Thus, scrutiny was considered necessary to determine if the empirical model and resolution accommodations for generating the calibration values of the softcopy rulers were indeed accurate for the intended viewing distances.

As presented earlier in this chapter, JND values can be derived from the statistics of non-unanimous paired comparisons. Therefore, it was determined to run a paired comparison study using six expert observers at Kodak to judge one scene (the "girl" scene) of the softcopy quality ruler sets at a viewing distance of 34 inches (Phillips and Shaw, 2011). The experimental conditions were set up following ISO 20462, though the ruler validation probe did not fulfill the standard requirement to use a minimum of 10 observers and 3 scenes. Table 5.4 contains the SQS$_2$ values with their associated ruler designation to be used for the ruler scenes judged at a viewing distance of 34 inches. The SQS$_2$ values are specific to the Canon EOS 1Ds Mark II D-SLR camera because that camera was used to capture the "girl" scene. As can be noted, the ruler images at the high quality end of the calibrated SQS$_2$, that is, those with low ruler designation numbers, have very similar JND values (less than 0.5 JND deltas).

A preliminary study indicated comparisons of all neighboring pairs was fatiguing due to difficulty of making evaluations more precise than one JND. Thus, it was determined

Table 5.4 Example ISO 20462 Part 3 SQS$_2$ JND values for scenes taken with a Canon EOS 1Ds Mark II D-SLR camera to be used for ruler images judged at a viewing distance of 34 inches (from supplemental material for ISO (2012)). Note the sub-JND spacings for the high-quality end of the calibrated scale (rulers with highest SQS$_2$ JND values).

Ruler Designation	SQS$_2$ value	Ruler Designation	SQS$_2$ value
0	32.09	16	21.08
1	31.88	17	20.01
2	31.45	18	18.91
3	30.94	19	17.78
4	30.42	20	16.65
5	29.88	21	15.48
6	29.34	22	14.28
7	28.77	23	13.06
8	28.14	24	11.80
9	27.48	25	10.51
10	26.73	26	9.18
11	25.86	27	7.81
12	24.98	28	6.41
13	24.06	29	4.96
14	23.10	30	3.47
15	22.11		

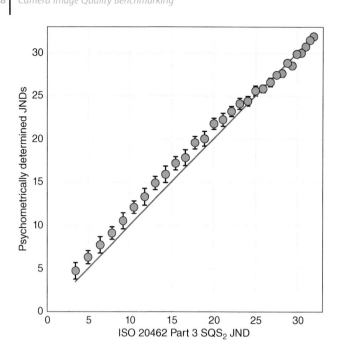

Figure 5.5 The relationship between the psychometrically determined JNDs using six expert judges and the ISO 20462 Part 3 calibrated SQS_2. The results are for the "girl" scene viewed at 34 inches as compared to the calibrated values (the solid line) for the same conditions, indicating that the modeled SQS_2 values are valid. Each data point for this experiment performed in Lab 2 represents $n = 6$ judgments and error bars are standard error. *Source:* Data for graph from Phillips and Shaw (2011). Reproduced with permission of Kodak.

that paired comparisons should be made from two subsets of images obtained by selecting only the odd-numbered ruler images for one set and only the even-numbered rulers for a second set. As such, the quality difference became more apparent, making the judgment less fatiguing and time consuming for the observers. Psychometrically based JNDs from the experimental data from the expert observers were compared to the SQS_2 calibrated values for the same viewing distance. Figure 5.5 shows that the psychometrically determined JNDs from the pooled observer data show good agreement with the calibrated values of the SQS_2, supporting their validation. While there is a quality overprediction of up to 1.84 JNDs for the empirically based JND values versus the SQS_2 values, expectations are that the discrepancy would dissipate with increasing the observer pool to 10 judges or more and increasing the number of ruler scenes assessed to 3 or more, following the ISO 20462 requirement for minimum judges and scenes.

The error bars in Figure 5.5 are standard errors for pooled judgments of the six observers. As can be seen, the standard error increases as the quality of the stimuli degrades, that is, as the SQS_2 values decrease. The error has an oscillation as an experimental artifact of sampling every other ruler for the odd-numbered and even-numbered sets describes above. However, the standard error plateaus at approximately 1 JND in the region of SQS_2 values of <16 for data from this limited number of observers and one scene.

More recently, a study was conducted in academia with 18 observers to compare SQS_2 values with paired comparison results for judging image quality of the same set of 10 scenes captured with 9 different smartphones (Jin *et al.*, 2017). SQS_2 JND values were obtained directly using the ISO 20462-3 method and z-scores were calculated from the paired comparison analysis. Both subjective methodologies were administered at a viewing distance of 34 inches. The correlation of these two approaches to quantifying overall image quality of the 90 images was very strong, with mean correlation of 0.89. This experiment also points to the validity of the SQS_2 JND scale used for the softcopy quality ruler methodology.

5.6.2.3 Quality Ruler Standard Deviation Trends

Further exploration of the data distribution of the quality ruler and alternative anchored paired comparison methods described in the Lab Cross-Comparisons section above was performed with results from four companies that ran experiments on the impact of geometric distortion (see Chapters 3 and 6) on overall image quality (Mauer and Clark, 2009). The participating companies were the same four companies as with the experiment for texture blur above, and, again, the same three companies followed the softcopy quality ruler approach while one company followed the alternative method. Four types of geometric distortion (barrel, pincushion, negative wave, and positive wave) were applied in increasing TV distortion levels up to 15% to six base images. Over 7000 individual subjective assessments from 60 observers were performed in obtaining the data in Figure 5.6. Each point represents the SQS_2 JND average of each company's observers for a given image. As was noted in the previous two subsections, observer variability increases with decreased quality of the stimuli. Again, the standard deviation is low at the high-quality end of the ruler scale and then begins to plateau at the mid-scale of the quality ruler. Note that the standard deviation varies between labs. Analysis indicates that Labs 2, 3, and 4 are statistically not different from each

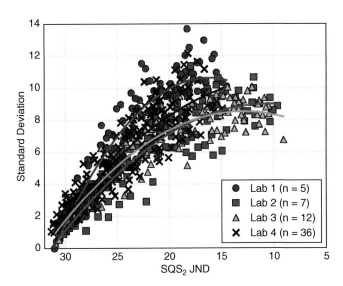

Figure 5.6 SQS_2 JND averages for each of four companies' labs. Labs 1, 2, and 3 followed the softcopy quality ruler approach while Lab 4 followed an alternative anchored paired comparison method. The standard deviation is plotted versus the SQS_2 JND average for each stimulus. Data sets are fitted to a second order polynomial fit. *Source*: Data for graph from Mauer and Clark (2009).

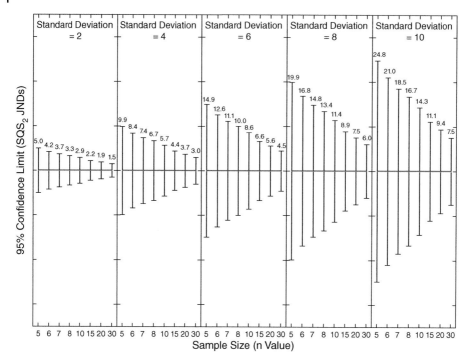

Figure 5.7 95% confidence limits (in units of SQS$_2$ JNDs) for varying sample sizes at given standard deviation level. *Source*: Data for graph from Mauer and Clark (2009).

other, whereas Lab 1 with the least amount of observers does differ. Addition of more observers could eliminate this Lab 1 difference, particularly when any observer in a small group, for example, $n = 5$, can strongly impact response.

Additional statistical modeling was performed on this data set to determine the impact of the value of the number of judgments per sample (i.e., the value of n) on the 95% confidence interval (CI), depending on the estimated standard deviation based on anticipated value of the SQS$_2$ JND scale for a given stimulus (Mauer and Clark, 2009). As can be seen in Figure 5.7, increasing the sample size greatly impacts the expected 95% CI for given standard deviation levels, particularly for higher standard deviation. The ISO 20462-3 standard requires a minimum of 10 observers and 3 scenes, or a sample size of 30 (10×3), for reportable SQS$_2$ JNDs (ISO, 2012). While Labs 1 and 2 had less than the recommended 10 observers, each observer in those labs judged geometric distortion levels applied to 6 scenes, so the sample size was 30 for Lab 1 and 42 for Lab 2. As can be observed from this sample size of 30 in Figure 5.7, the maximum presented in the figure, the 95% CI increases from 1.5 to 7.5 as the standard deviation of the observer data increases from 2 to 10. This experimental data supports the ISO quality ruler requirements as well as the standard's comment that the preferred experimental design is preferably 20 observers and 6 scenes, that is, a sample size of 120, which would reduce the 95% CI and strengthen conclusions from the data.

The implication from Figure 5.6 is that the standard deviation will be increasingly higher as the quality of the image stimuli decreases, which therefore relates to the standard deviation levels presented in Figure 5.7. Remember from Figure 5.3 that the intervallic categories of *excellent* to *not worth keeping* are associated with values on the SQS$_2$ JND scale. Thus, if the image set to be evaluated is expected to include images with image quality at the mid-scale or worse of the SQS$_2$ JND scale, that is, *good, fair, poor*, and *not worth keeping*, then additional observers or scenes should be added to narrow the 95% CI of the results.

These three studies above regarding statistical characteristics of subjective data with quality ruler methods indicate that anchored scaling for psychometric evaluation does provide an important tool for conveniently obtaining subjective results. Conclusions from psychometrics are based on statistical distributions. Therefore, understanding the fundamental statistical characteristics of anchored scaling presented above are important in determining experimental parameters that will influence the level of confidence in the conclusions.

5.6.2.4 Observer Impact

As was briefly mentioned in the Image Quality-Related Demonstration subsection, including replicates in a subjective experiment can provide insight regarding consistency of the observer pool's responses. Experience with the quality ruler method has led to published observations regarding discarding of observer data based on standard deviation and bias of replicates, using test stimuli that are images taken directly from the quality rulers themselves, known as *null images* (Keelan *et al.*, 2011). The experimenters withdrew observer data for those whose standard deviation were greater than 2.5 from the known SQS$_2$ JND values of the rulers used for this task of matching null images from the upper and mid-scale. This culling was performed in order to limit the analysis to observers who could closely match the null image with its exact match in the ruler set. The assumption was that these observers should also be providing a similar caliber of matching to the non-null experimental stimuli. Regardless of a specific standard deviation cut-off, inclusion of replicate judgments of null images and evaluation of the accuracy for given observers will provide insights regarding each observer's consistency and accuracy, which should be understood when choosing which observers to include in the data analysis pool.

The Keelan *et al.* (2011) paper also shows how the sensitivity varies from observer to observer for the same set of stimuli. In their study, the stimuli varied in noise level, which was then compared to the sharpness levels of the quality rulers to determine an SQS$_2$ JND value. For some observers, the slope of the SQS$_2$ versus the noise level was steep, indicating that presence of noise strongly impacted these observers' assessment of overall image quality. However, other observers were much less responsive to noise, even to the degree of it having little impact on their overall quality assessment. Other attributes have also shown observer sensitivity differences (Keelan, 2002a). Thus, observer-to-observer differences will have an impact, necessitating multiple observers to achieve a "standard observer" for a given study.

As has been noted in the literature, subjective evaluation will differ between experienced observers, those associated with the task at hand, and non-expert or naïve observers (Keelan, 2002a; Engeldrum, 2000). For some attributes, experienced observers may become oversensitized and provide evaluation differing substantially from naïve observers. Or, the experienced observers may have seen artifacts often enough to the point of becoming desensitized. From a benchmarking standpoint, the implications of this will discussed further in Chapter 9. However, the experience level of the observers should be noted in the analysis report.

In a study in academia comparing the advantages and disadvantages of single stimulus and quality ruler methods, researchers found that the latter method provided several advantages: smaller 95% confidence intervals, smaller RMSE (root mean square error) for linear regression, higher correlation with the objective measurement of the photographic attribute being varied, higher consistency between independent experiments over time, more robustness to range effects, faster assessment, and high significance of the psychophysical results due to the method's calibrated SQS JND scale (Redi *et al.*, 2010). A key disadvantage of the quality ruler method was noted as the effort and time needed to generate the rulers, specifically because the authors generated their own rulers following the whole empirical procedure described in Keelan (2002a) rather than following the approach incorporating the camera and display MTF curves or use of the existing quality rulers (since the study was run prior to the ISO softcopy rulers becoming available publicly). The authors also identified a possible disadvantage of the quality ruler method as suspected overestimation of quality, though this was in reference to the single stimulus results and not a known set of quality ratings.

For the single stimulus method in this study, the observers assigned an overall quality value of 0 (worst quality) to 100 (best quality) to the test scene. For the quality ruler method, a single scene of a sailboat varying in sharpness was determined to have the highest confidence in ratings obtained in preliminary screening experiments, and thus was used for the ruler itself. The observers compared the test scene to this univariate quality ruler shown in decreasing overall quality until the test image was judged better than one ruler image, but worse than its adjacent ruler image on the SQS JND scale. For each method, 18 observers judged a total of 21 unique scenes, including originals and each scene blurred at five different levels with a circular-symmetric 2D Gaussian kernel, thus totaling 126 images. A followup experiment included judgments by 10 observers, all different from the first experiment, of each method using 50 images. Findings described above were similar for each experiment.

This was the first independent study of the softcopy quality ruler approach published by Jin *et al.* (2009) and provided insights into the usage of such a method for subjective evaluation. Note that the test scenes in the Redi *et al.* (2010) study were all different from the ruler scene, so observers needed to extrapolate between overall quality appearance in one scene to that of another scene. The task was simplified because the attribute of blur being varied in the test scenes was applied by the same means as the univariate attribute of blur of the rulers. However, the observers were asked to judge based on overall quality, not sharpness per se. The viewing and display conditions varied from the subsequent revisions of the ISO method. However, the experiments were performed with the same conditions as used for the sailboat quality ruler generated for their experiments. Thus, the Redi *et al.* (2010) anchored scaling approach is valid and self-consistent. The advantages noted by the researchers for the softcopy quality ruler provided an independent validation of what was determined in industrial research during a similar time period. In addition, the experiment is an example of how mismatched scenes between a ruler scene and test scenes can be utilized for obtaining psychophysical results.

Another study in academia provides insights directly tied to the ISO 20462-3 method, most significantly regarding means of addressing pragmatic concerns of utilizing the quality rulers and statistical evaluation of ruler applications (Persson, 2014). By the time this research was being performed, the ISO standard had been officially revised to incorporate the softcopy approach and the standard ruler images had become available. As such, 3 rulers were chosen from the available 21 to be utilized in this research. Unlike the Redi *et al.* (2010) research, this research used test stimuli with scenes matched to the quality rulers in order to simplify the judgment process, and the attributes of noise and noise reduction levels were being studied, rather than the ruler attribute of sharpness.

The lab chose to use a display with a pixel pitch of 0.2331 mm versus the pixel pitch of 0.25 mm used for the preparation of the ISO standard. In order to understand necessary modifications to the viewing conditions, a derivation of the two theoretical display MTFs were made to compare to the measured MTF curves provided in Jin and Keelan (2010). With an assumption that both displays have perfectly square pixels of equal width and height, the theoretical MTFs were calculated to be very similar for the two displays and also approximated the measured MTFs well. Because of these findings, it was determined that the ISO SQS_2 JND calibration values could be utilized as long as the viewing distance was modified accordingly for the pixel pitch differences. For the pixel pitch ratio of the two displays, it was determined that the 34-inch SQS_2 JND calibration values could be used if the lab's display were viewed at 31.7 inches.

A preliminary study with four observers was run to ensure that the lab setup was operating as expected and that the viewing modification was appropriate. Twenty levels of Gaussian noise were added to ruler image number 3 of one of the ISO ruler scenes. A plot of the averaged SQS_2 JNDs versus the log of the object metric of noise showed that the data fit the form of Weber's Law with a coefficient of determination of 0.926, ensuring a successful protocol.

The final study had 47 observers, including both experienced and inexperienced observers. A set of noise test images were generated by applying five levels of Gaussian noise to ruler image number 3 of three different ISO quality ruler scenes. For each of the five noise levels, four levels of noise reduction were applied to generate the test images with noise reduction. In total, including a group of the five levels of Gaussian noise without noise reduction and the four levels of noise reduction applied to each, there were 25 variations × 3 scenes = 75 test images generated for the study. In addition, for 30 observers, the original image of ruler number 3 from each of the three ruler scenes was included as a null image to verify that observers could accurately and reproducibly match a ruler scene with itself.

Results from the final study corroborated many of the findings from industry described in the Quality Ruler Insights subsections above. Firstly, for the 30 observers who judged the null images, the standard deviation of the deviation from the correct match was calculated as in Keelan *et al.* (2011). Three of the 30 observers were identified to have standard deviations significantly higher than 2.5 and were removed from the study, indicating that vetting using null images could identify outlying observers. Secondly, in probing the normal approximation using the non-parametric bootstrap method, the total study distributions indicated the normal distribution did indeed describe the data set well (as assumed in the industrial studies). Thirdly, the size of confidence intervals increased as the quality of the test images decreased in a similar manner to the industrial studies (Phillips and Shaw, 2011; Mauer and Clark, 2009). Additional data analysis indicated that the confidence intervals increased most as the noise levels increased while confidence intervals decreased as the noise reduction increased. The latter trend follows from the former trend because, as the noise reduction increased, the inherent noise level dropped. Persson (2014) points out that this implies that the more a test image resembles the set of ruler images (i.e., nearly noiseless and varying univariately in sharpness), the easier the images are to judge and thus observers' results are more similar. This observation corresponds to the fact that appearance matching of sharpness, as done with the null images, is easier than value judgment of differing attributes. Fourthly, the confidence intervals decreased as the number of observers included in the pooling increased, which was also shown by Mauer and Clark (2009) in Figure 5.7. Fifthly, using Welch's *t*-test, the results of the experienced versus inexperienced observers were found to be statistically different. Specifically, images with the highest presence of noise were judged to be better by the experienced observers, indicating a difference in how the groups interpreted the impact of noise on overall quality.

Thus, for the quality ruler method, academic research findings support the findings in industry. The quality ruler method provides an example of anchored scaling that can be used to obtain reliable and repeatable subjective data. More importantly for image quality camera benchmarking, the quality ruler results for orthogonal attributes can be studied over time and retain relevancy because the rulers remain constant. Ultimately, the SQS$_2$ JND ruler results can be combined to generate multivariate overall quality predictions. This will be discussed further after presenting a practical example below.

5.6.4 Practical Example

Here is a practical example of using a quality ruler for psychometrics. Say, for example, a tester would like to understand how noise level and noise reduction impact overall image quality in consumer photographs using the ISO 20462 Part 3 method (ISO, 2012). This means that the impact of the noise in various scene content such as sky, skintone, and shadow regions needs to be assessed as part of the study rather than solely evaluating the impact of noise on a flat field such as patches in an OECF chart. In order to obtain the understanding of the impact of noise and noise reduction, a variety of scenes should be evaluated—the more scenes included in the study, the more the tester will understand the impact of the noise. In fact, variability due to scene content can vary more widely than observer variability (Keelan *et al.*, 2011). Thus, the importance of defining the scope and intent of the study cannot be understated with respect to the strategic selection of scene content.

When using the quality ruler, test scenes should be selected to mimic those available in the DRS in order to simplify the judging task for the observers. Scenes for this practical example should have content with low spatial frequency (see Chapter 6), which allows for better assessment of noise level as well as fine detail for assessing the impact of noise reduction. For example, the ruler scene with a mansion contains a region of blue sky with minimal frequency modulation as well as fine detail in the mansion's brick walls, foliage in the vines growing on the walls, and roof material. Note there is also a region of shadow underneath the entrance porch roof that can allow for better assessment of noise in darker areas. As noted in the ISO 20462-3 standard (ISO, 2012), there need to be a minimum of three scenes included in the study. This provides statistical significance as well as breadth of understanding the impact of noise and noise reduction.

For simulations, the recommendation is to use the second or third quality level of a ruler set as the base image to which the alterations are applied (in "Aptina Softcopy

Quality Ruler User's Manual" included in supporting files for ISO 204462-3 (ISO, 2012)). Then, the judging task is a comparison of very similar content. For camera benchmarking applications, the scene content of the test image to be photographed can be chosen such that it is similar to a given ruler scene. However, the earlier discussion in the chapter regarding cross-modal psychophysics should provide perspective regarding the fact that it is entirely possible, though more challenging, to judge a test scene for overall quality against a ruler scene which is very different in appearance. Remember that human observers are able to compare stimuli with *different* sensory modalities as described in the Cross-Modal Psychophysics section earlier in this chapter. Thus, the scene captured for benchmarking does not have to be identical to the ruler scene when probing the visual modality in isolation.

The experimental instructions are critical to guide the observer to assess the quality impact of interest. As stated earlier in this chapter, the instructions should indicate that the observers are to judge overall image quality and not compare a given attribute. However, the attributes in the test images do need to be part of the consideration of the overall quality. For this practical example, the instructions should clearly state that the test stimuli will vary in noise and/or texture detail. The instructions should also state that the ruler references vary in sharpness only. However, the instructions must state that the task is to judge the overall quality when comparing the test images (with varying noise and/or texture detail) to the ruler series (varying in sharpness). Sample instructions follow with permission by the author.

> "Welcome to the Visual Lab!
>
> You will now participate in a test evaluating image quality. It is important that you judge the OVERALL QUALITY of the WHOLE image—not some particular attribute or a particular area of the image.
>
> A pair of images will be presented on the monitor. The image on the left is labeled 'Ruler image' and the image on the right is labeled 'Test image.' The sharpness of the ruler image is varied by moving the slider bar. For each test image on the right, you are supposed to adjust the ruler image on the left so that there is a match in the overall quality between the two images.
>
> There are 78 test images for you to judge, varying in noise and/or texture detail. Your response will be recorded when you press the 'Next' button. If you want to re-evaluate a previous test image that is not currently on the screen, press the 'Back' button.
>
> Before the actual test, there will be a training session. Play with the slider to get used to how it works. When you are ready to start the actual test, press 'Exit Trial' and the test starts.
>
> Please remember that there are no right or wrong answers—image quality is defined by observer perception, and the purpose of this test is to find out what YOUR perception is of these test images. Good luck! :)" (Persson, 2014)

A training session, as mentioned in the sample instructions, will prepare the observer to be able to provide better assessments. This training should provide example test scenes with instructions guiding the observer to note relevant portions of the image when making an overall image quality assessment. For example, if a test scene has blue sky as a significant component of the image, the observer could be instructed to assess the noise level in that region and its impact on the overall image quality of the scene. If a test scene has a closeup portrait, the observer could be instructed to look at the hair strands and texture in the clothing to determine the amount of detail present in the test image and how that impacts overall quality when comparing to the quality of the ruler images.

Once the observer has finished the training with example test images, the formal test would begin. Note that the results from the training session are not used in the experimental analysis. The order of presenting the test images should be randomized for all observers to avoid start-up phenomena or fatigue influences for specific images at the start or end of the test, respectively. As noted in the ISO 20462-3 standard (ISO, 2012), the experiment should include a minimum of 10 observers and 3 different test scenes, with preferably 20 observers and 6 scenes. As described in previous sections of this chapter, the larger the sample size, n, the more reproducible and conclusive the results will become. However, there is a limit to the amount of time an observer can spend make assessments before fatigue and boredom can start to impact the results. Typically, results from the studies in CPIQ averaged about 15 seconds for an observer to make a ruler selection for a test image. Therefore, a total of approximately 200 images could be judged in an hour-long session, representing a long assessment period. Reducing the number of samples will naturally reduce the judging session duration and be less taxing on the observers. Keep in mind that there will be observers who will judge at rates slower than expected, so buffer time between scheduled observers is a pragmatic component of obtaining observer results.

For each scene having the same level of the treatment, that is, the various noise and noise reduction levels in this practical example, the resultant SQS_2 JND values should be averaged across observers. The mean value represents the subjective assessment for the impact of the given noise or noise reduction treatment level on overall quality. Calculating responses of individual observers across the range of treatment for pooled scenes would provide understanding regarding the observer value judgments on the tradeoff between noise and noise reduction compared to the ruler sharpness on overall image quality. In a typical set of 10 observers, the sensitivity will vary from those who tend to discount the appearance of noise and noise reduction on overall image quality, while other observers are measurably sensitive to the impact of noise and noise reduction levels. The responses could also be pooled for all observers across the range of treatment for individual scenes. This would provide understanding regarding the scene impact on observer assessment of overall quality. For example, scenes with large flat fields such as sky or shadow regions tend to accentuate noise levels compared to a scene filled with mid- and high-frequency spatial components (see Chapter 6) such as grass. Thus, the quality of the former type of scenes should be more impacted by noise than the latter. But, noise reduction strength tends to reduce mid-frequencies with low contrast, particularly texture, of original scenes. Therefore, image quality of scenes with dominant mid-frequency components should be more impacted by noise reduction levels.

Note that the judgments in this example should always be intent on comparing the impact of noise and noise reduction on overall image quality, and not the threshold detection levels of noise and noise reduction. This means that test scenes for which observers are able to detect noise or the impact of noise reduction are evaluated not for the simple presence of these attributes, but instead for the impact on the overall quality. In parallel, the quality level of the ruler scene should be judged for its level of sharpness and adjusted by the observer until the quality is deemed equal to the quality (not sharpness) of the test scene. The observers should determine what tradeoff in the level of sharpness in the ruler image conveys the same overall quality level as that generated by the noise or noise reduction impact of the test scene.

A final note relates to the psychometric aspect of the experimental design and analysis. In order to ensure that the perceptual measurement is using accurate "meters," that is, observers who provide reliable and accurate assessment, observer judgments should be scrutinized. As mentioned previously, incorporating replicates of null images (actual ruler images) in the test image set allows the analysis of both precision and accuracy. For example, having the observers judge five replicates of ruler scenes in the midrange and upper quality levels of the SQS_2, such as rulers 2 and 21 in Table 5.4 for a viewing distance of 34 inches, provides images with known values, so deviation from the reference values can be used to determine the "calibration" of the human "meters" used for the psychometrics. A clue regarding how to determine a failure threshold is stated in the "Aptina Softcopy Quality Ruler User's Manual" included in supporting files for the ISO 20462-3 standard (ISO, 2012) in which the recommendation is to discard data for observers having an average difference from the reference SQS_2 JND value and/or the standard deviation of the difference for the replicates "larger than a few JNDs." A more specific determination is stated in Keelan *et al.* (2011): data from observers were removed from the analysis because their standard deviations from the known calibrated values were significantly larger than 2.5 SQS_2 JNDs and noticeably worse than other observers.

5.6.5 Quality Ruler Applications to Image Quality Benchmarking

A significant portion of this chapter has focused on the quality ruler method because of the advantages this subjective method offers from a repeatability, consistency, and ease-of-use perspective. However, more important to the topic of this book, the method provides a systematic and formalized approach to generating a psychophysically based benchmarking process. The quality ruler results are obtained in SQS_2 overall quality JND values and these have been shown to be combinable using a variable-power Minkowski metric such that summation of individual attributes predicts the overall quality of the camera (Keelan, 2002b). For example, individual image quality attributes of a given camera, for example, lens geometric distortion, noise, and color saturation, can be judged using the quality ruler and subsequently the individual JND values can be added with the Minkowski metric to obtain an overall quality prediction. This process is repeated for additional cameras of interest, and then the resultant overall quality predictions for each camera can be compared in order to benchmark the devices.

The formula for this multivariate formalism is stated as follows:

$$\Delta Q_m = \left[\sum_i (\Delta Q_i)^{n_m} \right]^{1/n_m} \tag{5.10}$$

where

$$n_m = 1 + 2\tanh\left(\frac{\Delta Q_{max}}{16.9}\right) \tag{5.11}$$

Here ΔQ_m is the predicted overall quality change in SQS_2 JND values, ΔQ_i is the quality change of an *i*th image quality attribute that is orthogonal in nature to other attributes, and n_m is the empirically determined variable power of the metric that provides suppression accommodation for the attribute with the most severe degradation, ΔQ_{max}.

This formalism will be discussed in greater detail in Chapter 9 regarding methods of benchmarking. But, with respect to subjective evaluation itself, the Minkowski summation approach allows direct combination of subjective experimental data from the quality ruler without having to convert to a separate scoring scale or needing to determine a weighting function for the individual attributes. Note that other methods such as forced-choice paired comparison can be converted into JNDs, and then this Minkowski metric can also be utilized.

Another application of the quality ruler is studying the mathematical fit between subjective quality and objective metrics for a given image quality attribute in order to determine an objective metric well correlated to SQS_2 JNDs. For example, a psychophysical study has been utilized to indicate that an objective metric based on the green channel intensity of a neutral patch in a color chart captured in lab conditions can be used to predict optimal exposure level quality in pictorial scenes (He *et al.*, 2016). Eight different real world scenes were selected for the experiment in which seven exposure levels from underexposed through overexposed were captured for each of the scenes. Twenty-four observers rated the images using the quality ruler to obtain mean SQS_2 JND values for each exposure level. The subjective data was compared to three objective metrics calculated for the given gray patch of the chart: the green channel intensity, luminance (Y), and lightness (L^*). The root mean square (RMS) value was calculated for these three objective metrics and a mathematical model that fitted the subjective experimental data. The objective metric with the best fit was the green channel intensity. The established mathematical model allows for prediction of the overall quality loss in SQS_2 JND units from measuring the green channel intensity of the neutral patch in the chart.

In a similar manner, the mathematical fit for other image quality attributes can be studied. Once the fit is established, the subjective impact determined from the quality ruler experiments can be predicted from images captured in lab conditions rather than needing to run additional psychophysical experiments. Thus, a collection of subjective and objective relationships for image quality attributes can be generated, enabling prediction of SQS_2 JND values for given objective metrics subsequently determined in lab conditions. The quality loss values in SQS_2 JNDs can then be combined using the Minkowski summation described above. Expansion of this concept of predicted SQS_2 JND values from established quality loss mathematical fits will occur in future chapters.

5.7 Subjective Video Quality

Contributed by Hugh Denman

While many aspects of objective assessment of video image quality overlap with still image quality, for example, by analyzing individual frames of video clips, subjective assessment of video image quality is more complex than that of still images described above for several reasons. Firstly, the evaluation of video clips, including live streaming video, is influenced by both the spatial and the temporal functions of the HVS. This contrasts with still image assessment, where only the spatial component is involved.

Secondly, systems for processing and transmitting digital video are more complex than those for still images; video encoding systems in particular exhibit time-varying, content-dependent artifacts whose visibility is also content-dependent. Thirdly, both audio quality and audio/visual synchronization impact video quality perception (Frater *et al.*, 2001; Watson and Sasse, 1996). This means that cross-modal influences are part of the evaluation of video image quality, adding more elements to be assessed.

5.7.1 Terminology

The terminology used in the literature describing subjective video quality testing is diverse and is distinct in some respects from that used for still image assessment. In industry, the terminology in common use is derived from the assessment standards developed by the International Telecommunication Union (ITU), and particularly the BT.500 recommendation (International Telecommunication Union, 2012), detailed below.

The videos to be assessed are called *clips* or *sequences*. Each clip depicts a *source*, abbreviated SRC, which is a specific scene or piece of footage. For each source there is generally an *original* clip, which is supposed to be entirely unimpaired, and a set of *processed video sequences*, or PVSs, which are generated by applying some processing or deliberate degradation to the original. In ITU terminology, the different treatments are referred to as *hypothetical reference circuits*, or HRCs, reflecting the origins of the ITU protocols in the sphere of video broadcast hardware. An HRC generally involves a video compressor and decoder, each operating with specified parameters, and may involve simulation of network transmission errors. The most common scenario is that the effects of the HRC are to be assessed by means of direct or implicit comparison of the PVSs with the original, for each source.

In the case of camera testing and benchmarking, there will not be an original clip as the source must be captured by each camera individually. Care must be taken to ensure that the resulting clips are as similar as possible.

The participants from whom quality ratings will be collected are interchangeably termed viewers, subjects, or observers.

As with all quality assessment exercises, it is important to determine whether the experiment is to establish worst-case performance, called critical condition in ITU terminology, or typical performance. This affects all aspects of the experiment design.

5.7.2 Observer Selection

As noted above in the discussion of still image quality evaluation, observer selection can have a significant influence on the experimental outcome. The distinction generally drawn is between expert and non-expert, or naïve, observers. Experts might constitute broadcast engineers, video editors or producers, or other professionals occupationally concerned with video quality. Non-expert observers are those drawn from some more general population, such as college students or non-video professionals. Generally, expert assessors could be expected to provide a critical-condition, or especially rigorous, assessment of quality. However, as noted in the previous sections, some video technology experts are attuned to particular artifacts and can miss those not pertaining to their domain of expertise (Corriveau, 2006). For example, a video compression engineer might be conditioned to scan video for blocking artifacts typical of imperfect compression, and might thereby miss a color shift.

In BT.500, the ITU highlights a consistency study they performed that indicates that systematic differences can occur between subjective results from different laboratories. The hypothesis presented to account for this is that the makeup of the observer groups influenced the result, even among non-expert observers. In other words, the categorization of observers into expert and non-expert is not sufficient to obtain repeatable results. A definitive approach to selecting observer groups so as to obtain repeatable results across different laboratories has not yet been established, though factors such as age, gender, affective state, and individual level of interest in the content have been identified as influential. Therefore, the ITU recommends that an experimental report contain a full description of the characteristics of the observers, such as age, gender, and employment category.

As with observers for still image quality evaluation, it is generally recommended that observers should be screened for visual acuity and color blindness. However, for non-critical assessment, some researchers deliberately forego this screening to obtain results more representative of the population at large (Moorthy *et al.*, 2012). Pinson points out that it is inappropriate for the experimenter to reveal to an otherwise unknowing observer that they have failed an acuity or color blindness test; instead the observer should proceed through the evaluation, and their results be subsequently discarded (Pinson *et al.*, 2015).

5.7.3 Viewing Setup

The viewing setup encompasses the display, ambient lighting, and positioning of observers. Good examples of viewing environment specification are found in the ITU BT.500 recommendation, which describes setups for both a *home* condition and a *lab* condition. The lab condition is the more stringent, and is intended to enable broadcasters to assess the quality of video in an absolute sense; this represents critical-condition performance. The home condition, by contrast, is intended to replicate a typical home television viewing environment, and enables characterization of video quality in the consumer context. A summary of these conditions appears in Table 5.5. Note that some aspects are specified as a range of acceptable values, and others are not specified

Table 5.5 Comparison of ITU BT.500 recommendations for viewing setup (International Telecommunication Union, 2012). The Lab condition is for stringent assessment, while the Home condition is slightly more critical than a typical home. *Source*: Reproduced with permission of ITU.

Observers' Viewing Condition	Lab	Home
Ratio of luminance of inactive screen to peak luminance at assessor	> 0.02	> 0.02
Maximum observation angle	30°	30°
Ratio of luminance of background behind monitor to peak luminance of picture	approx. 0.15	Unspecified
Chromaticity of background	D65	Unspecified
Room illumination	Low	200 lux
Viewing distance	Unspecified	> 4 times screen height

at all for the home condition. A particular assessment experiment should fix and report specific details for all aspects of the setup.

In some experimental setups, multiple observers will be observing the videos simultaneously. Thus, the distance to the screen and the viewing angle will differ slightly for all observers in the pool. As well as ensuring that the seating is placed so that all observers are within the acceptable limits for the chosen condition, the experimenter should carefully check that the display is consistent for all observer positions. Color appearance, in particular, can be affected even by small off-axis viewing angles for some LCD displays. Display selection is considered in further detail below.

5.7.4 Video Display and Playback

There are three critical aspects of presentation of the clips for assessment: the display, the playback system, and compression of the clips for compatibility with the playback system, if required.

The display technology is most commonly LCD. Some researchers persisted in using CRT displays for critical assessment for years after the LCD was established as the dominant consumer technology (Seshadrinathan *et al.*, 2010a), but broadcast-quality LCD monitors are now generally available. As mentioned above, the display should be inspected for color consistency for the range of viewing angles to be used in the study, and calibrated for color accuracy. If consumer television hardware is being used, any facilities for color enhancement or motion interpolation should be disabled for the test. The display should be set to a refresh rate that is a multiple of the video frame rate, as otherwise the display duration of each video frame will vary.

The size of the display has long been recognized as influencing quality perception for images and video, modeled in JND terms as early as 1988 (Barten, 1988). To a first approximation, this effect is due to variations with field size in the contrast sensitivity function (see Chapter 7) at low spatial frequencies (Barten, 1999). However, even when the distance to the display is kept constant in the relative sense, for example six screen heights (6H), different screen sizes influence perceived quality (Sugama *et al.*, 2005). More recently, the scene content has been shown to influence whether a larger display improves or degrades the perceptual quality (Chu *et al.*, 2013).

The playback system must be capable of sustaining smooth, uninterrupted playback of the video clips. Current consumer technologies such as DVD or Blu-ray disc are quite reliable in this regard. Computer software is more versatile, but can require extra care to ensure smooth playback. While the test is running, the system should not be running any background services (such as backup, indexing, or virus scanning). Playback of high resolution video, especially when using little or no compression, will require a dedicated, high-bandwidth storage system, for example, multiple hard drives in a RAID (redundant array of independent disks) configuration or SSD (solid-state drive) storage. The most reliable approach for smooth playback on computer is to play the sequence entirely from RAM. At high resolutions, RAM limitations constrain the length of clip to tens of seconds for most computers, but this is still adequate for most subjective quality testing scenarios. The cost of a computer system suitable for video quality assessment using playback of uncompressed sequences has remained at roughly $10 000 for the past decade or so (Pinson *et al.*, 2015). This reflects how the growth in video resolution has kept pace with the dramatically increased capabilities of computer hardware over that period.

Playback of video using a computer can exhibit the problem of *tearing*, in which parts of two successive frames are displayed simultaneously, one in each half of the display. This effect will only last for one refresh cycle of the monitor, but can nevertheless be very noticeable, for example when panning shots are shown on a large screen. To avoid this, video playback software should implement a *double buffering* or *vertical sync* feature. This synchronizes the playback with the display so that each individual frame is displayed smoothly, without tearing.

Wherever possible, the playback system should be chosen so that re-compression of the clips for assessment is not necessary. In some cases, this may be unavoidable; compression is required for DVD, Blu-ray, or delivery over the Internet. Even when using a computer for playback, the video codec being assessed may not be supported with the playback software, and so clips may need to be transcoded (converted from one digital format to another) for compatibility. If recompression is necessary, the highest possible encoding settings should be used. Because the quality degradation introduced by video compression is highly non-linear, it should not be assumed that recompression for playback will affect all clips equally: subtle artifacts in the sources could be masked or accentuated by the re-compression. Therefore, assessment using a playback system involving recompression is only reliable for coarse quality judgments, as far as the intrinsic quality of the video clips themselves is concerned. Of course, if the playback system has been deliberately chosen to reflect the use case for assessment, more detailed inferences are valid.

When only the video performance of a system is to be assessed, no audio should be included. As noted previously in this chapter, the quality of an audio signal, when present, can influence the perception of video quality.

5.7.5 Clip Selection

Clip selection is a crucial aspect of experimental design when undertaking the subjective assessment of the video quality of a camera. There are two scenarios to consider. The first is the situation where it is possible and sufficient to access the video processing engine in isolation, for example where an engineer is choosing a video compression system as part of an exercise in modeling or optimizing the performance of a proposed camera design. In this case, a library of digital reference sequences can be used. These reference clips should be flawless, exhibiting no compression artifacts or other distortions. The clips for assessment are prepared by processing the reference sequences with each proposed video subsystem. The reference clips themselves can be used within the experiment as a basis for comparative evaluation.

More relevant to the topic of this book, the second scenario is where the camera must be assessed as a black box. Here, the clips for assessment are obtained by capturing some test scenes using the camera itself. This precludes the possibility of isolating the impact of the video processing from the influence of the optics, color correction, and other components. The test scene should involve repeatable, controlled moving elements and changing illumination, so that repeated captures are highly similar. This facilitates comparison of different video processing settings within a camera, or comparison of different cameras. It may still be possible to obtain a reference clip, by capturing the test scene using a camera whose performance greatly exceeds the cameras to be assessed. For example, when evaluating mobile phone cameras, a broadcast-quality studio camera could be used to obtain a reference capture of the scene. The physical requirements for a video test scene are discussed further in Chapter 6.

The length of the video clips is an important parameter. As outlined above, the playback system may impose limits on the length of the clips used. Even where this is not a constraint, most subjective video quality assessment experiments use clips of between 8 and 15 seconds in length (Pinson *et al.*, 2015), with recent research suggesting that a clip length of 5 seconds gives an optimal trade-off of stable subjective assessment versus efficiency (Moss *et al.*, 2015). Short clips are preferred for the assessment of quality degradations, as it is challenging for the typical observer to integrate assessment of a sequence of longer length into a single evaluation. The biases at work include a recency effect, whereby the quality of the last 10–20 seconds of footage has a disproportionate influence on the overall quality percept; a negative-peak effect, such that the depth of a quality degradation is more influential than its duration; and a forgiveness effect resulting in small quality degradations being disregarded after some time (Pearson, 1998). These effects do not influence perception of short sequences, in the sense that there is not a significant discrepancy between the mean of the instantaneous perception of quality and the observer's overall, after-the-fact assessment (Barkowsky *et al.*, 2007; Seshadrinathan and Bovik, 2011). In longer clips, the principal danger is that an artifact occurring near the end of the sequence will disproportionately influence the quality rating, due to recency bias.

Perhaps the most important consideration in clip selection is whether the goal is to assess critical performance or typical performance. Both requirements present challenges. Critical performance assessment is complicated by the fact that the performance of a particular video processing or compression algorithm is strongly dependent on the content of the video sequence, due to the non-linearity of the operations involved. A particular system may only introduce artifacts when processing material containing fast motion, or when processing animation rather than live-action footage. The specific conditions that can trigger quality degradations differ between video processing systems. These factors make it challenging to prepare a set of short extracts that will exercise all the video processing systems under assessment equally and comprehensively, without resulting in an impractically large set of clips. This is particularly the case when benchmarking cameras, where the source clips must be captured under reproducible conditions rather than extracted from a library. A critical set should include examples of large motions such as pans and tracking shots; fast subject motion, for example sports footage in mid-shot or medium close up; non-rigid motion such as flames or water rapids; scenes captured under low illumination levels; scenes with rapidly varying illumination conditions, for example containing flash photography; and scenes with slowly varying illumination, in which lights are slowly dimmed or faded between colors (Pinson *et al.*, 2013).

For typical performance assessment, the aim is to present a set of video clips such that the quality ratings of the clips overall reflects users' perception of the quality of the video camera. This requires some knowledge of the *videospace*, the analog of the photospace for still imagery (Säämänen *et al.*, 2010). The videospace is determined by the consumer segment and the particular use case in question: users of mobile phone cameras, for instance, often make video recordings at sports events, musical concerts, and social events in the home, while cameras for video conferencing applications will be used predominantly for indoor, relatively static scenes. These different scenarios pose distinct challenges for the video processing in the camera. For example, skintone enhancement will improve performance for indoor social events but not concerts. Thus, the chosen clips should have broad coverage of the videospace. This entails practical difficulties if a set of cameras must be benchmarked, as sufficiently similar footage must be recorded in each context with each camera. For wide shots of large-scale scenes, such as concerts and sports events, multiple cameras can be set side-by-side to record simultaneously. In this case, the discrepancy in position is small relative to the distance to the scene, so the clips can still be subjectively compared. There are more challenges to obtain sufficiently similar recordings with multiple cameras at indoor social events.

Even given some corpus of footage representative of the videospace, it is effectively impossible to extract a set of short clips that will provide overall representative performance. This is for two reasons. Firstly, the subjective impact of a momentary quality degradation can be quite different within a short clip compared to within a long clip, due to the temporal biases mentioned above. The second reason is that video quality degradations are content-dependent, hence time-varying, and also codec dependent. So within a body of footage representing the "indoor social event" context, say, there will be particular moments where a given video codec might introduce compression artifacts. It would clearly not be representative to extract these difficult moments for assessment in isolation. Extracting a set that contains such moments to a representative degree presupposes an understanding of their subjective impact, which is what the experiment is trying to establish. The problem is compounded by the consideration that the set of clips would need to give representative performance over a range of codecs and compression settings.

Therefore, despite the considerations suggesting that short clips be used described earlier in this section, typical video performance is most accurately assessed using longer clips. In the non-camera context, extracts from programs for broadcast or feature films are typically used. For consumer cameras, complete recordings up to several minutes in length are suitable; the difficulty here is that the footage is often not as compelling as video for broadcast. Note that where the customer videospace naturally includes short videos, as is the case with mobile cameras, these should be included to a representative degree. The issue is not one of video length directly, but rather of the difficulty of the selection of representative extracts.

The final consideration in clip selection is that the affective content of the video to be assessed can influence subjective quality ratings, and that the effect varies from observer to observer. For example, an observer's quality rating for feature film clips is generally higher for clips of interest to the observer (Kortum and Sullivan, 2010), and degraded footage of football games is given a higher subjective quality rating by football fans than by non-fans (Palhais *et al.*, 2012). This effect is difficult to quantify and account for statistically. In larger subjective experiments, particularly those using multiple pools of observers, stability may be improved by seeking to use clips without specific emotional resonance, or by having observers rate the desirability or interest level of each clip and discarding ratings for clips rated at the extremes of desirability for that observer.

5.7.6 Presentation Protocols

The presentation protocol, as described in this section, governs how the clips for assessment are shown to the participants. The choice of a particular presentation protocol from various options governs the trade-off between experiment efficiency and the sensitivity and statistical stability of the results. An efficient presentation protocol is one that yields as many clip ratings as possible per session. The more clips that can be rated, the more sources can be used, which broadens the coverage, and also the number of different processing parameters (HRCs) that can be assessed, which yields fine-grained conclusions. An efficient protocol is desirable because observer time is typically the limiting resource in subjective video quality experiments, due to both recommended limits on session length and total observer time, detailed below, and the practical availability of observers for experiments. Militating against efficiency is the incorporation of redundancy in the protocol, which may improve the consistency of observer ratings or the sensitivity of the experiment.

Video quality assessment can be undertaken using *interactive* or *non-interactive* presentation. In non-interactive presentation, the clips for assessment are displayed in a predetermined sequence. Multiple observers can participate in each session, each watching the same display, provided that the viewing conditions are satisfied for each observer.

As in still image assessment, non-interactive presentation protocols for video quality assessment can be broadly categorized into *single stimulus* and *double-stimulus* variants. In single-stimulus experiments, each clip is assessed in isolation. Short clips are typically presented twice in succession, with a mid-gray field shown for a few seconds to separate presentations. Longer clips are shown once. After the presentations of the clip for assessment, 20 or 30 seconds are allowed for each participant to record their rating.

Double-stimulus experiments involve the assessment of pairs of clips. Each pair may be presented using simultaneous, side-by-side playback, or in sequence. Simultaneous presentation is more efficient, but for higher resolutions the display may not accommodate side-by-side playback. Although not yet formally assessed in the literature, the authors' experience indicates that a split-screen presentation can be used to yield actionable results. The comparison is more natural when the screen is divided vertically, into left and right halves, rather than horizontally. Each half of the display should show the same portion of the source, rather than the left half of the display showing the left half of the source and similarly for the right, as this gives a better basis for comparison. The halves should be separated with a black vertical line of a few pixels' width. Where the playback system does not support split-screen presentation, split-screen videos must be prepared by the experimenter. It is crucial that these split-screen sequences be prepared using lossless compression or as uncompressed video, to avoid the introduction of artifacts.

In interactive presentation, a single observer uses computer software to select clips for playback, repeat clips, and toggle the display between clips within a pair during playback. The SAMVIQ (Subjective Assessment of Multimedia Video Quality) method described below in the ITU Recommendations section formalizes one such approach. This has a cost in efficiency as multiple observers cannot participate simultaneously, and also because each observer may take an arbitrarily long time in rating each clip. Furthermore, the variation in results between observers may increase, as these methods have a higher dependency on the level of observer engagement than non-interactive methods. The advantages are that the interactive nature of the experiment may help to sustain observer engagement, and that domain experts undertaking critical evaluation can take more time to bring their skills to bear on the task.

If an original, uncorrupted clip is available for each source, this will be incorporated into the presentation protocol as a reference, or basis for comparison. For single-stimulus presentation, the reference is simply included, unlabeled, as one of the clips to be rated. This is called a hidden-reference (HR) approach, and enables the experimenter to infer the comparative rating of each clip against the reference. In double-stimulus presentation, the reference will usually be one member of every pair. Where the observers are not informed which member is the reference, it is important to randomize the order of each pair.

In all presentation protocols, it is important to randomize the order of assessments, so as to minimize order dependent effects.

To avoid observer fatigue, evaluation sessions should be limited to a maximum duration of 30 minutes; some researchers recommend a maximum of 20 minutes (Pinson *et al.*, 2015). An observer may participate in multiple sessions, but the total time per observer should be limited to about one hour, to ensure sustained engagement with the experiment. Issues of fatigue and engagement can be mitigated by ensuring a varied selection of source clips.

As with still image quality assessment, some practice evaluations should be performed at the start of each session to familiarize the observer with the task and the evaluation interface, for example five practice evaluations on the observer's first session and three for subsequent sessions. The results of these practice assessments should not be included in the analysis.

5.7.7 Assessment Methods

The three most commonly used methods for video assessment are category scaling, continuous scaling, and functional assessment. Category scaling can be used if suitable categories are available for the dimension being assessed. If adjectival categories are not available or not desirable for the dimension being assessed, numerical categories can be used instead. These two variants are logically termed adjectival categorical scaling and numerical categorical scaling in BT.500 (International Telecommunication Union, 2012).

Continuous scaling involves assigning a numerical value from some range, typically 0 100, to each stimulus. The observer may explicitly record the number, or make their assessment by making a mark on a continuous vertical scale, which will then be encoded as a numerical value. Where appropriate, the scale is divided into segments and each segment is labeled with a category rating, to make the intended range of the ratings explicit to the observers. For the assessment of longer clips, observers can express their instantaneous perception of quality by updating the position of a mechanical slider continuously throughout the presentation; the position of the slider is sampled by the experiment software and interpreted on the continuous scale.

In some contexts, it may be appropriate to assess the stimulus quality on the basis of the ease with which some functional task, typically related to information extraction, is performed. Such functional tasks include, for example, counting the number of some specific target object present in the stimulus, or reading some text depicted in the scene. In this case, the stimulus is assessed by rating the performance of this task on the stimulus, in terms of accuracy or speed.

The BT.500 recommendation provides three category scales for absolute quality, impairment level, and comparative quality. The scale for each of these is sometimes

Table 5.6 Subjective scales used for rating quality or impairment levels as recommended in ITU BT.500 (International Telecommunication Union, 2012). *Source*: Reproduced with permission of ITU.

Level	Quality Scale	Impairment Scale
5	Excellent	Imperceptible
4	Good	Perceptible but not annoying
3	Fair	Slightly annoying
2	Poor	Annoying
1	Bad	Very annoying

referred to as the *absolute category rating* (ACR) scale. For the first two scales, an observer assigns a level of quality or impairment, respectively, depending on the type of attribute. Table 5.6 contains the specific levels of the quality and impairment scales from which the observer selects. Assessment of an attribute that has a positive influence on image quality, such as color reproduction, would be assigned a quality level, for example, 4 = "Good," whereas an attribute that degrades image quality, such as noise, would be assigned an impairment level, for example, 2 = "Annoying." For the third scale, the comparison scale, the observer compares the test video clip to a reference clip and then assigns a level from the comparison scale, for example, 0 = "The same." Table 5.7 contains the seven levels used in the comparison scale. An odd number of levels is used for each of these scales. As mentioned earlier in Section 5.3 regarding the category scaling methodology, this ensures that there is a central value, which makes evaluation easier for the observer. Other task-specific scales may be used for specific applications, for example with categories for text legibility or image usefulness.

The authors have found that it is important to provide the observers with anchoring by presenting examples and definitions of extremes; this assists with the ease and repeatability of subjective evaluation using these scales. While not a formal anchored scaling approach, this hybrid approach of category scaling with end anchors is useful.

Table 5.7 Subjective scale used for rating the comparison of a test clip to a reference clip as recommended in ITU BT.500 (International Telecommunication Union, 2012). *Source*: Reproduced with permission of ITU.

Level	Comparison Scale
−3	Much worse
−2	Worse
−1	Slightly worse
0	The same
+1	Slightly better
+2	Better
+3	Much better

Research using both simulated user preferences and published data suggests that using fewer categories increases the variation of the final ratings, and that variation in ratings over observers is highest toward the middle of the scale (Winkler, 2009). On the other hand, single-stimulus experiments conducted with a five-level categorical scale provide high efficiency and maximize cognitive ease for participants (Tominaga *et al.*, 2010), and using finer-grained rating scales has not been shown to significantly improve data accuracy (Pinson *et al.*, 2015).

5.7.8 Interpreting Results

Once the experiment is complete, the data must be screened to eliminate ratings supposed to be invalid. The results can then be analyzed to provide conclusions and actionable insights.

Screening the data entails detecting observers whose ratings differ from the other observers' ratings to a degree that suggests that their response is entirely atypical, or that they have misunderstood the assessment task. This can be carried out on the basis of agreement, wherein the criterion is that ratings are numerically identical, or association, which requires that observers' ratings follow the same tendency (Pinson *et al.*, 2015). Association is sufficient for quality inferences.

BT.500 describes a screening method in which each observer is assessed for acceptability based on confidence intervals derived from the ratings of the other observers. This is an observer-agreement, rather than observer-association, test. For observer-association assessment, the Pearson correlation or the Spearman rank-order correlation are commonly used (Seshadrinathan *et al.*, 2010b). The Pearson correlation test is valid only for normally distributed data derived using a ratio scale; if these criteria are not met, the Spearman correlation should be used. In either case, a correlation threshold must be chosen on the basis of the data arising from the experiment itself.

In the simplest approach, the observers' ratings for each HRC are averaged to provide a Mean Opinion Score (MOS) for each treatment. This has most commonly been derived from experiments using numerical quality categories ranging from 1 (lowest quality) to 5 (highest quality). Where a reference clip has been incorporated, each observer's rating is related to his or her rating of the reference. This may be done by the observers themselves in an explicit-reference, double-stimulus experiment, or by the experimenter after the fact in a hidden-reference experiment. These relative scores should then be scaled and centered about a common mean to obtain a standard score, or z-score (Seshadrinathan *et al.*, 2010a). The average of the resulting z-scores are Difference Mean Opinion Scores, or DMOS (note that DMOS is also used as an acronym for Degradation Mean Opinion Score). This conversion to z-scores mitigates the differences due to each observer's biases, expectations of quality, strength of opinion, and individual preference for the sources, as these factors are reflected equally in their assessment of the reference and the PVS. Thus, this approach increases the validity of pooling or averaging the ratings of multiple observers. Note that the z-scores should be computed for ratings within each session separately, as user expectations may vary over time.

Mean opinion scores are used very frequently, informally, within industrial assessments. However, it is important to recognize that computation of an average is only valid for ratings on an interval or ratio scale. More generally, commonly used statistical methods such as Student's t-test and Analysis of Variance (ANOVA) (Boslaugh, 2012) require that the data be normally distributed, which implies the use of an interval or ratio scale. It is possible to rescale data obtained using an ordinal scale to an interval scale (Harwell and Gatti, 2001). If the data do not pass a Normal distribution criterion, such as the χ^2 test, analysis must be based on median rather than mean scores. For example, the Mann-Whitney U-test (Corder and Foreman, 2014) should be used instead of Student's t-test.

These statistical methods are used to determine whether a difference in the MOS or DMOS scores of different HRCs is statistically significant. When such a difference is established, it can be used, for example, as part of a cost-benefit model to determine whether investment in a more computationally expensive coding scheme or a higher encoding bitrate is justified.

For experiments using longer sequences, it is recommended to collect ratings continually throughout the presentation to counter the effect of temporal biases, as described above. This yields a considerable quantity of data for analysis, which may be difficult to summarize to form an actionable insight. BT.500 suggests converting these ratings to histograms describing the probability distribution of different quality levels for each HRC. In some contexts, an after-the-fact overall rating may be more useful, for example, where a consumer's general impression of quality is sought, and certainly these single ratings are more amenable to analysis than continuous assessment data.

5.7.9   ITU Recommendations

The material above outlines the general considerations for subjective video quality assessment experiments. In industry, these experiments are most often conducted using the formalized methodologies established by the ITU. The key documents describing these approaches include the following:

- ITU-R Rec. BT.500 (2012), Methodology for the Subjective Assessment of the Quality of Television Pictures
- ITU-T Rec. P.910 (2008), Subjective Video Quality Assessment Methods for Multimedia Applications
- ITU-R Rec. BT.1788 (2007), Subjective Assessment of Multimedia Video Quality (SAMVIQ)

The prefix BT indicates that the document pertains to "Broadcasting service, Television," while the P series of recommendations describe "Terminals and subjective and objective assessment methods." BT.500, already mentioned in several contexts above, targets the assessment of broadcast television, and involves tightly controlled lighting and display conditions. P.910 was developed for multimedia digital video, assuming lower bitrates than those used for broadcast television, and requires less stringent viewing conditions to match the diverse conditions in which Internet video is consumed.

BT.1788 provides an in-depth description of a method for the *subjective assessment of multimedia video quality* (SAMVIQ). This is an interactive-presentation method, using on-demand video playback via computer. By contrast, the methods of BT.500 are suitable for non-interactive presentation on a television with a video playback device such as a DVD player.

There are other ITU documents providing media-specific details, for example, for testing HDTV, standard definition TV, and stereoscopic TV. The ITU document "ITU-R Rec

BT.1082 (1990), Studies Toward the Unification of Picture Assessment Methodology," though dated, provides informative background and context for the problems of subjective video assessment.

Of the specific approaches for quality assessment described by the ITU, three from BT.500 are the best known. These are *double-stimulus impairment scale* (DSIS), *double-stimulus continuous quality scale* (DSCQS), and *simultaneous double stimulus for continuous evaluation* (SDSCE). These and some other notable methods are described below.

5.7.9.1 The Double-Stimulus Impairment Scale Method

In the double-stimulus impairment scale method, the presentation of the stimuli is serial: the reference, a flat mid-gray, the test condition, and then a repeat of the mid-gray. The video clips used are about 10 seconds in length. The mid-gray is presented for three seconds' duration, which allows the observer a short break to process the two stimuli. This sequence may be repeated if the experiment time permits.

As indicated in the method name, the scale for this method is the ITU-R impairment scale described in Table 5.6. The method is recommended for the assessment of clearly visible degradations.

This method is also described in P.910, where it is designated *degradation category rating*, or DCR.

5.7.9.2 The Double-Stimulus Continuous Quality Scale Method

The double-stimulus continuous-quality scale method is similar to DSIS, with the difference that the order of the reference and test clips is chosen by the experimenter, and varies for each assessment. The first video is designated "A" and the second is designated "B." Thus, the observer does not know which of the two is the reference.

Observers grade both A and B on a form with continuous vertical scales for each video. The scale consists of a vertical line split into equal-sized segments, each labeled in accordance with the ITU-R quality scale shown in Table 5.6. The "Excellent" rating is at the top of the scale, and the "Bad" rating at the bottom of the scale. The observers make their assessment by placing a mark at the point corresponding to their judgment on the quality scale. The pairs are observed twice, with the observers making their assessment on the second viewing. Where only a single observer is participating in each session, they are allowed to switch between A and B at will before making an assessment, rather than being presented with the clips in strict sequence.

The interpretation of DSCQS ratings consists of quantifying the difference in the vertical scale between the reference and test conditions—note that the individual scores should not be interpreted as an absolute rating. This methodology is more sensitive to small differences in quality than DSIS.

5.7.9.3 The Simultaneous Double-Stimulus for Continuous Evaluation Method

The simultaneous double-stimulus for continuous evaluation method is an approach suitable for evaluating arbitrarily long video sequences. As such, it enables assessment of a representative corpus of video in which content-dependent performance degradation will occur to a proportionate degree. In this method, the reference and the video to be assessed are presented side-by-side, with synchronized playback. Participants are given an electronic voting device, which contains a slider with continuous motion over a 10 cm range of positions. Participants are instructed to continuously compare the reference and test videos, and to continuously update the slider position to reflect their judgment, with one extreme corresponding to perfect fidelity, and the other corresponding to a completely degraded signal. The experiment software will sample the slider position twice per second, and record this series of assessments for subsequent analysis.

This method was developed for the use of the Motion Picture Experts Group (MPEG), to enable the assessment of video coding at very low bitrates. It can be applied in any context where the quality degradation is expected to be time-varying or content-dependent.

5.7.9.4   The Absolute Category Rating Method

The most straightforward approach to clip rating is *absolute category rating* (ACR). Here, observers directly rate each video using the quality categories of Table 5.6. It is recommended that stimuli are presented with a three-second mid-gray adaptation field preceding the video, and a ten-second mid-gray post-exposure field following the video. If a hidden reference is incorporated, the method is termed ACR-HR.

This method has the weakness that the results can be strongly influenced by the order of presentation. In BT.500, a variant designated *single stimulus with multiple repetition* (SSMR) is identified, which addresses this issue. In this approach, first a training pass is presented, containing every stimulus exactly once, to permit the users to calibrate their ratings. Two assessment passes follow, each also containing every stimulus, and the observer's rating is derived from the mean of the two assessment passes. The order of stimuli is randomized within each pass. This greatly reduces the influence of stimulus order on the results.

5.7.9.5   The Single Stimulus Continuous Quality Evaluation Method

The *single stimulus continuous quality evaluation* (SSCQE) method described in BT.500 uses the same approach and apparatus of SDSCE, with the difference that only one stimulus is presented for evaluation, and the rating is interpreted as an absolute quality assessment.

5.7.9.6   The Subjective Assessment of Multimedia Video Quality Method

In this SAMVIQ approach, all treatments of a given sequence are available for playback on computer via an interactive software interface. The users can select, pause, and replay the clips at will, and review and revise their scores until they are ready to move to the next sequence. Typically, the set of treatments includes a reference clip, explicitly identified as such, and also the same reference clip presented as one of the clips for assessment (a hidden reference).

The observer's assessments are recorded on a continuous scale with linearly spaced markers labeled with the categories from the quality, or ACR, scale. SAMVIQ is unusual in that it is a single-stimulus method, strictly speaking, but because the observer can use the software to perform a direct comparison with a labeled reference, it has some of the characteristics of a double-stimulus method. It is the only method that permits observers to compare any pair of treatments they wish. The observers can refine their judgments until they have an internally consistent set of ratings. For example, an observer may choose to ensure that if a treatment A has been rated above a treatment B, and B has been rated above C, then A is rated above C. This deliberation on the part of the observer can result in longer experiment running times.

5.7.9.7 ITU Methodology Comparison

A comparison of the methodologies is useful to consider. Of all video quality assessment methods, ACR is the most efficient. Provided that the experiment is conducted with care, results are consistent across different groups of test observers (Huynh-Thu and Ghanbari, 2005). Of the double stimulus methods, DSIS (also known as DCR) is the fastest, and it does not appear to be less accurate than DSCQS or SAMVIQ (Tominaga *et al.*, 2010). It is limited to the assessment of degradations; quality-enhancing treatments can not be assessed using this scale. It is best suited to coarse degradations. It is difficult to distinguish small degrees of unsharpness, for example, with this method.

DSCQS seeks to improve accuracy by presenting each treatment twice. However, such an improvement in accuracy has not been established in the literature. Furthermore, rating on a continuous scale is a more strenuous cognitive task than category rating with five labels. However, the method is sufficiently general to assess both degradations and enhancements, and is sensitive to small differences in quality.

As described above, SDSCE has the advantage that long video sequences can be assessed, and thus can give an assessment more representative of typical performance. Naturally, experiments using this method take longer to run, and also result in a much larger amount of data that need to be analyzed to extract summary results.

The SAMVIQ method can yield precision equivalent to ACR with fewer observers (Huynh-Thu and Ghanbari, 2005). As it is an interactive method, each participant must be able to complete a session individually.

5.7.10 Other Sources

Subjective video quality assessment is an active area of research, and experimental design is developing from being intuition-driven and informed by experience toward a growing body of established best practices for efficient, statistically robust methods. The overview articles by Pinson *et al.* (2015), Chen *et al.* (2015), and Corriveau (2006) are excellent sources for further details and references. Redi *et al.* (2015) have presented a comprehensive account of how a deeper understanding of viewers has informed quality assessment experiments.

Clips suitable for subjective experiments are available from the Consumer Digital Video Library, provided by the U.S. Institute for Telecommunication Sciences (Pinson, 2013). The Video Quality Experts Group (VQEG) has conducted numerous studies of subjective video quality, using DSCQS and ACR-HR, and makes available a library of video clips processed with different levels of distortion, along with subjective ratings of these clips (Brunnström *et al.*, 2009). The Laboratory for Image & Video Engineering at the University of Texas also provides a database of clips suitable for subjective quality experiments, along with previously obtained subjective ratings (Seshadrinathan *et al.*, 2010a).

Summary of this Chapter

- The field of psychophysics was established in the early 19th century, having formed the foundation for subjective image quality evaluation.
- Subjective evaluation involves having observers make perceptual judgments about some physical property, such as weight, length, temperature, and so on.

- It is a well-established fact from research and experience that it is possible to make reliable, reproducible subjective assessments of image quality when carried forth in a controlled manner.
- Measurement scales used in quantifying subjective evaluation include nominal, ordinal, interval, and ratio scales.
- Methodologies for evaluating observer responses include rank order, category scaling, acceptability scaling, anchored scaling, forced-choice comparison, and magnitude estimation.
- Observers are able to reliably perform cross-modal psychophysical assessments, that is, making perceptual judgments comparing two differing sensory modalities such as the human visual and auditory systems.
- Psychophysical studies of still image quality are typically limited to evaluation using visual sensation. Thus, they are fundamentally less complex for an observer than cross-modal assessments.
- Psychophysical studies of video image quality are often cross-modal, employing the visual and auditory modalities.
- Thurstonian scaling is often used for paired comparison (and triplet comparison) studies to generate an interval scale for a given image quality attribute.
- Use of quality rulers is a specific example of applying an anchored scaling methodology to run subjective evaluations.
- Quality ruler experiments show that as overall image quality declines, observers diverge on what constitutes poor image quality. Conversely, observer variability decreases for what is considered excellent image quality.
- Increasing the number of observers and the number of stimuli in an experiment will decrease the 95% confidence limits and provide less variable predictions of image quality.
- The type of observer, for example, experienced or naïve, will impact subjective results.
- The type of stimuli will impact subjective results. In video quality evaluation, a recency bias, that is, disproportionate influence of the last segment of footage, needs to be considered.
- Double-stimulus impairment scale (DSIS), double-stimulus continuous quality-scale (DSCQS), and simultaneous double stimulus for continuous evaluation (SDSCE) are the best known video quality assessment methods.
- Absolute category rating (ACR) scales for video quality measurement include quality, impairment, and comparison scales.
- Assessment of video quality typically includes the mean opinion score (MOS) of the ACR scales.
- The goal of the subjective study, whether for still or video image quality, must be considered in order to determine items such as appropriate method, stimuli, test instructions, and analysis options when designing the study to ensure intended answers are obtained.

References

Barkowsky, M., Eskofier, B., Bitto, R., Bialkowski, J., and Kaup, A. (2007) Perceptually motivated spatial and temporal integration of pixel based video quality measures, in *Welcome to Mobile Content Quality of Experience*, ACM, New York, NY, USA, MobConQoE '07, pp. **4**:1–7.

Barten, P.G.J. (1988) Effect of picture size and definition on perceived image quality, in *Conference Record of the 1988 International Display Research Conference*, San Diego, CA, USA, pp. 142–145.

Barten, P.G.J. (1999) *Contrast Sensitivity of the Human Eye and Its Effects on Image Quality*, SPIE Press, Bellingham, WA, USA.

Bartleson, C.J. and Grum, F. (1984) Visual measurements, in *Optical Radiation Measurements*, vol. **5**, Academic Press, Cambridge, MA, USA.

Boslaugh, S. (2012) *Statistics in a Nutshell*, O'Reilly Media, Sebastopol, CA, USA.

Brunnström, K., Hands, D., Speranza, F., and Webster, A. (2009) VQEG validation and ITU standardization of objective perceptual video quality metrics [standards in a nutshell]. *IEEE Signal Process. Mag.*, **26**, 96–101.

Chen, Y., Wu, K., and Zhang, Q. (2015) From QoS to QoE: A tutorial on video quality assessment. *IEEE Commun. Surv. Tutorials*, **17**, 1126–1165.

Chu, W.T., Chen, Y.K., and Chen, K.T. (2013) Size does matter: How image size affects aesthetic perception?, in *Proceedings of the 21st ACM International Conference on Multimedia*, ACM, New York, NY, USA, MM '13, pp. 53–62.

Corder, G.W. and Foreman, D.I. (2014) *Nonparametric Statistics: A Step-by-Step Approach*, John Wiley & Sons, Inc., Hoboken, NJ, USA, 2nd edn.

Corriveau, P. (2006) Video quality testing, in *Digital Video Image Quality and Perceptual Coding* (eds H.R. Wu and K.R. Rao), Boca Raton, CRC Press, Boca Raton, FL, USA, chap. 4.

Engeldrum, P.G. (2000) *Psychometric Scaling: A Toolkit for Imaging Systems Development*, Imcotek Press, Winchester, MA, USA.

Frater, M.R., Arnold, J.F., and Vahedian, A. (2001) Impact of audio on subjective assessment of video quality in videoconferencing applications. *IEEE Trans. Circuits Syst. Video Technol.*, **11**, 1059–1062.

Gescheider, G.A. (1997) *Psychophysics: The Fundamentals*, Lawrence Erlbaum Associates, Mahwah, NJ, USA, 3rd edn.

Gilbert, A.N., Martin, R., and Kemp, S.E. (1996) Cross-modal correspondence between vision and olfaction: The color of smells. *Am. J. Psych.*, **109**, 335–351.

Harwell, M.R. and Gatti, G.G. (2001) Rescaling ordinal data to interval data in educational research. *Rev. Educ. Res.*, **71**, 105–131.

He, Z., Jin, E.W., and Ni, Y. (2016) Development of a perceptually calibrated objective metric for exposure. *Electronic Imaging: Image Quality and System Performance*, **2016**, 1–4.

Huynh-Thu, Q. and Ghanbari, M. (2005) A comparison of subjective video quality assessment methods for low-bit rate and low-resolution video, in *Proceedings of the IASTED International Conference*, Honolulu, HI, USA, pp. 69–76.

I3A (2007) Camera Phone Image Quality – Phase 1 – Fundamentals and review of considered test methods. IEEE.

I3A (2009) Camera Phone Image Quality – Phase 2 – Subjective Evaluation Methodology. IEEE.

International Telecommunication Union (2012) Recommendation 500-13: Methodology for the subjective assessment of the quality of television pictures, ITU-R Rec. BT.500.

ISO (2005a) ISO 20462-1:2005 Photography – Psychophysical experimental methods for estimating image quality – Part 1: Overview of psychophysical elements. ISO.

ISO (2005b) ISO 20462-2:2005 Photography – Psychophysical experimental methods for estimating image quality – Part 2: Triplet comparison method. ISO.

ISO (2012) ISO 20462-3:2012 Photography – Psychophysical experimental methods for estimating image quality – Part 3: Quality ruler method. ISO.

Jin, E., Phillips, J., Farnand, S., Belska, M., Tran, V., Chang, E., Wang, Y., and Tseng, B. (2017) Towards the development of the IEEE P1858 CPIQ standard - a validation study. *Electronic Imaging: Image Quality and System Performance*, **2017**, 88–94.

Jin, E.W. and Keelan, B.W. (2010) Slider-adjusted softcopy ruler for calibrated image quality assessment. *J. Electron. Imaging*, **19** (1), 011009.

Jin, E.W., Keelan, B.W., Chen, J., Phillips, J.B., and Chen, Y. (2009) Soft-copy quality ruler method: implementation and validation. *Proc. SPIE*, **7242**, 724206.

Keelan, B.W. (2002a) *Handbook of Image Quality – Characterization and Prediction*, Marcel Dekker, New York, USA.

Keelan, B.W. (2002b) Predicting multivariate image quality from individual perceptual attributes. *Proc. IS&T PICS Conference*, pp. 82–87.

Keelan, B.W., Jin, E.W., and Prokushkin, S. (2011) Development of a perceptually calibrated objective metric of noise. *Proc. SPIE*, **7867**, 786708.

Kortum, P.T. and Sullivan, M. (2010) The effect of content desirability on subjective video quality ratings. *Human Factors: The Journal of the Human Factors and Ergonomics Society*, **52**, 105–118.

Marks, L.E. (1974) On associations of light and sound: The mediation of brightness, pitch, and loudness. *Am. J. Psych.*, **87**, 173–188.

Mauer, T.A. and Clark, J.H. (2009) Lens geometric distortion cross-company results. Unpublished research.

Moorthy, A.K., Choi, L.K., Bovik, A.C., and de Veciana, G. (2012) Video quality assessment on mobile devices: Subjective, behavioral and objective studies. *IEEE J. Sel. Top. Signal Process.*, **6**, 652–671.

Moss, F.M., Wang, K., Zhang, F., Baddeley, R., and Bull, D.R. (2015) On the optimal presentation duration for subjective video quality assessment. *IEEE Trans. Circuits Syst. Video Technol.*, **26**, 1977–1987.

Palhais, J., Cruz, R.S., and Nunes, M.S. (2012) *Mobile Networks and Management: Third International ICST Conference, MONAMI 2011, Aveiro, Portugal, September 21-23, 2011, Revised Selected Papers*, Springer Berlin Heidelberg, Berlin, Heidelberg, chap. Quality of Experience Assessment in Internet TV, pp. 261–274.

Pearson, D.E. (1998) Viewer response to time-varying video quality. *Proc. SPIE*, **3299**, 16–25.

Persson, M. (2014) *Subjective Image Quality Evaluation Using the Softcopy Quality Ruler Method*, Master's thesis, Lund University, Sweden.

Phillips, J.B., Coppola, S.M., Jin, E.W., Chen, Y., Clark, J.H., and Mauer, T.A. (2009) Correlating objective and subjective evaluation of texture appearance with applications to camera phone imaging. *Proc. SPIE*, **7242**, 724207.

Phillips, J.B. and Shaw, B. (2011) Kodak validation of ISO 20462 Part 3 SQS$_2$ values. Unpublished research.

Pinson, M.H. (2013) The consumer digital video library [best of the web]. *IEEE Signal Process Mag.*, **30**, 172–174.

Pinson, M.H., Barkowsky, M., and Callet, P.L. (2013) Selecting scenes for 2D and 3D subjective video quality tests. *EURASIP J. Image Video Process.*, **2013**, 1–12.

Pinson, M.H., Janowski, L., and Papir, Z. (2015) Video quality assessment: Subjective testing of entertainment scenes. *IEEE Signal Process. Mag.*, **32**, 101–114.

Redi, J., Liu, H., Alers, H., Zunino, R., and Heynderickx, I. (2010) Comparing subjective image quality measurement methods for the creation of public databases. *Proc. SPIE*, **7529**, 752903.

Redi, J.A., Zhu, Y., Ridder, H., and Heynderickx, I. (2015) *Visual Signal Quality Assessment: Quality of Experience (QoE)*, Springer International Publishing, Heidelberg, Germany, chap. How Passive Image Viewers Became Active Multimedia Users, pp. 31–72.

Rice, T.M. and Faulkner, T.W. (1983) The use of photographic space in the development of the disc photographic system. *J. Appl. Photogr. Eng.*, **9**, 52–57.

Säämänen, T., Virtanen, T., and Nyman, G. (2010) Videospace: classification of video through shooting context information. *Proc. SPIE*, **7529**, 752906.

Seshadrinathan, K. and Bovik, A.C. (2011) Temporal hysteresis model of time varying subjective video quality, in *2011 IEEE International Conference on Acoustics, Speech and Signal Processing (ICASSP)*, Prague, Czech Republic, pp. 1153–1156.

Seshadrinathan, K., Soundararajan, R., Bovik, A.C., and Cormack, L.K. (2010a) Study of subjective and objective quality assessment of video. *IEEE Trans. Image Process.*, **19**, 1427–1441.

Seshadrinathan, K., Soundararajan, R., Bovik, A.C., and Cormack, L.K. (2010b) A subjective study to evaluate video quality assessment algorithms. *Proc. SPIE*, **7527**, 75270H.

Stevens, S.S. (1946) On the theory of scales of measurement. *Science*, **103**, 677–680.

Stevens, S.S. (1957) On the psychophysical law. *Psychol. Rev.*, **64**, 153–181.

Stevens, S.S. (1975) *Psychophysics: Introduction to Its Perceptual, Neural and Social Prospects*, John Wiley & Sons Ltd, Chichester, UK.

Sugama, Y., Yoshida, T., Hamamoto, T., Hangai, S., Boon, C.S., and Kato, S. (2005) A comparison of subjective picture quality with objective measure using subjective spatial frequency, in *Proceedings of the 2005 IEEE International Conference on Multimedia and Expo, ICME*, Amsterdam, The Netherlands, pp. 1262–1265.

Thurstone, L.L. (1927) A law of comparative judgment. *Psychol. Rev.*, **34**, 273–286.

Tominaga, T., Hayashi, T., Okamoto, J., and Takahashi, A. (2010) Performance comparisons of subjective quality assessment methods for mobile video, in *Second International Workshop on Quality of Multimedia Experience (QoMEX)*, Trondheim, Norway, pp. 82–87.

Trotter, C. (1878) Note on 'Fechner's Law'. *J. Physiol.*, **1**, 60–65.

Watson, A. and Sasse, M.A. (1996) Evaluating audio and video quality in low-cost multimedia conferencing systems. *Interacting with Computers*, **8**, 255–275.

Winkler, S. (2009) On the properties of subjective ratings in video quality experiments, in *International Workshop on Quality of Multimedia Experience*, San Diego, CA, USA, pp. 139–144.

Zwick, D.F. (1984) Psychometric scaling of terms used in category scales of image quality attributes. *Proc. Symposium of Photogr. and Electr. Image Quality*, pp. 46–55.

6

Objective Image Quality Assessment—Theory and Practice

Chapters 2 and 3 made the point that it is possible to describe image quality to a fair degree of accuracy through a limited set of image quality attributes. In this chapter, methods for the quantitative assessment of those attributes will be presented. These *objective image quality metrics* form the basis of the benchmarking methodology described in this book. It is important that such metrics be subjectively correlated, or that there exist methods to transform them into results that do correlate with perception. Chapter 7 will describe methods to accomplish this.

The definition of an objective metric in this context is that it is independent of human judgment. This is in contrast with subjective methods, which by definition rely on the judgment of the experimental observers. If performed well, however, both methods are able to produce repeatable and accurate results.

With these definitions, it may seem difficult to find a correlation between results from objective and subjective experiments. Nonetheless, subjective studies, if correctly performed, can produce results that are representative of the average of a population. If it is possible to model the overall properties of a human population, there should in principle be no problem to find objective metrics that indeed do reflect how an average person perceives specific attributes.

The goal of this chapter is to describe objective metrics whose results are, or can be, correlated to subjective findings in order to be used for camera image quality benchmarking. In many cases, however, these metrics will have a wider scope than just producing visually correlated results. They may also be used to provide information about the characteristics of the capturing device, which might give clues to the design considerations made as well as settings of image tuning parameters.

The objective metrics described in this chapter, in order of appearance, are:

- Exposure and tone
- Dynamic range
- Color
- Shading
- Geometric distortion
- Stray light
- Sharpness and resolution
- Texture blur
- Noise
- Color fringing

Camera Image Quality Benchmarking, First Edition. Jonathan B. Phillips and Henrik Eliasson.
© 2018 John Wiley & Sons Ltd. Published 2018 by John Wiley & Sons Ltd.
Companion website: www.wiley.com/go/benchak

- Image defects
- Video quality metrics

Of these metrics, sharpness and resolution will be given a proportionally larger space compared to the other metrics. The reason for this is that we are introducing the spatial frequency concept and the modulation transfer function (MTF) when discussing this metric. An understanding of spatial frequency is important for other metrics relying on the scale of measurement, such as texture blur and noise. The MTF is important not only for sharpness and resolution, but also for texture blur. A working knowledge of the MTF is also necessary to understand the subjective image quality ruler and the SQS, as discussed already in Chapter 5.

6.1 Exposure and Tone

In one sense, *tone reproduction* may be regarded as the ability of a camera to render all areas of an image as optimally as possible with respect to overall signal level. A well-balanced tone reproduction then means that it should be possible to distinguish details in both highlights and shadows and all the midtones in between, if such details are present in the original scene. In reality, however, due to artistic considerations or otherwise, this may not be desirable in every situation. For instance, if the scene contains specular highlights, there might not be any use in trying to adjust exposure and tone in order not to have them blow out. Also, if a strong light source is present in the scene, it may not be useful to adjust the exposure such that the filament inside the lamp is visible. Thus, the optimal exposure will depend on the scene, that is, two different scenes with the same average luminance may need completely different exposure levels in order to obtain a pleasing image. For this reason, several different exposure methodologies have been developed during the past years, such as center-weighted exposure, spot metering, matrix-based methods, and so on. With digital photography, even more tools are at the photographer's disposal, for example, real-time histogram analysis, but even so, the basic concepts have not changed as a result of the transition from chemistry-based to digital photography. Therefore, consulting the classic works on this topic from the past century, in particular the three-book series by Ansel Adams (Adams, 1995a,b,c), is highly recommended.

6.1.1 Exposure Index and ISO Sensitivity

Since it is possible to relate the camera output level to a particular input luminance, it is also possible to aid the camera user in adjusting the camera settings for reasonably accurate exposure. The purpose of the ISO 12232 ISO sensitivity standard is to provide such an aid, by specifying a *recommended exposure index* (REI) and a *standard output sensitivity* (SOS) (ISO, 2006), as well as ISO sensitivities with respect to noise and saturation.

Technically, the term *exposure* refers to the amount of light energy per unit area at the camera focal plane. Exposure is therefore expressed in units of lux seconds. The *exposure index* provides a link between the parameters affecting the exposure, and this relationship is usually expressed through the equation

$$\frac{N^2}{t} = \frac{LI_E}{k} \tag{6.1}$$

where N is the f-number, t the exposure time, L the average scene luminance, I_E the exposure index, and k a calibration constant. Consequently, for a constant scene luminance, we obtain:

$$\frac{N^2}{tI_E} = \text{constant} \tag{6.2}$$

This provides freedom for the photographer to choose the most appropriate combination of parameters for a specific scene. For instance, in sports photography it is often important to freeze motion, which requires a short exposure time. In order to maintain correct exposure, either the f-number must be decreased or the exposure index increased if the exposure time is decreased.

Formally, the exposure index in ISO 12232 is defined as

$$I_E = \frac{10}{H} \tag{6.3}$$

where H is the focal plane exposure required to obtain correct exposure according to one of the criteria described in the ISO 12232 standard. This is the base quantity used for specifying the sensitivity or *ISO speed* using certain criteria. The ISO speed of a camera can be determined according to different conditions such as the exposure required to reach saturation, or the exposure required to reach a certain output value, or even the exposure needed to reach a certain signal to noise value. These criteria are described in the ISO 12232 standard and result in a number of different methods to calculate the ISO sensitivity.

In the saturation based ISO speed definition, the exposure index is calculated for a focal plane exposure just below saturation of the image sensor. Thus, image highlights in a typical scene will be just below the maximum possible camera signal level, which typically yields the lowest noise image.

The more recent standard output sensitivity defines the exposure criterion to be to obtain an average luminance, L_{std}, that, in the case of an 8-bit image, yields an image value of 118. For an sRGB image, this corresponds to a scene reflectance of 18%, by virtue of Eq. (6.15).

For either of the two definitions above, the exposure index can be changed by varying the gain of the image sensor. Adjusting the gain will change the output signal, therefore compensating for inadequate focal plane exposure. For instance, if the current exposure is only filling pixels up to half of the full well capacity, and the ADC is only using half of its range, increasing the gain by a factor of two will provide a signal that just saturates the sensor ADC, effectively increasing the exposure index by a factor of two. Of course, this will decrease the signal to noise ratio for reasons explained in Chapter 4.

6.1.2 Opto-Electronic Conversion Function

Even though varying preferences make it difficult to establish what is meant by a good tone reproduction, it is still possible to measure. The verification of tone reproduction through objective testing can be made using the ISO 14524 standard (ISO, 2009). This standard describes the steps necessary to determine the *opto-electronic conversion function* (OECF), which is a measure of the relationship between the focal plane exposure and camera output levels. Figure 6.1 shows a typical measured OECF.

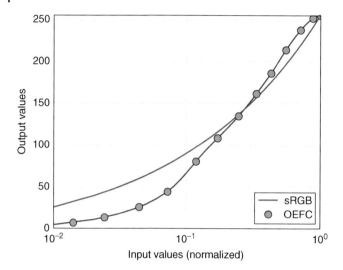

Figure 6.1 Example OECF. The red curve shows the transfer curve of the sRGB color space, as described by Eq. (6.15).

The OECF can be measured with or without a lens. In the former case, the camera sensor is directly illuminated by light of varying intensity. In the latter case, a test chart with patches of varying reflectance or transmittance is used.

6.1.3 Practical Considerations

Modern image processing in cameras often includes *local tone mapping* as part of high dynamic range (HDR) functionality. This means that different parts of the image may have different OECFs, which certainly makes the OECF concept less relevant. Also, the SOS may be affected by this processing, since local manipulation of the tone map will lead to different relations between the mid-gray 18% reflectance patch and what digital image code value this actually corresponds to.

The issue with stray light (or flare) is also important to consider when measuring the OECF. High levels of flare, introducing an offset in the image, will affect the OECF at the low end where the input signal level is low. This is mostly true in the case when a lens and a test chart are used to determine the OECF. Since a substantial part of any flare present will typically originate from the lens, a direct measurement without a lens may eliminate most of the flare, thus giving a more accurate estimate of the OECF. However, since some type of camera lens must be used when capturing images, this more accurate OECF measurement may not exactly characterize the digital images captured by the camera.

6.2 Dynamic Range

The ability to capture a certain range of luminance (radiance) values of a scene is known as the dynamic range of a camera. As discussed in Chapter 4, the dynamic range of an image sensor may be defined as the ratio of the full well capacity to the temporal dark noise. More strictly, one can define a digital camera dynamic range metric as the ratio of the maximum unclipped signal to the lowest possible signal, defined as being obtained when the signal to noise ratio is equal to 1.

For a dynamic range metric to be meaningful, it should refer to the scene content and therefore be defined in terms of the scene luminance. The ISO 15739 standard describes a methodology for determining the dynamic range. In short, it makes use of the incremental gain as described in Section 6.9.3 to obtain an estimate of the scene-referred noise and signal obtained by capturing images of a test chart with patches of varying brightness. The lowest signal can be found by measuring the temporal noise in a dark patch, which ideally should be dark enough so the noise is dominated by the dark noise.

6.3   Color

As stated in the beginning of this chapter, results obtained from an objective image quality metric are independent of human judgment. However, when trying to obtain an objective assessment of the color attribute, this will no longer hold true. Merely the fact that colors are perceived almost the same under a wide range of light sources gives a hint that the processing done by the human visual system has a big influence on how we perceive color. This can be further corroborated by various color appearance phenomena, such as simultaneous contrast (demonstrated in Chapter 1), the tendency to perceive more saturated colors as being brighter (Helmholz–Kohlrausch effect), and many others (Fairchild, 2013).

Nevertheless, given that we are able to model the human visual system with reasonable accuracy, it should still be possible to find metrics that do describe the color analysis capabilities of various devices, such as a digital camera. In such a description of human color vision, the ability to partially or completely *discount the illuminant* plays an important role. Therefore, a color metric also needs to incorporate a description of the light source illuminating the scene, in addition to the human visual system and the scene properties. The material in this section is mainly built on the information found in Fairchild (2013), Hunt (2004), Reinhard *et al.* (2008), and Wyszecki and Stiles (1982).

6.3.1   Light Sources

Our eyes are sensitive to electromagnetic radiation in the wavelength range from approximately 390 nm to around 700 nm. In this range, light of pure wavelengths (monochromatic light) will be perceived as going from red through orange, yellow, green, blue, and violet as the wavelength is decreased from the upper limit to the lower bound. This is visualized in Figure 6.2. In reality, purely monochromatic light sources do not exist, since an infinitely thin distribution of wavelengths will contain zero energy. Therefore, all real light sources will have some distribution of light across a range of wavelengths. Such distributions can still be close to monochromatic, such as in a laser or, to a lesser extent, a light emitting diode (LED). Some examples of wavelength distributions of light sources are shown in Figure 6.3.

In nature, light sources are mostly broadband and ranging over the entire visible spectrum and beyond. A fundamental class of sources is the *black body*.[1] From physics,

---

1 A perfect black body will absorb any incident radiation. Since a body in thermal equilibrium must emit the same amount of energy as it absorbs, a heated black body will then become a source of radiation.

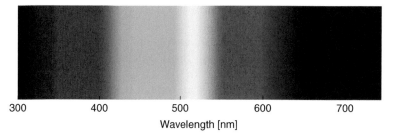

300   400   500   600   700

Wavelength [nm]

Figure 6.2 The visual spectrum. Note that due to limitations of the reproducing medium, the colors corresponding to particular wavelengths only serve as an approximate illustration of the actual color.

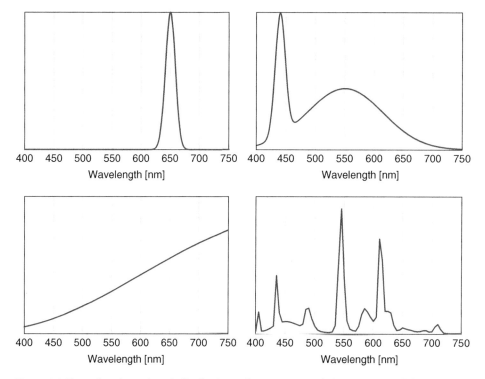

Figure 6.3 Examples of wavelength distributions of some common light sources. Top left: red LED; top right: white LED; bottom left: halogen light; bottom right: fluorescent light.

we know that any body having a temperature above absolute zero will emit electromagnetic radiation with a wavelength and temperature dependent radiance (in $\mathrm{Wsr^{-1}m^{-3}}$), $S(\lambda, T)$, given by *Planck's radiation law* (Reif, 1985):

$$S(\lambda, T) = \frac{2\pi hc^2 \lambda^{-5}}{\exp\left(\frac{hc}{\lambda kT}\right) - 1} \tag{6.4}$$

where h is Planck's constant, c the speed of light in vacuum, and k Boltzmann's constant. Figure 6.4 shows the wavelength distribution of this function for various temperatures. As seen from this figure, the peak of the function shifts toward shorter wavelengths

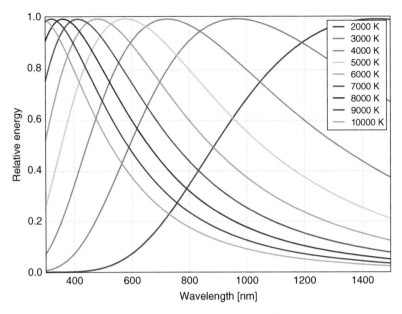

Figure 6.4 The wavelength distribution of the radiance of black bodies with varying temperatures. The plots are normalized to unity peak value.

as the temperature is increased. Taking the derivative of this expression in order to calculate the position of the peak (expressed in nm), leads to *Wien's displacement law*:

$$\lambda_{max} \approx \frac{2.898 \times 10^6}{T} \tag{6.5}$$

This explains why, for instance, a piece of iron being worked on by a blacksmith will shift from red through yellow to white as it gets hotter. Thus, the temperature of the body is directly related to its color and therefore the term *color temperature* is regularly used to characterize a light source.

A black body is a construct not readily encountered in nature. An often used approximation is a cavity in a wall. Light entering the cavity will bounce around inside and for each bounce lose some of its intensity, so that after a large number of bounces a negligible amount will be left to escape. In practice, objects may come more or less close to a pure black body. The property describing how far away an object is from a true black body is the *emissivity*, $\epsilon(\lambda)$, of the body. This is a number between zero and one, which is also wavelength dependent. An ideal black body has an emissivity of 1 over the entire wavelength range. Any other source, or *gray body*, will have an emissivity smaller than 1 for some wavelengths.

It is possible to extend the color temperature concept to arbitrary sources, with spectral power distributions that may be quite different from a black body, for example, fluorescents. In this case, the term *correlated color temperature* (CCT) is used.

There are many different types of light sources. In addition to various gray bodies, artificial sources such as lasers, LEDs, and fluorescent tubes are abundant. This is becoming increasingly apparent as the traditional light bulbs in our homes are being replaced by such sources in order to save energy, since light bulbs waste large amounts of energy in the infrared region. The International Commission on Illumination, CIE (Commission Internationale de l'Éclairage), has standardized a smaller set of *standard illuminants* (CIE, 2004; Wyszecki and Stiles, 1982; Hunt and Pointer, 2011; Reinhard *et al.*, 2008). The most commonly encountered are:

- Daylight series D
- Incandescent illuminant A
- Fluorescent sources F1-F12
- Uniform illuminant E

It is important to make the distinction between an *illuminant* and a *light source*. Illuminants are theoretical light sources that may not be possible to realize in real life. Examples are the daylight D series of illuminants. On the market, it is possible to purchase daylight sources, but these will only be approximations of the D illuminants and may be made up of, for example, fluorescent tubes or filtered tungsten lights with varying degrees of accuracy. A light source, on the other hand, denotes a source that is physically available.

The CIE D illuminants can have CCTs in the range 4000 to 25 000 K. Formulas for calculating the spectral power distribution in this range have been standardized by the CIE (CIE, 2013) and can also be found in standard works such as Wyszecki and Stiles (1982). In practice, the D50 and D65 illuminants are the most common ones. The CCTs of these illuminants are 5003 and 6504 K, respectively.

An incandescent source, illuminant A, is also available in the CIE data set. This is a black body radiator with a color temperature of 2856 K. This illuminant is frequently encountered in work involving color.

The set of fluorescent sources covers the CCT range from 2940 K (F4, warm white fluorescent) to 6500 K (F7, daylight fluorescent). Some of them are also known under their trade names, such as TL84 (F11).

The last illuminant in the list is the uniform illuminant E. This has a flat spectrum, that is, the spectral power density is constant across the whole wavelength range. Figure 6.5 shows example plots of the spectral power distributions of some of the illuminants discussed above.

6.3.2 Scene

The objects in the scene that are illuminated by the light source will, depending on the material, either absorb, transmit, or reflect the incident radiation. Due to conservation of energy, the following relationship is expected to hold between the (wavelength-dependent) absorbance, α, transmittance, τ, and reflectance, ρ:

$$\alpha(\lambda) + \tau(\lambda) + \rho(\lambda) = 1 \tag{6.6}$$

Therefore, if two of these quantities are known, the third can be calculated using Eq. (6.6), unless the material is fluorescent. Fluorescence is more common than might be thought at first. In order to make clothing appear brighter and with more brilliant colors, optical brighteners are often added to detergents. These compounds exploit fluorescence, absorbing ultraviolet light and re-emitting this in the visible region. Also for paper, brighteners are often added in order to make the paper appear whiter.

The sensation of color will be related to the distribution of wavelengths reflected back toward the observer, and is thus given by the combination of the wavelength distribution

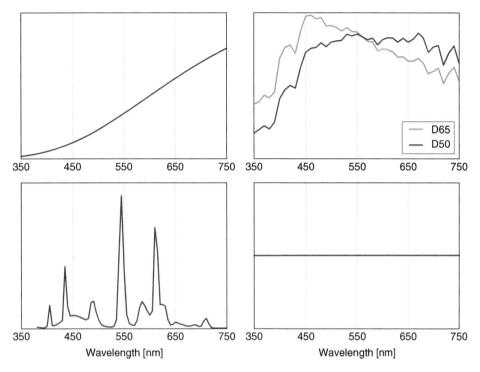

Figure 6.5 CIE illuminant power spectral distributions. Top left: A; top right: CIE D50 and D65; bottom left: CIE F11; bottom right: CIE E.

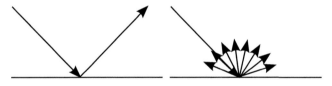

Figure 6.6 Reflections from a surface. Left: specular reflection; right: diffuse reflection.

of the reflectance of the object and the light source. What will also affect the appearance of the object is the angular distribution of the light reflected. The two opposite extremes are referred to as *specular* and *Lambertian* reflections. The first case represents a perfect mirror. Here, all the light coming from one direction will be reflected in only one distinct direction, illustrated in Figure 6.6. The second case is a perfectly diffuse surface, scattering the light in all possible directions uniformly. In this case, the distribution of light follows the *Lambert cosine law*, which states that the intensity of light will fall off as the cosine of the angle to the normal of the surface. Since the projected area will also fall off with the cosine of the angle, the net result is that the radiance (luminance) will be constant as a function of viewing angle. See Chapter 4 for definitions of radiometric and photometric quantities. This is the complete opposite behavior to specular reflection, where the reflected light will only be visible in one direction. In reality, no perfect Lambertian surface exists, even though it is often a good approximation. To properly

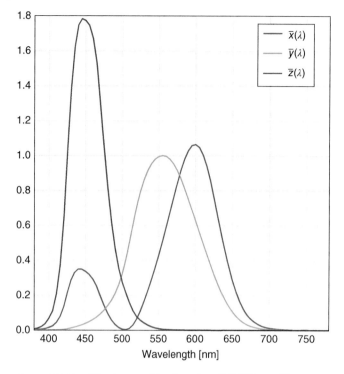

Figure 6.7 The CIE color matching functions $\bar{x}(\lambda)$, $\bar{y}(\lambda)$, and $\bar{z}(\lambda)$.

account for the surface reflectance properties of real materials, a quantity known as the *bidirectional reflectance distribution function* (BRDF) is defined (Reinhard *et al.*, 2008).

6.3.3 Observer

As has already been stressed, the measurement of color cannot be made without taking the human observer into account. Therefore, a model of the human visual system as it applies to color has to be established. The *de facto* standard for color measurements using such a model is based on the CIE *standard observer*, which specifies a set of *color matching functions* (CMFs) (CIE, 2004), shown in Figure 6.7. Using these functions, it is possible to express any spectral stimulus, such as the scene radiance, irradiance of a light source, and so on, as three coordinates in the *XYZ color space*. The reason that three coordinates are used is based on the fact that the human vision is *trichromatic*, that is, has three types of photoreceptors (the cones) that are active in color vision. Given an arbitrary spectral stimulus, $I(\lambda)$, the CIEXYZ coordinates[2] are calculated as follows:

$$X = \int_{380 \text{ nm}}^{780 \text{ nm}} I(\lambda)\bar{x}(\lambda)d\lambda$$

$$Y = \int_{380 \text{ nm}}^{780 \text{ nm}} I(\lambda)\bar{y}(\lambda)d\lambda$$

---

2 At this point, it should be mentioned that there exist two types of CIEXYZ coordinates, one for a 2° visual field, and one for a 10° field. All references to CIEXYZ values in this book are based on the 2° observer.

$$Z = \int_{380 \text{ nm}}^{780 \text{ nm}} I(\lambda)\bar{z}(\lambda)d\lambda \qquad (6.7)$$

Since it is difficult to visualize colors in a three dimensional space, the following reduction in dimensionality can be made. First, the XYZ coordinates are normalized through

$$x = \frac{X}{X + Y + Z}$$
$$y = \frac{Y}{X + Y + Z}$$
$$z = \frac{Z}{X + Y + Z} \qquad (6.8)$$

With this choice of normalization, $x + y + z = 1$, and therefore, the z coordinate can be determined if the two other coordinates are known. These two coordinates, x and y, are known as the CIE *chromaticity coordinates*. The normalization makes it possible to express the chromaticities of, for example, light sources and illuminants in a nonambiguous way. Table 6.1 summarizes the chromaticities of the standard CIE illuminants discussed above. Chromaticity coordinates are often also plotted in a *chromaticity diagram*, such as the one shown in Figure 6.8. The horseshoe-shaped outline in the diagram is a plot of xy values for all pure wavelengths from 380 to 780 nm. This is a visualization of the *gamut*, that is, the set of all visible colors of the standard observer.

Since the xy coordinates do not contain any intensity information, the Y coordinate is often supplied together with the chromaticity coordinates, giving an xyY triplet.

Table 6.1 CIE *xy* chromaticities and correlated color temperatures of CIE standard illuminants.

Illuminant	(x, y)	CCT (K)
A	(0.4476, 0.4074)	2856
D50	(0.3457, 0.3585)	5003
D65	(0.3127, 0.3290)	6504
D75	(0.2990, 0.3148)	7504
F1	(0.3131, 0.3373)	6430
F2	(0.3721, 0.3753)	4230
F3	(0.4091, 0.3943)	3450
F4	(0.4402, 0.4033)	2940
F5	(0.3138, 0.3453)	6350
F6	(0.3779, 0.3884)	4150
F7	(0.3129, 0.3293)	6500
F8	(0.3459, 0.3588)	5000
F9	(0.3742, 0.3728)	4150
F10	(0.3461, 0.3599)	5000
F11	(0.3805, 0.3771)	4000
F12	(0.4370, 0.4044)	3000
E	(0.3333, 0.3333)	5454

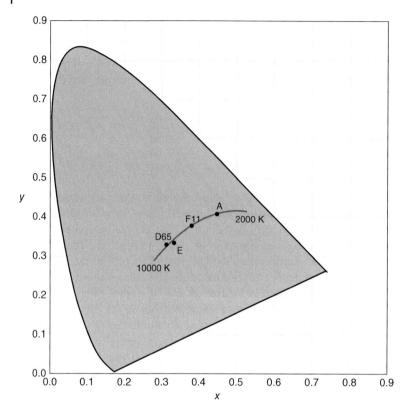

Figure 6.8 Example of a CIE xy chromaticity diagram. Included in the diagram are chromaticities of black bodies between 2 000 and 10 000 K (Planckian locus, red line), as well as a selection of CIE standard illuminants.

The backward transform to X and Z values are

$$X = \frac{x}{y}Y$$

$$Z = \frac{1 - x - y}{y}Y \tag{6.9}$$

6.3.4 Basic Color Metrics

One drawback of the XYZ system is that it is not *perceptually uniform*, that is, a constant distance between two coordinates at different positions within the XYZ space will not be perceived as a constant difference in color. For this reason, other color spaces have been developed that show better uniformity across the space. One important space is the *Uniform Chromaticity Scale* (UCS), which can be derived from the XYZ values according to

$$u' = \frac{4x}{-2x + 12y + 3}$$

$$v' = \frac{9y}{-2x + 12y + 3} \tag{6.10}$$

Even though the UCS shows much better perceptual uniformity, it still does not account for other important aspects of human color vision that are critical for color work. The CIELAB and CIELUV systems take the description of colors one step further by also including chromatic adaptation and the nonlinear response of the HVS. Chromatic adaptation relates to the ability of the human visual system to distinguish the same color under a wide range of different illuminations. As seen from Table 6.1 and Figure 6.7, the chromaticities of the CIE illuminants span quite a large space. Even so, the color red, for instance, will be perceived as red under any of the above illuminants. This must be taken into account by any description that aims to model human color vision with reasonable accuracy.

The CIELUV color space can be regarded as an extension of the UCS and the translation into this space is made according to the following formulas:

$$
L^* = \begin{cases} \left(\frac{29}{3}\right)^3 \frac{Y}{Y_n}, & \frac{Y}{Y_n} \leq \left(\frac{6}{29}\right)^3 \\ 116\left(\frac{Y}{Y_n}\right)^{1/3} - 16, & \frac{Y}{Y_n} > \left(\frac{6}{29}\right)^3 \end{cases}
$$

$$
u^* = 13L^*(u' - u'_n)
$$
$$
v^* = 13L^*(v' - v'_n) \tag{6.11}
$$

Here, Y, u', and v' are the luminance and chromaticities of the color stimulus, while Y_n, u'_n, and v'_n are the luminance and chromaticities that would result if a perfectly white diffuser could be placed in the scene, thus representing the *white point* of the scene, which, in essence, is a representation of the light source illuminating the scene. The cube root dependence of the L^* coordinate with respect to the luminance Y models the non-linear relation between the luminance and the perceived relative brightness in the scene.

A color space competing with CIELUV is CIELAB. Both color spaces were presented by the CIE at the same time, in 1976, since they at that time were deemed to be equally valid. In later years, however, CIELAB has become the more popular of the two due to, for example, the better ability of CIELAB to predict color differences (Fairchild, 2013). The calculation of the L^* lightness value and the two color opponent channels a^* and b^* is done in the following way:

$$
\begin{pmatrix} L^* \\ a^* \\ b^* \end{pmatrix} = \begin{pmatrix} 0 & 116 & 0 & -16 \\ 500 & -500 & 0 & 0 \\ 0 & 200 & -200 & 0 \end{pmatrix} \begin{pmatrix} f\left(\frac{X}{X_n}\right) \\ f\left(\frac{Y}{Y_n}\right) \\ f\left(\frac{Z}{Z_n}\right) \\ 1 \end{pmatrix} \tag{6.12}
$$

where

$$
f(r) = \begin{cases} \frac{\left(\frac{29}{3}\right)^3 r + 16}{116}, & r \leq \left(\frac{6}{29}\right)^3 \\ r^{1/3}, & r > \left(\frac{6}{29}\right)^3 \end{cases} \tag{6.13}
$$

In addition to the X, Y, and Z values of the color stimulus, the values X_n, Y_n, and Z_n of the white point are also needed.

With a color space that is perceptually uniform, one can calculate differences between colors. In this way it is possible to quantify how close a particular color is to other colors. Thus, the color reproduction of a camera can be assessed by calculating how far the camera-captured colors are from the original scene colors. This is done by determining a *delta E* value, ΔE.[3] Using the CIELAB space and assuming it is completely perceptually uniform, we can calculate a color difference as the Euclidean distance between two color samples, known as the CIE 1976 ΔE metric:

$$\Delta E_{76} = \sqrt{(L_1^* - L_2^*)^2 + (a_1^* - a_2^*)^2 + (b_1^* - b_2^*)^2} \tag{6.14}$$

Unfortunately, the CIELAB space is not perfectly uniform with respect to human perception, implying that there will not be complete correspondence between the calculated and perceived color differences over the entire color space. For this reason, the simple color difference has been improved in recent years to include, for example, the ΔE_{94} and CIEDE2000 (ΔE_{00}) color difference metrics.

The calculated color differences for the ΔE_{94} and ΔE_{00} norms are made such that a value of 1 corresponds approximately to a just noticeable color difference. The CIEDE2000 metric, while providing more accurate results, in particular for saturated blue colors and near-neutral values (Reinhard *et al.*, 2008), is quite complicated compared to the ΔE_{94} metric. For this reason, ΔE_{94} may be preferred in most cases (Fairchild, 2013).

6.3.5 RGB Color Spaces

As discussed in the section on noise, the raw image sensor signals are typically linearly transformed using white balance weights and a color matrix in order to produce acceptable color. This will transform the sensor signals into an RGB color space suitable for display on a monitor. The most common such space is the sRGB color space, defined in the IEC 61966-2-1 standard (IEC, 1999).

An RGB color space usually has an associated *gamma curve*. An old cathode ray tube (CRT) type display had a nonlinear relationship between the voltage signal applied and the intensity of the emitted light. This relationship can be modeled quite well by a power law, $y = x^\gamma$. In most RGB color spaces this is accounted for by including a nonlinear transformation. In the sRGB case, the transformation from linear RGB values normalized to the range between zero and one, $\{R, G, B\}$, to nonlinear values, $\{R', G', B'\}$, is

$$\{R', G', B'\} = \begin{cases} 12.92\{R, G, B\}, & \{R, G, B\} \le 0.0031308 \\ 1.055\{R, G, B\}^{1/2.4} - 0.055, & \{R, G, B\} > 0.0031308 \end{cases} \tag{6.15}$$

The inverse transformation is

$$\{R, G, B\} = \begin{cases} \frac{\{R', G', B'\}}{12.92}, & \{R', G', B'\} \le 0.04045 \\ \left(\frac{\{R', G', B'\} + 0.055}{1.055}\right)^{2.4}, & \{R', G', B'\} > 0.04045 \end{cases} \tag{6.16}$$

---

3 The "E" comes from the German word *Empfindung*, meaning "sensation."

To transform sRGB values to CIELAB, the following steps are performed:

1) Normalize the RGB data so that the maximum digital code value of 255 gets mapped to a value of 1.
2) Linearize the RGB data using Eq. (6.16).
3) Transform linear RGB data to the XYZ space.
4) Transform XYZ data to CIELAB using Eqs. (6.12) and (6.13). The white point values should be set equal to $X_w = 0.95043$, $Y_w = 1$, and $Z_w = 1.0888$, corresponding to a D65 white point.

Since the CIELAB color space is dependent on the illuminant, it is very important not to confuse illuminants in color space conversions. As an example, if we have measured a test chart under illuminant A, it might be argued that the white point of A should be used in the calculations of the CIELAB values. If we are going to compare colors in an actual scene, this is correct. If evaluating the color reproduction of a camera, however, this situation is no longer true. In this case, the images have been *rendered* into an RGB color space with its own white point (e.g., D65). Then, completely white, defined as $R = G = B = 1$, is implicitly assumed to correspond to the white point of the color space. Therefore, when measuring color differences from measurements in images, the CIELAB values should be calculated using the white point of the RGB color space used to render the images, even if the actual image was captured using a different light source.

6.4 Shading

A shading metric needs to handle luminance as well as color shading. For both cases, the measurement is typically made by capturing an image of a sufficiently uniform surface, illuminated by a set of well defined light sources. The resulting images are analyzed by dividing them into a regular grid. In order to obtain a metric that is reasonably well correlated with human vision, the RGB values of the image should be converted to a more perceptually correlated color space, such as CIELAB. Then, for the luminance shading metric, differences in L^* values between the image portions within in the grid are calculated. For the reported metric, both the maximum and the RMS deviation may be reported. This could be useful since even though average values are calculated for relatively large parts of the image, a large local variation might skew the result and produce a metric that is less correlated to perception.

To calculate the color shading, the CIELAB values obtained above are utilized, but in this case, the maximum (or RMS) value of the chrominance variations

$$\Delta C_i^* = \sqrt{(a_i^* - a_{mid}^*)^2 + (b_i^* - b_{mid}^*)^2} \tag{6.17}$$

is calculated, where the index i refers to the average values of the individual grid segments. For a detailed description of the method prescribed by CPIQ, upon which the above description is based, refer to I3A (2009a) and IEEE (2017), and also the ISO 17957 standard (ISO, 2015b).

In a mobile phone camera, the largest contributor to the color shading is most likely the IR cut filter, especially if it is of the reflective type. Since the passband on the long wavelength side changes with angle, the amount of IR present in the scene will affect the color shading to quite a considerable degree. For this reason, it is necessary to perform measurements of color shading using a variety of light sources, both those containing large amounts of IR, such as incandescent sources, as well as sources with low amounts, such as fluorescents. Also camera parameters, such as optical zoom position, may affect the result to a considerable degree.

6.5 Geometric Distortion

The effects of distortion have already been discussed in Chapters 3 and 4. To measure distortion, test charts like those shown in Figure 6.9 are used. The principle in both cases is the same: find the positions of the dots or the intersections of the lines and compare those to the position at which they would have been in an undistorted image (I3A, 2009c; IEEE, 2017). The optical distortion, D, is then calculated as

$$D = \frac{r' - r}{r} \qquad (6.18)$$

where r' is the measured radial distance from the center of the image, and r the radial distance in the undistorted case, see Figure 6.10. The result of the measurement will be a curve such as the one shown in Figure 6.11.

Another distortion metric that is often encountered is the *TV distortion*. The definition is illustrated in Figure 6.12. The same test charts used for the optical distortion measurement can also be used to measure TV distortion. The TV distortion metric gives one number, calculated as

$$D_{TV} = \frac{\Delta H}{H} \qquad (6.19)$$

where ΔH is usually calculated at the corners of the image. It should be noted that the TV distortion metric is differently defined in some standards, where instead of dividing by the full image height, half the height is used, thus providing a result that is twice as large as the definition above. Another important aspect of TV distortion that needs to be stressed is the fact that for some distortion types, the distortion can be severely

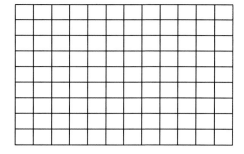
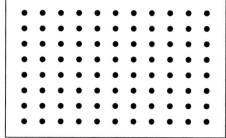

Figure 6.9 **Example distortion charts.**

Figure 6.10 Definition of the optical distortion metric. Dashed lines correspond to the undistorted case.

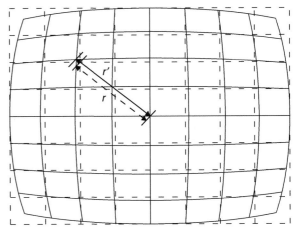

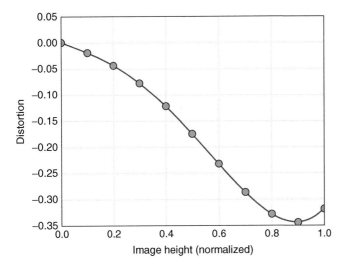

Figure 6.11 Presentation of optical distortion.

Figure 6.12 Definition of the TV distortion metric.

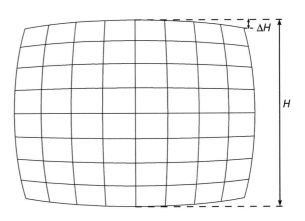

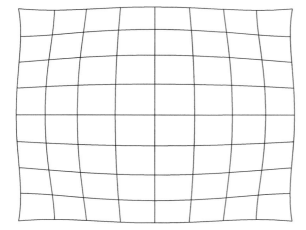

Figure 6.13 Distortion chart corresponding to the data presented in Figure 6.11. The height difference, ΔH, at the corners is clearly very close to zero in this case, leading to negligible TV distortion. However, the distortion is clearly noticeable.

underestimated, or even zero. This is illustrated in Figure 6.13, where the deviation ΔH will be measured to be zero, while clearly the distortion is quite noticeable. Therefore, TV distortion as a metric should be avoided as much as possible in favor of the more reliable optical distortion metric.

6.5.1 Practical Considerations

In the choice between a dot chart and grid chart for measuring optical distortion, it is tempting to use a dot chart since it might be easier to detect the dots with higher accuracy for that type of chart. However, the same reason that potentially increases the accuracy of detecting the dots on a subpixel level, namely the comparatively large size of the dots, may in some cases decrease the accuracy. The reason for this is that for severe distortion, as well as for very high amounts of other optical aberrations, the shape of the dots may be changed, which means that the center of gravity is also changed. This could potentially lead to a less accurate estimation of the dot center and thereby reduce the measurement accuracy.

6.6 Stray Light

In Chapter 3, the effects of stray light were discussed. Basically, two types of stray light may be identified, that is, ghosting and flare. Ghosting shows up as localized artifacts that can change appearance as well as position depending on the location of strong light sources both inside and outside the field of view of the camera. Flare, on the other hand, manifests itself as an offset in the image intensity, leading to a general reduction of contrast in the image. The former type is very difficult to quantify due to its localized nature and strong dependence on the location of light sources. Since flare is associated with an overall decrease in contrast, it should be possible, at least in principle, to measure. In fact, the ISO standard 9358 (ISO, 1994) describes a method to measure the *veiling glare index* (VGI). The definition of veiling glare is similar to that of flare, as described in this book.

In ISO 9358, VGI is defined for the lens only, not the complete camera. Still, it is possible to adapt the method so that it can be used for a camera system. The measurement is

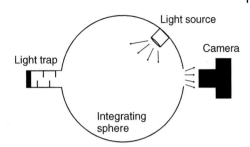

Figure 6.14 The setup used in the veiling glare measurement.

performed as follows. An *integrating sphere* is used, having a light source that illuminates the interior of the sphere. An integrating sphere is a hollow sphere covered on the inside with a white diffusely reflecting material. If a light source is pointed inside the sphere, the whole interior will be uniformly illuminated. One opening is made in the sphere, where the camera/lens to be tested is located. At the opposite end of the sphere, a *light trap* is placed. This light trap can be a tube with baffles and covered on the inside with a black non-reflecting material to ensure that a negligible amount of light can escape. The whole setup is shown in Figure 6.14. The camera captures an image of the interior of the sphere with the light trap placed in the center of the field of view. It is very important not to overexpose the white parts, since the image values of those parts in the captured image will be used to calculate the VGI, according to

$$\text{VGI} = \frac{L_B}{L_W} \times 100 \qquad (6.20)$$

where L_W are averaged image values calculated from the white parts in close vicinity to the light trap, and L_B is the average image value calculated inside the light trap. The resulting value is expressed as a percentage. Typical values range from fractions of one percent for very good systems to several percent for a poorly performing camera.

Flare may also be characterized using estimations of the MTF (Williams and Burns, 2004). The presence of flare may introduce artifacts leading to long tails of the PSF, which can be picked up by, for example, the slanted edge algorithm. An offset in the MTF measurement will lead to a delta function at zero spatial frequency. Therefore, if the point where the low frequency response crosses the vertical axis can be estimated, a flare metric may be constructed.

6.6.1 Practical Considerations

Since the VGI is calculated as a ratio of bright to dark image values, it is very sensitive to any changes in the dark area (the light trap). Therefore, if the image is not linear, for example, an sRGB image, the tone curve can have a big impact on the result.

6.7 Sharpness and Resolution

The pixel resolution of a camera, more correctly called the number of effective pixels on the image sensor (ANSI, 2015), has commonly been related to how sharp images produced by that particular camera can be. That the pixel resolution of an image capturing device does not necessarily correlate with perceived sharpness was pointed out already in Chapter 3. In order to understand the reasons behind the distinctions between sharpness, pixel resolution, and limiting resolution, the concept of the *modulation transfer function* (MTF) must be firmly understood.

In essence, the MTF describes the *spatial frequency response* (SFR) of an optical system. The spatial frequency can be thought of as the level of detail in the scene: the smaller details, the higher spatial frequency and the larger structures, the lower spatial frequency. In the time domain, the same concept is expressed using the relationship between temporal frequency, f, measured in Hz, and the period time, T, in seconds:

$$f = \frac{1}{T}$$

In the spatial domain, the relationship between the spatial frequency in one dimension, v, and the distance in that dimension, x, consequently becomes

$$v = \frac{1}{x}$$

The unit of spatial frequency must certainly be correlated with its corresponding spatial unit, so that if the distance is measured in, for example, millimeters, the spatial frequency is expressed in cycles per millimeter or line pairs per millimeter (lpm, or lp/mm). The latter is usually used in optical measurements.

Graphically, one can use bar patterns, as shown in Figure 6.15, to illustrate the spatial frequency concept. In order to find the limiting resolution of a particular camera, one can now simply capture an image of the target in this figure and determine the narrowest pattern that can just barely be resolved. The spatial frequency at that point may then be used as a measure of the sharpness, and is often referred to as the *limiting resolution* of the camera under test. As already pointed out, such a metric does not correlate very well with the perceived sharpness. This is illustrated in Figure 6.16, where the right image clearly has a higher resolution, while the left image presents an overall sharper impression. Obviously, the capability of the camera to reproduce low and intermediate spatial frequencies is very important for the sharpness performance. To properly account for this we will use linear systems theory, which will lead up to the definition of the MTF, already mentioned in the beginning of this section.

Generally, both the lens and the image sensor can be regarded as linear systems, and consequently the signal emanating from the camera sensor when capturing an image

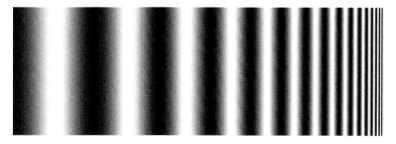

Figure 6.15 Graphical example of structure with varying spatial frequency content. Low spatial frequencies are found at the left end of the image and high spatial frequencies at the right end.

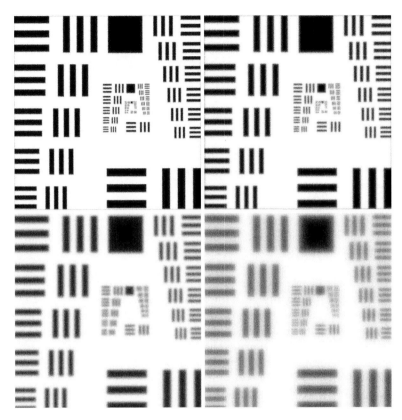

Figure 6.16 Images showing the distinction between sharpness and resolution. The left upper image is clearly sharper than the right upper image. However, in the zoomed in parts, shown in the bottom row, the right image shows more detail in fine structures.

from the lens, $g(x, y)$, can be described by the convolution of the input signal, $f(x, y)$, and the point spread function, $h(x, y)$:

$$g(x) = f(x, y) * h(x, y) = \int_{-\infty}^{\infty} f(x', y')h(x - x', y - y')dx' dy' \qquad (6.21)$$

From the convolution theorem, we know that the convolution of two functions in the spatial domain is equivalent to the multiplication of their Fourier transforms in the spatial frequency domain. The Fourier transform of the point spread function (PSF, defined in Chapter 4) is known as the *optical transfer function* (OTF). Since the point spread function provides information about the spatial modification of input signals, the OTF will describe how different spatial frequencies will be modified by the system. Mathematically, the relation between the OTF, $H(v, \mu)$, and the point spread function is written as

$$H(v, \mu) = \int_{-\infty}^{\infty} \int_{-\infty}^{\infty} h(x, y) \exp[-2\pi i(vx + \mu y)] \, dx dy \qquad (6.22)$$

This implies that the OTF is complex, that is:

$$H(v, \mu) = M(v, \mu) \exp[i\Theta(v, \mu)] \qquad (6.23)$$

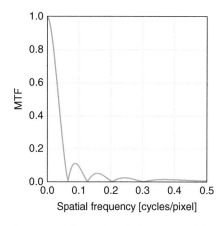

Figure 6.17 Illustration of phase reversal due to negative OTF values. Left: MTF of defocus blur; right: blurred image due to defocus.

where $M(v, \mu)$ describes how the magnitude of the input signal will be modified by the system and $\Theta(v, \mu)$ how the phase of the signal will change. For symmetric PSFs for which no translation of the input will occur (i.e., the PSF is an *even* function), the OTF will be real and the only possible values for the phase function $\Theta(v, \mu)$ are either zero or multiples of π. For odd multiples, the OTF will be negative, in which case a *phase reversal* of the signal will occur. This is the case for, for example, defocus, which has a PSF that can be described by the cylinder function

$$C(x, y) = \begin{cases} (\pi d)^{-2} & \sqrt{x^2 + y^2} \leq d \\ 0 & \sqrt{x^2 + y^2} > d \end{cases}$$

The OTF corresponding to this point spread function is

$$H(v, \mu) = \frac{2J_1(a\sqrt{v^2 + \mu^2})}{a\sqrt{v^2 + \mu^2}} \tag{6.24}$$

where $J_1(r)$ is the Bessel function of the first kind (Arfken, 1985). This function is shown in Figure 6.17 together with an image of a Siemens star target where the phase reversal can be clearly seen when moving the gaze from edge to center, thereby observing increasingly higher spatial frequencies.

The OTF is a two dimensional quantity and therefore not entirely easy to visualize. For this reason, one or more one dimensional slices in some appropriately chosen directions are usually presented. Such a one dimensional representation of the OTF is generally referred to as the *modulation transfer function*. The reason for its name will become clear from the following example. Imagine an input signal given by the function

$$f(x, y) = a + b \cos(2\pi kx) \tag{6.25}$$

The output of our optical system is described by Eq. (6.21):

$$g(x, y) = \int_{-\infty}^{\infty} \int_{-\infty}^{\infty} [a + b \cos(2\pi kx')] h(x - x', y - y') \, dx' dy'$$

This can be simplified to

$$g(x, y) = \int_{-\infty}^{\infty} [a + b \cos (2\pi k x')] \, \ell \, (x - x') \, dx' \tag{6.26}$$

where

$$\ell(x) = \int_{-\infty}^{\infty} h(x, y) dy \tag{6.27}$$

is known as the *line spread function* (LSF). By imposing that

$$\int_{-\infty}^{\infty} \ell(x) \, dx = 1$$

and also using the commutativity property of convolution, we can write

$$g(x, y) = a + b \int_{-\infty}^{\infty} \ell(x') \cos (2\pi k(x - x')) \, dx'$$

$$= a + b \left[\cos 2\pi k x \int_{-\infty}^{\infty} \ell(x) \cos 2\pi k x' \, dx' \right.$$

$$\left. + \sin 2\pi k x \int_{-\infty}^{\infty} \ell(x) \sin 2\pi k x' \, dx' \right] \tag{6.28}$$

The Fourier transform of the LSF is equal to

$$L(v) = \int_{-\infty}^{\infty} \ell(x) \exp(-2\pi i v x) \, dx$$

$$= \int_{-\infty}^{\infty} \ell(x) \cos(2\pi v x) \, dx - i \int_{-\infty}^{\infty} \ell(x) \sin(2\pi v x) \, dx$$

Being a complex quantity, we may write this as

$$L(v) = |L(v)| \exp(iP(v)) = |L(v)|[\cos(P(v)) + i \sin(P(v))] \tag{6.29}$$

Making the substitution $k = v$ and noting that the integrals in Eq. (6.28) can be identified as the real and imaginary parts of Eq. (6.29), yields

$$g(x, y) = a + b|L(v)|[\cos 2\pi v x \cos(P(v)) + \sin 2\pi v x \sin(P(v))]$$

$$= a + b|L(v)| \cos(2\pi v x - P(v))$$

Thus, compared to the input signal in Eq. (6.25), the amplitude of the output will be modified by the function $|L(v)|$ and the phase by the function $P(v)$, known as the *phase transfer function* (PTF). This is illustrated in Figure 6.18. Let us now define the *modulation* as the ratio of the amplitude to the average, that is,

$$\text{Modulation}[f] = \frac{(\max[f] - \min[f])/2}{(\max[f] + \min[f])/2}$$

$$= \frac{\max[f] - \min[f]}{\max[f] + \min[f]} \tag{6.30}$$

From Figure 6.18, this gives us

$$\text{Modulation}[f] = \frac{b}{a}$$

$$\text{Modulation}[g] = |L(v)| \frac{b}{a}$$

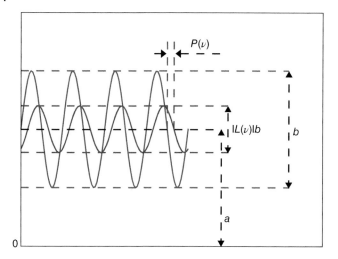

Figure 6.18 Change in amplitude and phase of a one-dimensional function passing through a system with an MTF given by $|L(v)|$ and a PTF $P(v)$.

The *modulation transfer* at a specific spatial frequency v is now defined as

$$\frac{\text{Modulation}\,[g]}{\text{Modulation}\,[f]} = \frac{|L(v)|b/a}{b/a} = |L(v)| \tag{6.31}$$

Consequently, the magnitude of the Fourier transform of the line spread function is the modulation transfer function, MTF, $m(v)$:

$$m(v) = \left| \int_{-\infty}^{\infty} \ell(x) \exp\left(-2\pi ivx\right) dx \right| \tag{6.32}$$

There are two properties of the MTF that are worth mentioning. The first tells us the value of the MTF at zero spatial frequency:

$$m(0) = \left| \int_{\infty}^{\infty} \ell(x) \exp\left(-2\pi i0x\right) dx \right| = \left| \int_{-\infty}^{\infty} \ell(x)\, dx \right| = 1 \tag{6.33}$$

The second relates the MTF to the OTF:

$$m(v) = \left| \int_{-\infty}^{\infty} \ell(x)\, \exp\left(-2\pi ivx\right) dx \right|$$

$$= \left| \int_{-\infty}^{\infty} \int_{-\infty}^{\infty} h(x,y)\, \exp\left(-2\pi ivx\right) dxdy \right|$$

$$= \left| \int_{-\infty}^{\infty} \int_{-\infty}^{\infty} h(x,y)\, \exp\left(-2\pi i(vx + 0y)\right) dxdy \right|$$

$$= |H(v,0)| \tag{6.34}$$

Therefore, the MTF is the magnitude of a slice of the OTF along one axis, and is equal to 1 at zero spatial frequency. For an optical system, the MTF will be a monotonically decreasing function, reaching zero at some cutoff frequency. When measured in the output image of a camera or on film, one might find MTF curves that exhibit a local maximum before decreasing and reaching zero at some higher spatial frequency value.

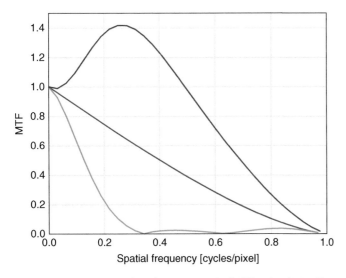

Figure 6.19 Some examples of MTF curves. Red: diffraction limited lens; green: defocused lens; blue: sharpening filter applied in the image processing.

This behavior can be due to, for example, sharpening filters in the digital case (Sonka *et al.*, 1999) and adjacency effects in the film case (Saxby, 2002). In Figure 6.19 some examples of MTF curves are shown.

The MTF provides a powerful tool to analyze sharpness in images. If we go back to the example in the beginning of this section, Figure 6.16, and study the MTF of the two systems that produced the images, it becomes immediately clear what is causing the discrepancies (see Figure 6.20). At low spatial frequencies, the left image has a lower MTF compared to the right, while at higher spatial frequencies the situation is reversed.

6.7.2 The Contrast Transfer Function

One way of obtaining the MTF is to measure the modulation of sinusoidal bar targets, varying in frequency. Since it is, relatively speaking, more challenging to produce such sinusoidal charts, one might try a simpler variant with sharp transitions between dark and bright areas. An example of such a chart is shown in Figure 6.21. A bar pattern with sharp transitions can be expressed in terms of harmonic functions through the Fourier series

$$f(x, y) = a + b\frac{4}{\pi} \sum_{n=0}^{N} \frac{(-1)^n \cos{(2\pi(2n + 1)kx)}}{2n + 1}$$

If we apply the same mathematical treatment to this expansion as for the sinusoidal pattern above, we find that the transfer function will be given by a new quantity, the *contrast transfer function* (CTF). The relation between the CTF, $c(v)$, and the MTF is (Coltman, 1954)

$$c(v) = \frac{4}{\pi} \left(m(v) - \frac{m(3v)}{3} + \frac{m(5v)}{5} - \cdots \right) \tag{6.35}$$

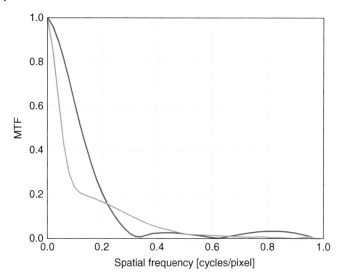

Figure 6.20 The distinction between sharpness and resolution explained by the MTF. The red curve is the MTF of the imaging system used to produce the left image in Figure 6.16, and the green curve represents the MTF corresponding to the right image in that figure.

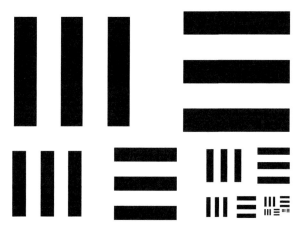

Figure 6.21 Example resolution chart.

It should be noted that the signs of the terms in this series are irregular. For the inverse relation, however, the signs are alternatingly positive and negative:

$$m(v) = \frac{\pi}{4} \left(c(v) + \frac{c(3v)}{3} - \frac{c(5v)}{5} + \ldots \right) \tag{6.36}$$

It is evidently somewhat cumbersome to translate between the CTF and MTF due to the fact that, in principle, an infinite number of harmonics have to be included in the calculation. However, as touched upon above, an optical system has a cutoff frequency, above which the MTF is zero. Therefore, in practice, a finite number of terms are involved in the calculation. Furthermore, the higher the spatial frequency, the less terms are needed.

6.7.3 Geometry in Optical Systems and the MTF

An image is two dimensional, and to fully capture the spatial frequency response of an optical system, the two dimensional OTF needs to be measured. Since the MTF is just a slice of the OTF in some direction, different MTF curves will be obtained for different orientations in the case when the point spread function is asymmetric. In particular, optical systems usually exhibit a radial symmetry, leading to a radial-tangential symmetry of the PSF. An example is shown in Figure 6.22. In order to obtain the highest and lowest MTF responses, one has to measure along the best and worst directions. In lens MTF measurements, as mentioned above, this corresponds to measuring in the tangential and sagittal (radial) directions. In digital images, which can exhibit some vertical-horizontal symmetry due to the layout of the pixel array, it has been customary to measure the MTF in the vertical and horizontal directions, using for instance the slanted edge method, which will be described in much more detail further on. Certainly, this will produce MTF curves which are different from the tangential and sagittal curves, as illustrated in Figure 6.23. Usually, the largest variations in sharpness are due to the lens. Therefore in most cases, it is recommended to measure the MTF in images in the tangential and sagittal directions.

As shown in Figure 6.22, the PSF of a lens may vary substantially across the field of view, thus violating the requirement on shift invariance needed to be able to measure the MTF. However, if the window chosen for the measurement is small enough, the variation of the PSF will be so small within this window that it, for all practical purposes, can be regarded to be constant. Still, the measurement region is an important factor when determining which method is best suited for measuring the MTF in a certain application.

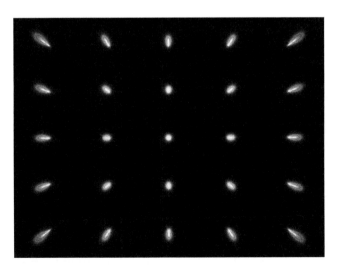

Figure 6.22 Position and orientation of point spread functions in the image produced by an example lens.

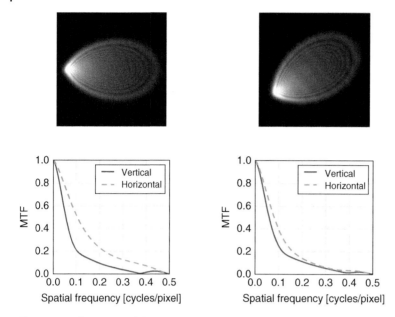

Figure 6.23 Illustration of discrepancies in MTF curves depending on orientation. The right PSF is a copy of the left, but rotated 30 degrees counterclockwise. The bottom graphs show MTFs calculated in the vertical and horizontal directions. The results are evidently different.

6.7.4 Sampling and Aliasing

Measuring the MTF in digital images requires the data to be sampled on some regular grid with spacing Δx. Mathematically, we may express this as

$$f_s(x) = f(x) \sum_{n=-\infty}^{\infty} \delta(x - n\Delta x) \qquad (6.37)$$

where $f_s(x)$ is the sampled signal, $f(x)$ the original signal, and $\delta(x)$ the Dirac delta function. Figure 6.24 shows an example of sampling of a harmonic signal. We see in this case that if we were to try to reconstruct the sampled signal by interpolation, the result would be another harmonic signal with a lower frequency. This is known as *aliasing* since the high frequency input signal takes on the appearance (i.e., alias) of a lower frequency signal when it is reconstructed. The Fourier transform of the sampled signal in Eq. (6.37), $F_s(\nu)$, can be shown to be

$$F_s(\nu) = \sum_{n=-\infty}^{\infty} F\left(\nu - \frac{n}{\Delta x}\right) \qquad (6.38)$$

where $F(\nu)$ is the Fourier transform of the original signal. The spectrum of the sampled signal is consequently the spectrum of the original signal repeated an infinite number of times at regular intervals $n/\Delta x$. Figure 6.25 shows the sampled spectrum for two situations, one where the bandwidth of the original system is larger than the *sampling frequency* $\nu_s = 1/\Delta x$, and one where it is lower. In the case where the bandwidth is too large or the sampling frequency is too low, higher frequencies will be "folded" back into lower frequencies, which will result in aliasing artifacts. To avoid this situation, either

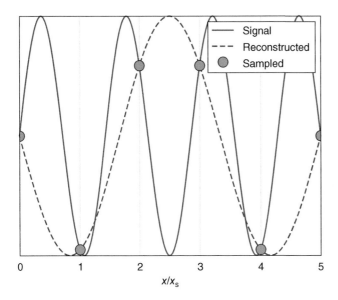

Figure 6.24 Example of aliasing. The dashed line represents the reconstructed signal. x_s is the sampling interval.

the sampling frequency needs to be higher, or the original signal must be filtered in order to remove frequency content above half the sampling frequency. This latter frequency is known as the *Nyquist frequency*. The latter case of course means that information will be lost in the sampling process. Formally, this is expressed by the *Sampling Theorem*[4] (Vollmerhausen and Driggers, 2000), which states that a band limited signal, $f(x)$, with no frequency content above the Nyquist frequency, $v_s/2$, can be perfectly reconstructed from samples taken at intervals $\Delta x = 1/v_s$ by the function

$$f(x) = \sum_{n=-\infty}^{\infty} f(n\Delta x) \frac{\sin\left[\pi\left(\frac{x}{\Delta x} - n\right)\right]}{\pi\left(\frac{x}{\Delta x} - n\right)}$$

$$= \sum_{n=-\infty}^{\infty} f\left(\frac{n}{v_s}\right) \text{sinc}\,(xv_s - n) \tag{6.39}$$

The artifacts resulting from aliasing usually appear as moiré patterns, and were discussed in Chapter 3.

When measuring the MTF, aliasing will lead to an overestimation of the MTF, especially in higher spatial frequency regions. As previously discussed, this can be overcome by oversampling the signal and techniques to do that will be discussed later in this chapter.

6.7.5 System MTF

In a camera, the lens is not the only component affecting the sharpness. As mentioned in Chapter 4, the size of the pixel itself limits the ability to resolve infinitely small objects,

---

4 In many cases involving either or all of the names Shannon, Nyquist, and Whittaker.

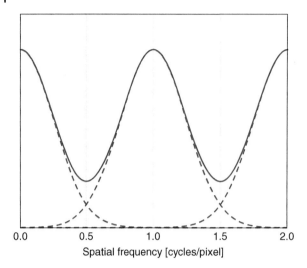

0.0   0.5   1.0   1.5   2.0

Spatial frequency [cycles/pixel]

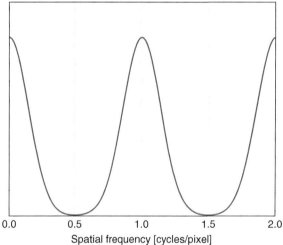

0.0   0.5   1.0   1.5   2.0

Spatial frequency [cycles/pixel]

Figure 6.25 Examples of a band limited signal that exhibits aliasing (upper curve), and without aliasing (lower curve).

resulting in a point spread function similar to the shape of the pixel. If the photodetector of the pixel is a square with side a, the PSF is

$$h(x, y) = \begin{cases} \frac{1}{a^2} & x, y \leq a \\ 0 & x, y > a \end{cases} \tag{6.40}$$

The Fourier transform, that is, OTF, of this function is

$$H(v, \mu) = \text{sinc}\,(av)\,\text{sinc}\,(a\mu) \tag{6.41}$$

Thus, the MTF of the pixel (assuming no crosstalk) is

$$m(v) = \text{sinc}\,(av) \tag{6.42}$$

However, in practice the PSF of the photodetector, even if square, can be modified by the PSF of the microlenses that are put on top of the photodetectors in order to increase the sensitivity of the image sensor (Parulski *et al.*, 1992).

In order to reduce color moiré effects due to insufficient sampling and color interpolation, some cameras may contain an *optical lowpass filter* (OLPF). Such a filter is often made up of a sandwich having several pieces of a birefringent material (Hecht, 1987), which splits the light passing through into several polarized beams that will be spread out by some distance d, proportional to the material thickness, from the original position in the vertical and horizontal directions. The MTF of such a filter is given by

$$m_{\text{OLPF}}(v) = \cos(2\pi v d) \tag{6.43}$$

Image signal processing can also reduce sharpness and contribute with MTFs from, for example, color interpolation, noise reduction, sharpening, and so on. For the example of simple bilinear color interpolation in Eq. (4.15), the MTF of the green interpolated pixels may be expressed as

$$m_G = \frac{1 + \cos(2\pi v)}{2} \tag{6.44}$$

Similar calculations can be performed for the red and blue channels. Since this calculation only applies to the green pixels in non-green locations, the Bayer pattern gives rise to a different response from pixel to pixel, and therefore the above expression is not completely correct (Yotam *et al.*, 2007). In order to take into account the phase dependency, more elaborate calculations need to be performed. The above derived expression is, however, still interesting for the qualitative understanding of the Bayer pattern MTF.

The resulting image is a chain of convolutions with the PSFs of the various components and the input signal. In the frequency domain, this corresponds as we know to a multiplication. Therefore, the total MTF of the camera may be obtained by simply multiplying all the individual MTFs from the lens, sensor, and so on. Figure 6.26 shows the result of combining the different MTFs together into the full, system MTF.

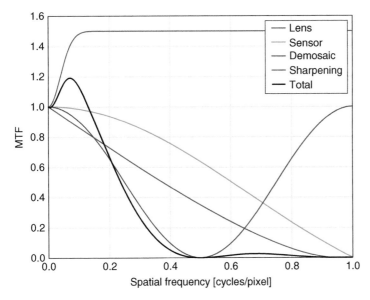

Figure 6.26 The system MTF is the combination of several MTFs from separate parts of the camera.

From the above discussion it should be clear that the MTF can be measured in a number of ways, at least in principle. The modulation could be measured directly on bar patterns of varying spatial frequency, either sinusoidal or discrete. The point spread function may be measured directly by imaging a small point and by a Fourier transform the MTF is found. Other techniques could also be envisioned, for instance using random targets (Brauers *et al.*, 2010) with known spectral content like white noise or other features. A variant of these techniques will be revisited when discussing measurements of texture blur.

The ISO 12233 standard (ISO, 2014) describes two methods for estimating the MTF by measuring the spatial frequency response (SFR). The distinction between MTF and SFR is made to emphasize the fact that the measured frequency response of a camera may change depending on the measurement conditions as reflected in the test chart used to perform the measurement. So in practice, a digital camera cannot be exactly characterized using a single MTF curve. Two SFR methods are described in the standard: edge-based SFR (E-SFR) and sine-based SFR (S-SFR), which is measured using a sinusoidal Siemens star test chart. Both of these methods will be described and discussed in the following sections.

As touched upon above, one way of measuring the MTF is to directly measure the PSF as captured by the camera. In practice, this method is made quite difficult by the fact that in order for the imaged point source to be as small as possible, its intensity will also drop. This will make the signal to noise ratio of the measurement very low, thus introducing large errors. Furthermore, the PSF will be sampled by the camera array, and this will potentially introduce aliasing into the measurement and also limit the resolution substantially.

The first problem can be dealt with by using a sharp edge as a target, for instance a knife edge, instead of a point source. The convolution of an ideally sharp edge with the PSF of the camera is

$$I_{edge}(x, y) = \int_{-\infty}^{\infty} \int_{-\infty}^{\infty} \Theta(x') h(x - x', y - y') dx' dy'$$

$$= \int_{-\infty}^{\infty} \Theta(x - x') \ell(x') dx'$$

$$= \int_{-\infty}^{x} \ell(x') dx' = \left[e(x') \right]_{-\infty}^{x} = e(x)$$

Here, $\Theta(x)$ is the Heaviside step function, defined as

$$\Theta(x) = \begin{cases} 0, & x \leq 0 \\ 1, & x > 0 \end{cases} \tag{6.45}$$

and $\ell(x)$ the LSF defined in Eq. (6.27). The result above implies that the LSF is the derivative of $e(x)$, the *edge spread function* (ESF):

$$\ell(x) = \frac{de(x)}{dx} \tag{6.46}$$

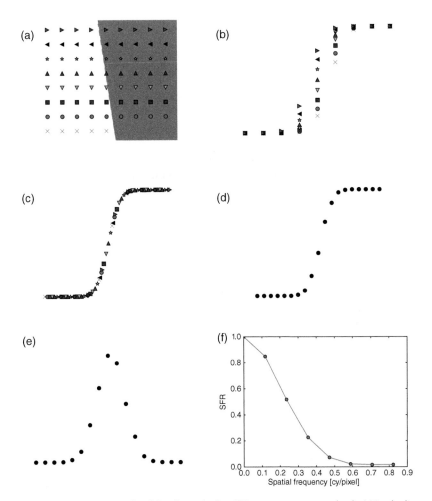

Figure 6.27 The principle of the slanted edge SFR measurement method. a) Line by line sampling of image values across the edge. b) Edge profiles of each line. c) Displaced edge profiles yielding an oversampled ESF. d) Binned ESF. e) Differentiated edge profile, yielding the LSF. f) SFR calculated from Fourier transform of LSF.

Therefore, it is possible to obtain the SFR by capturing an image of a sharp edge, then taking the derivative of a cross section of the edge followed by a Fourier transform. However, it is possible to improve this even further by slanting the edge with respect to the pixel array (Reichenbach *et al.*, 1991; Fischer and Holm, 1994). Then scans in the direction almost perpendicular to the edge will be slightly displaced, see Figure 6.27. By scanning the edge profile for each row in the image followed by a displacement of the profile corresponding to the shift in position relative to the other edge profiles and then superposing the resulting profiles on top of each other, an oversampled edge profile is obtained. In this way a much more accurate assessment of the SFR can be made. In the ISO 12233 standard (ISO, 2014), the edge profile samples thus obtained are averaged within bins that are 1/4 pixel wide. Consequently, a four times oversampling is accomplished (Williams, 1998; Burns, 2000; Williams and Burns, 2001; Burns and Williams, 2002; Williams, 2003).

In summary, the algorithm to calculate the edge SFR in ISO 12233 is as follows:

1) For each line in the edge image, estimate the location of the edge.
2) Using the edge locations, estimate the edge direction and position using linear regression.
3) Displace each line according to the position found by calculating its position on the line obtained in the previous step.
4) Place each displaced line on top of each other to form an oversampled edge profile.
5) Average edge values within bins of size 1/4 pixel and resample the edge profile on the resulting grid.
6) Calculate the derivative of the supersampled edge to get the line spread function by convolving the edge function with a finite difference filter.
7) Calculate the discrete Fourier transform of the LSF. The SFR is the absolute value of the Fourier transform.

There are various ways to improve this algorithm. One is to extend it to also allow for measurements at other angles than along the vertical and horizontal directions (Reichenbach *et al.*, 1991; Kohm, 2004; Ojanen and Tervonen, 2009; Masaoka *et al.*, 2014), thereby accommodating true tangential and sagittal measurements. To accomplish this, one way is to replace items 3 and 4 with a rotation of the pixel coordinate space (Ojanen and Tervonen, 2009). Assuming that the position of the edge, represented by an arbitrary point (x_e, y_e) at the edge, and its orientation, represented by the angle θ_e, are known, the distance, r, to the edge from each pixel coordinate, (x, y), inside the cropped image along the scanning direction, θ_s, becomes

$$r = \frac{(y - y_e) \cos \theta_e - (x - x_e) \sin \theta_e}{\sin (\theta_s - \theta_e)} \tag{6.47}$$

These distances can now be paired with their respective image values and then sorted so that the positions are in ascending order to form the raw ESF corresponding to Figure 6.27c. Then, the ESF values within 1/4 pixel bins can be averaged to form the final supersampled ESF profile.

Two more things should be mentioned regarding the binning and differentiation procedures. Firstly, the binning is equivalent to performing a lowpass filtering operation with an averaging filter $1/L$ pixels in width, where $L = 4$ in this case. The frequency response of such a filter is

$$H_{\text{bin}}(v) = L \int_{-1/2L}^{1/2L} \exp(-2\pi i v x)\, dx = L \left[\frac{\exp(-2\pi i v x)}{-2\pi i v} \right]_{-1/2L}^{1/2L}$$

$$= \frac{\sin (\pi v/L)}{\pi v/L} = \text{sinc}\,(v/L) \tag{6.48}$$

with the spatial frequency v in units of cycles per pixel. Therefore, it is possible to compensate for the binning by dividing the SFR values by this function. In practice, however, this correction is quite small and usually only amplifies noise at high spatial frequencies which means that it can be excluded in most cases.

Secondly, the finite difference filter $(0.5\ 0 - 0.5)$ used to approximate the derivative will also introduce a frequency response that should be compensated for (Burns, 2000).

The discrete Fourier transform of this filter becomes

$$H_{\text{filter}}(v) = 0.5|1 - \exp(4\pi iv)| = 0.5\sqrt{2(1 - \cos(4\pi v))} = \sin(2\pi v) \qquad (6.49)$$

The derivative of a function $f(x)$ has the Fourier transform

$$F'(v) = \int_{-\infty}^{\infty} \frac{df(x)}{dx} \exp(-2\pi ivx)\, dx$$

This can be integrated by parts, which yields

$$F'(v) = -2\pi ivF(v)$$

where $F(v)$ is the Fourier transform of $f(x)$. Therefore, the derivative acts like a "filter" in the Fourier domain. In order to compensate for the finite difference filter approximating the derivative, we should divide the calculated SFR by the absolute value of the spatial frequency response of this filter and then multiply by the absolute value of the frequency response of the derivative, that is, $2\pi v$. In other words, the compensation due to the finite difference filter is made by dividing the calculated SFR by the function

$$h_{\text{comp}} = \frac{\sin(2\pi v/L)}{2\pi v/L} = \text{sinc}(2v/L) \qquad (6.50)$$

where we have also taken into account the oversampling by the factor L in order to express the spatial frequency in cycles per pixel.

Item 2 in the description of the ISO 12233 methodology prescribes that the edge should be fitted to a straight line. In some cases, for instance if the lens suffers from substantial amounts of optical distortion, a straight line fit may not produce an accurate enough result. To improve the measurement, it is possible to instead use a higher order polynomial to obtain a better fit. This should be done with care, however, since it might make the algorithm more sensitive to noise in the image. If a test chart with edges rotated in order to enable tangential and sagittal measurements is used, this issue becomes less problematic. When using near-horizontal and near-vertical edges at off center positions in the image, excessive distortion may affect the shapes of those edges substantially. However, if the edges are oriented to be near-parallel or perpendicular to a line through the edge centers and the center of the image, optical distortion will influence the edge geometry to a lesser extent. The reason for this is that in a well centered system, the effect of distortion is to displace points in the radial direction. Therefore, an edge that is oriented perpendicularly to the radius will then only be displaced and not distorted. In a real situation, there will, however, certainly still be some distortion of the edge, but may still be less than for a vertical or horizontal edge.

6.7.8 Sine Modulated Siemens Star SFR

As an alternative to measuring the edge SFR, one can directly measure the modulation in a test chart having sine modulated starburst patterns with varying spatial frequency. An elegant way to do such a measurement is to use a sine modulated Siemens star pattern (Loebich *et al.*, 2007), shown in Figure 6.28. This method (S-SFR) is also described in ISO 12233 as an alternative way of measuring SFR. In comparison with the slanted edge SFR (E-SFR), there are cases where different results are obtained using sinusoidally varying features instead of sharp edges. This may, for example, be because adaptive image processing algorithms for color interpolation, noise reduction, and sharpening are likely to treat such features differently.

Figure 6.28 A sine modulated Siemens star test pattern.

Using this particular starburst pattern test chart, it is possible to accommodate a wide range of spatial frequencies within a smaller part of the image compared to other charts employing sinusoidal or bar pattern features. Furthermore, it is also possible to measure the SFR in different directions, thereby naturally being able to assess the SFR in, for example, tangential and sagittal directions.

The measurement algorithm as described in ISO 12233 and elsewhere (Loebich *et al.*, 2007) is outlined in the following. The Siemens star is divided into a number of segments. Within each segment, the distance from the center of the star to the periphery is determined. In order to correct for geometric distortion, this step is necessary.

Within each segment, the radius from center to circumference is divided into several sub-radii. Each distance from center to every individual sub-radius represents one spatial frequency. The spatial frequency, v_n, of sub-radius r_n is given by

$$v_n = \frac{N_p}{2\pi r_n} \tag{6.51}$$

where N_p is the number of sinusoidal cycles around the periphery of the star. For each segment and radius, the pixels that are nearest to the circular segment are chosen and their intensity values are determined as a function of angle. The sinusoidal pattern that emerges is fitted to the function

$$y(\theta) = a + b \sin(\theta + \theta_0) \tag{6.52}$$

With the coefficients a and b determined, the modulation is calculated using Eq. (6.30). By repeating this calculation for each radius that we previously decided to use, we can plot the obtained modulation values as a function of spatial frequency, thus forming an estimate of the MTF. Observe that this estimate has not been normalized, since the true MTF is found by dividing the measured modulation by the input modulation, according to Eq. (6.30). Since the input modulation should not change as a function of spatial frequency, one could in principle calculate this value from the modulation of two large patches, one having the same gray value as the darkest value of the Siemens star, and the other having the same value as the brightest part of the Siemens star. Issues related to linearization and normalization of sinusoidal Siemens star measurements are described by Artmann (2016). It should also be mentioned that the determination of the center of the Siemens star is quite important—if a misregistration is made, the resulting SFR values may be underestimated (Birch and Griffin, 2015).

6.7.9   Comparing Edge SFR and Sine Modulated Siemens SFR

The motivation behind introducing the Siemens star measurement alongside the slanted edge technique may at least be traced back to measurements performed on test patterns captured on film (Cadou, 1985). In this case, adjacency effects, which are nonlinear, caused MTF measurements on sine-modulated targets to differ from measurements performed on edge targets. In a digital camera, adaptive color interpolation, noise filtering, and sharpening algorithms will introduce similar behavior, providing different responses depending on the local contrast in the image. Therefore, in a digital image, a sharp edge will likely induce a different response compared to the more slowly varying sinusoidal patterns. As a result, in some cases it can be helpful to measure and report both the E-SFR and the S-SFR.

From a technical measurement point of view, there are a few important facts to keep in mind when deciding which method should be used for a particular situation. First, the sinusoidal pattern that makes up the Siemens star takes up quite a large part of the camera field of view. If the lens of the camera that is being tested has large variations in sharpness across the field of view, this will mean that the Siemens star method will not be measuring the same PSF for low and high spatial frequencies. Thus, the result will be less accurate compared to the slanted edge method, which uses a much smaller spatial extent.

Second, the fact that there is a gap between the lowest spatial frequency and zero frequency makes it difficult to determine the correct scaling of the curve and therefore introduces inaccuracies when trying to estimate the MTF. Also here, the slanted edge method has an advantage, since spatial frequencies all the way down to zero are obtained.

Furthermore, at least one study shows discrepancies between the slanted edge and Siemens star methods applied to raw images, where there is no image processing that could affect the results (Williams *et al.*, 2008). A possible explanation to this behavior could be a combination of less accurate results for the Siemens star method close to the Nyquist frequency combined with difficulties in normalizing the curve at zero spatial frequency. However, no published study has to the knowledge of the authors thus far followed up on the cited work to further explain this behavior.

The observation that the slanted edge method shows different behavior compared to the Siemens star method, especially for highly sharpened images (Artmann, 2015), may more reflect the fact that image processing algorithms in the camera treat the two test chart types differently. There is no conclusive reason to believe that the slanted edge method would generally provide more inaccurate data for those cases, and therefore the results of the measurement are likely to reflect the actual response. Furthermore, the impression of sharpness in an image is closely related to the rendering of sharp edges, which further motivates the use of the slanted edge algorithm in methods aiming to provide a good correlation with human vision.

6.7.10 Practical Considerations

As pointed out previously, the E-SFR and S-SFR measurements require linear data. In chapter 4, we described the fact that the output from a camera is typically nonlinear, since a gamma curve is usually applied to the image data. Also, more complicated operations, such as local tone mapping operators might also have been applied. Therefore, SFR measurements as described in this section should not be performed directly on the image, unless it is known that the camera indeed produces linear data. The ISO 12233 standard prescribes a linearization step by measuring the OECF of the camera and uses this to make the linearization. This type of characterization might be tricky in the case when local tone mapping has been applied to the image data. In this case it is not certain that the image signal for the edge or Siemens star has been treated the same as the data in the test chart used to determine the OECF. Another approach that might be taken is to perform the linearization by simply inverting the gamma curve (IEEE, 2017). Most cameras apply the sRGB gamma curve, described by Eq. (4.16). The inverse of this function is given by

$$X_{in} = \begin{cases} \left(\frac{X_{out}+0.055}{1.055}\right)^{2.4} & X_{out} > 0.04045 \\ \frac{X_{out}}{12.92} & X_{out} \leq 0.04045 \end{cases} \tag{6.53}$$

In addition, if the contrast of the edge or Siemens star is low, the signal region within which the SFR calculation is performed may be small enough to be very close to linear (Burns, 2005). By lowering the contrast, issues with clipping in the low or high signal parts of the image can also be avoided. However, if the contrast is lowered too much, the signal to noise ratio of the measurement may be too low to generate an accurate result. Therefore, it might be useful to perform measurements over a range of chart contrasts.

6.8 Texture Blur

Figure 4.26 shows the effects of adaptive noise filtering on the output image. Such filtering clearly manages to retain the sharpness impression by preserving the edges in the images. However, this is at the expense of blurring of fine detail in low contrast textured parts. Such an artifact is difficult to capture using the standard MTF measurement methods, particularly the slanted edge method.

In order to characterize this very important artifact, much activity has been spent developing methods for measuring texture blur. In recent years, two new metrics have arisen to approach this problem, the *dead leaves* method (Cao et al., 2009, 2010) and kurtosis measurements on white noise targets (Artmann and Wueller, 2009). As indicated in a subjective validation study (Phillips and Christoffel, 2010), the dead leaves

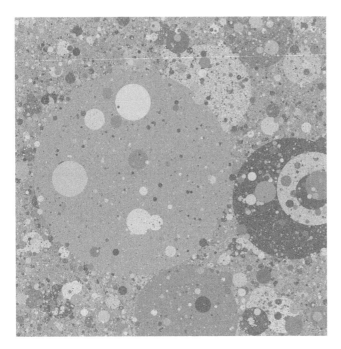

Figure 6.29 A dead leaves test pattern.

method appears more robust and has been the most widely adopted method of the two. For this reason, the following discussion will focus on this method.

The denomination *dead leaves* comes from the (fairly distant) resemblance of the test target used for the experiment to randomly scattered leaves of varying sizes. Figure 6.29 shows an example of the typical pattern found in the dead leaves test chart. This type of test chart has some very appealing properties: it is insensitive to scaling, rotations and translations, as well as exposure and contrast. This simplifies the testing procedure significantly. It also resembles texture features found in natural scenes, which makes it possible to assess the texture degradation by direct visual inspection of captured images.

The principle of the dead leaves measurement relies on the fact that the statistics of the test chart are well known. The spatial frequency response of the chart is therefore also known. Furthermore, the functional form of the spatial frequency response can be very closely modeled by a power law, according to

$$I_{\text{DL}}(m, n) = \begin{cases} N^4 \langle i \rangle^2 & m^2 + n^2 = 0 \\ \frac{c(N)}{(m^2+n^2)^{\eta/2}} & m^2 + n^2 \neq 0 \end{cases} \tag{6.54}$$

with

$$c(N) = \frac{\text{Var}(i)N^4}{\sum_{m,n} \frac{1}{(m^2+n^2)^{\eta/2}}} \tag{6.55}$$

Here, the discrete Fourier transform of the image i of size $N \times N$ is I_{DL}, for spatial frequency coordinates m, n. The variance of the image values is $\text{Var}(i)$, the average is $\langle i \rangle$, and the exponent η is approximately 1.857 (Cao *et al.*, 2009; McElvain *et al.*, 2010).

From Eq. (6.21) and the convolution theorem, we know that the optical transfer function, $H(\nu, \mu)$, describes the connection between the Fourier transform of the input and output signals ($I_{in}(\nu, \mu)$ and $I_{out}(\nu, \mu)$, respectively) in the following way:

$$I_{out}(\nu, \mu) = I_{in}(\nu, \mu)H(\nu, \mu) \qquad (6.56)$$

Therefore, the OTF of the system is found by calculating the discrete Fourier transform of the captured image, $I_{out}(m, n)$, and dividing it by the spectrum of the test chart, which is known from Eq. (6.54), that is,

$$H(m, n) = \frac{I_{out}(m, n)}{I_{DL}(m, n)} \qquad (6.57)$$

The MTF is now obtained by assuming radial symmetry of $H(m, n)$ and averaging the modulus, $|H(m, n)|$, in the angular direction around circles with radii corresponding to the spatial frequencies of interest.

6.8.1 Chart Construction

In order to obtain the desired statistical properties of the dead leaves test chart, the chart should be constructed as follows. The distribution of circle radii should follow a $1/r^3$ function, where the upper and lower bounds, r_{min} and r_{max}, are chosen such that the smallest sizes are considerably less than the pixels in the captured image and the largest size should be considerably larger than r_{min}, but not larger than the image itself.

The generated circles are drawn from the above distribution and are placed randomly in the image in such numbers that the entire image area is covered. This typically involves several millions of circles. The gray levels of the circles are drawn from a uniform distribution ranging from relative intensity 0.25 up to 0.75. For a thorough description of the chart generating procedure, the reader is advised to consult the paper by McElvain *et al.* (2010).

6.8.2 Practical Considerations

The dead leaves method has been shown to be quite noise sensitive (McElvain *et al.*, 2010). This has the effect of amplifying high spatial frequencies, and thus overestimating the MTF/SFR, which could lead to incorrect conclusions about the texture response of the camera being measured. As a remedy, it was suggested that the power spectral density of a uniform gray patch with the same average gray level as the dead leaves features could be used to compensate for the exaggerated high frequency response in the following way:

$$H(m) = \sqrt{\frac{|I_{out}(m)|^2 - |I_{gray}(m)|^2}{|I_{DL}(m)|^2}} \qquad (6.58)$$

where $|I_{out}(m)|^2$, $|I_{gray}(m)|^2$, and $|I_{DL}(m)|^2$ are the power spectral densities of the measured image, gray image, and dead leaves chart, respectively.

One should be somewhat cautious with this compensation, however, since it is unlikely that adaptive noise reduction algorithms will treat the uniform gray patch in the same way as the dead leaves features. Thus, the spatial frequency content of the two may be quite different at high spatial frequencies.

6.8.3   Alternative Methods

An extension to the dead leaves model has been proposed where the original gray level target is modified to become colored (Artmann and Wueller, 2012). The motivation behind this step was that it was sometimes found that there was a difference in the appearance of texture in real life scenes containing, for example, grass or other colored features, and the standard dead leaves texture chart. This work has been expanded in later works (Kirk *et al.*, 2014; Artmann, 2015).

Even though the dead leaves model is appealing, the sensitivity to noise may make it less reliable in many situations. As an alternative method, the sinusoidal Siemens star SFR method might be considered to be used to characterize texture blur, as has been suggested for an ISO standard (Artmann, 2016). By varying the sinusoidal modulation level of the test target, one might anticipate different behavior of the image signal processing algorithms that most likely will be reflected in the S-SFR data.

Another way of dealing with the difference in treatment of the dead leaves chart and uniform areas is to calculate the cross correlation spectrum between input and output (Kirk *et al.*, 2014). The main disadvantage of this approach is that it becomes a full reference method, and the accuracy is therefore highly dependent on the spatial similarities between the input scene and output image.

6.9   Noise

Most noise in an image from a digital camera originates from sources within the image sensor, as discussed in Chapter 4. By noise we generally mean random fluctuations of a signal, which makes the estimation of the desired signal values more difficult to obtain. In an image, this will lead to masking of certain image features, and artifacts not present in the original scene will be introduced. To estimate the magnitude of the noise, statistical methods need to be employed. The signal may then be represented by the theoretical mean value, and the noise by the variance. As an estimate of the impact of the noise, the *signal to noise ratio* (SNR) can be defined as the ratio

$$SNR = \frac{\mu}{\sigma} \tag{6.59}$$

where μ is the mean value and the square root of the variance, $\sigma = \sqrt{\text{Var}(X)}$, is usually referred to as the *standard deviation* of the signal. In imaging, the SNR is typically obtained by calculating the sample mean and standard deviation of a uniform patch in a test chart. Since the image signal gets modified on its way through the ISP, so does the noise. Sharpening, noise filtering, tone mapping, and similar operations, will all amplify or attenuate the noise and even change the appearance of the noise with respect to color and spatial features. Therefore, the simple SNR estimate in Eq. (6.59) will often not provide an accurate representation of the appearance of noise in an image. The following subsections will discuss some of the aspects of noise appearance and how to perform a more detailed assessment.

6.9.1   Noise and Color

In principle, the noise from the different color channels in the raw image produced by the image sensor should be statistically independent. However, there are situations where the assumption of statistical independence between color channels will not hold. For instance, the structure of the pixels in the sensor may introduce correlations between pixels. More important, however, is the correlation being introduced by the color interpolation and color correction.

Color correction typically implies a linear transformation of the color signal using a 3×3 *color correction matrix* (CCM), although higher order matrices and 3D lookup tables can be used. The combined effect of white balancing and color correction is described by the equation

$$\left(R' \; G' \; B' \right) = \left(R \; G \; B \right) \begin{pmatrix} \rho_R & 0 & 0 \\ 0 & \rho_G & 0 \\ 0 & 0 & \rho_B \end{pmatrix} M \tag{6.60}$$

where R', G', and B' are the transformed red, green, and blue color values, and ρ_R, ρ_G, and ρ_B the white balance weights for red, green, and blue, respectively. The matrix M is the CCM:

$$M = \begin{pmatrix} m_{11} & m_{12} & m_{13} \\ m_{21} & m_{22} & m_{23} \\ m_{31} & m_{32} & m_{33} \end{pmatrix} \tag{6.61}$$

We can calculate the variances of the transformed red, green, and blue signals:

$$\begin{aligned} \mathrm{Var}(R') =\ & \rho_R^2 m_{11}^2 \mathrm{Var}(R) + \rho_G^2 m_{21}^2 \mathrm{Var}(G) + \rho_B^2 m_{31}^2 \mathrm{Var}(B) \\ & + 2\rho_R\rho_G m_{11}m_{21}\mathrm{Cov}(R,G) \\ & + 2\rho_R\rho_B m_{11}m_{31}\mathrm{Cov}(R,B) \\ & + 2\rho_G\rho_B m_{21}m_{31}\mathrm{Cov}(G,B) \\ \mathrm{Var}(G') =\ & \rho_R^2 m_{12}^2 \mathrm{Var}(R) + \rho_G^2 m_{22}^2 \mathrm{Var}(G) + \rho_B^2 m_{32}^2 \mathrm{Var}(B) \\ & + 2\rho_R\rho_G m_{12}m_{22}\mathrm{Cov}(R,G) \\ & + 2\rho_R\rho_B m_{12}m_{32}\mathrm{Cov}(R,B) \\ & + 2\rho_G\rho_B m_{22}m_{32}\mathrm{Cov}(G,B) \\ \mathrm{Var}(B') =\ & \rho_R^2 m_{13}^2 \mathrm{Var}(R) + \rho_G^2 m_{23}^2 \mathrm{Var}(G) + \rho_B^2 m_{33}^2 \mathrm{Var}(B) \\ & + 2\rho_R\rho_G m_{13}m_{23}\mathrm{Cov}(R,G) \\ & + 2\rho_R\rho_B m_{13}m_{33}\mathrm{Cov}(R,B) \\ & + 2\rho_G\rho_B m_{23}m_{33}\mathrm{Cov}(G,B) \end{aligned}$$

This is more compactly written using the *covariance matrix*

$$C = \begin{pmatrix} \mathrm{Var}(R) & \mathrm{Cov}(R,G) & \mathrm{Cov}(R,B) \\ \mathrm{Cov}(R,G) & \mathrm{Var}(G) & \mathrm{Cov}(B,G) \\ \mathrm{Cov}(R,B) & \mathrm{Cov}(B,G) & \mathrm{Var}(B) \end{pmatrix} \tag{6.62}$$

so that the covariance matrix of the transformed red, green, and blue values becomes

$$C' = M^T \begin{pmatrix} \rho_R & 0 & 0 \\ 0 & \rho_G & 0 \\ 0 & 0 & \rho_b \end{pmatrix} C \begin{pmatrix} \rho_R & 0 & 0 \\ 0 & \rho_G & 0 \\ 0 & 0 & \rho_b \end{pmatrix} M \tag{6.63}$$

with M^T being the transpose of matrix M.

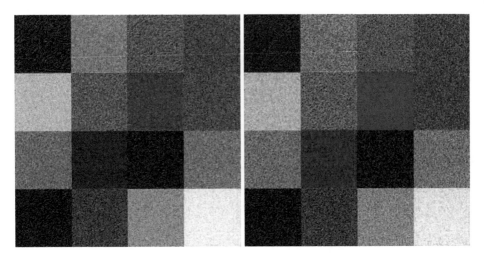

Figure 6.30 Amplification of noise due to the color correction matrix. Left: without CCM; right: with CCM. Note how noise coloration becomes more prominent when a CCM is applied.

For the case when the raw red, green, and blue noise values are uncorrelated, the covariances are all zero, **C** becomes a diagonal matrix, and consequently

$$\mathrm{Var}(R') = \rho_R^2 m_{11}^2 \mathrm{Var}(R) + \rho_G^2 m_{21}^2 \mathrm{Var}(G) + \rho_B^2 m_{31}^2 \mathrm{Var}(B)$$
$$\mathrm{Var}(G') = \rho_R^2 m_{12}^2 \mathrm{Var}(R) + \rho_G^2 m_{22}^2 \mathrm{Var}(G) + \rho_B^2 m_{32}^2 \mathrm{Var}(B)$$
$$\mathrm{Var}(B') = \rho_R^2 m_{13}^2 \mathrm{Var}(R) + \rho_G^2 m_{23}^2 \mathrm{Var}(G) + \rho_B^2 m_{33}^2 \mathrm{Var}(B) \tag{6.64}$$

Since all variables in these expressions are greater than or equal to zero, the noise values of the transformed signals will be increased. Furthermore, even if the original color signals are uncorrelated, the noise of the transformed data will not be, at least as long as the CCM contains off-diagonal elements. Therefore, applying a color correction matrix to the data will amplify the noise (Dillon *et al.*, 1978) and introduce correlations between the color channels. This is shown in Figure 6.30. These two effects will clearly change the appearance of the noise substantially. The appearance of colored noise compared to pure luminance noise will be discussed more in Chapter 7.

6.9.2 Spatial Frequency Dependence

Apart from correlation between color channels, there could also be correlation within the noise data itself, that is, a *spatial correlation*. Such correlations might also change the appearance of the noise drastically (Buzzi and Guichard, 2005; Baxter *et al.*, 2014). We may quantify this through the *autocorrelation function* (ACF), which describes the "similarity" between a particular noise sample and surrounding samples in space or time. The *Wiener–Khintchine theorem* (Easton, 2010) states that the ACF is equal to the Fourier transform of the *noise power spectrum* (NPS). The NPS, $N(v, \mu)$, is calculated as the squared Fourier transform of the noise samples, $n(x, y)$:

$$N(v, \mu) = \left| \int_{-\infty}^{\infty} \int_{-\infty}^{\infty} n(x, y) \exp\left(-2\pi i (vx + \mu y)\right) dx dy \right|^2 \tag{6.65}$$

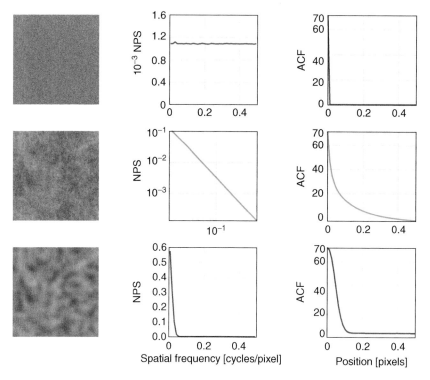

Figure 6.31 Examples of noise with different power spectra and autocorrelation functions. Top: white Gaussian noise; middle: $1/v$ noise; bottom: noise filtered with a Gaussian convolution kernel. Left: noise samples; middle: noise power spectrum; right: autocorrelation function. Note that all images have the same variance.

For discretely sampled noise, as in, for example, a digital image, the NPS is typically calculated using the fast Fourier transform (FFT) algorithm (FFTW, 2015).

The integral of the NPS gives the variance of the noise. This means that the zero input value of the ACF is also the variance of the noise. Figure 6.31 shows some samples from noisy images together with their noise power spectra and autocorrelation functions. All three images have the same noise variance as can be seen from the zero input value of the ACF. This clearly illustrates how different the noise may appear depending on the NPS, and consequently also the ACF.

To make the plots in Figure 6.31, 100 noisy image samples of size 256×256 pixels were generated. The noise was generated by drawing samples from a random number generator with a Gaussian distribution, thus having a flat spatial frequency characteristic. For each sample image, the NPS was calculated using the FFT algorithm. All two-dimensional noise power spectra were then averaged into one NPS. Since the NPS in this case had a circular symmetry, the data were averaged along the radial direction to obtain a one-dimensional curve. This curve was subsequently binned into equally sized spatial frequency values. In order to obtain smooth curves, this type of averaging is necessary when calculating the NPS.

The top image shows uncorrelated, white Gaussian noise as obtained from the original, unfiltered images. Since the noise has no spatial correlation, the ACF only shows one peak at zero spatial frequency, and then all subsequent samples are zero. In the middle figure, the white noise images were filtered using a spatial frequency domain filter with a $1/v$ frequency dependence. As seen in the image, the noise characteristic has clearly changed, even though the variance is still the same as for the Gaussian noise. The broadening of the ACF clearly shows that the degree of correlation of the noise samples has increased. In the bottom image, the Gaussian noise was filtered by a Gaussian kernel. Once again, this gives an appearance quite different from the original noise image. Also in this case, we can see that the ACF has broadened, reflecting the increased spatial correlation.

6.9.3 Signal to Noise Measurements in Nonlinear Systems and Noise Component Analysis

Apart from the factors discussed above, the noise in a camera is also dependent on the tonal characteristics. As was described in Chapter 4, the output image is modified by a gamma curve. In addition, a tone curve, typically "S"-shaped, may be applied to suppress dark regions and increase highlights in order to make the image appear more pleasing. The consequence of this is to make the relation between input scene radiances and output image values nonlinear. This is expressed through the OECF, discussed in section 6.1.2. Such processing will certainly affect the signal to noise ratio as a function of signal value. To account for this, the noise and signal can be transformed using the inverse OECF and thus refer the signal to noise ratio back to the input of the camera. Figure 6.32 illustrates this concept. As can be seen, the slope of the OECF for a particular signal level will determine the noise amplitude. Therefore, at a certain input luminance level, the SNR can be determined as

$$\text{SNR} = \frac{L}{\sigma(L)/g(L)} = \frac{Lg(L)}{\sigma(L)} \tag{6.66}$$

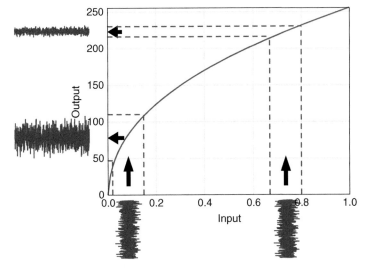

Figure 6.32 Illustration of how noise gets modified in a camera system with nonlinear signal transfer characteristics. A higher slope, as in the shadow region, will amplify the noise considerably, while in the highlights the slope is lower, leading to diminished noise.

where L is the input luminance, $g(L)$ the slope of the OECF at this luminance level, and $\sigma(L)$ the standard deviation of the values in the output image for an image signal value $I(L)$. The ISO 15739 standard (ISO, 2013) uses this concept and describes a methodology to measure noise in digital cameras taking these effects into account. The factor $g(L)$ is in the standard referred to as the *incremental gain*.

The ISO 15739 standard also makes estimates of noise for temporal as well as fixed pattern noise (FPN). The FPN can be found by making repeated measurements of the same uniform patch, thus obtaining a large number of images differing only in temporal noise. The standard deviation of the average of the images will, for a sufficiently large quantity of images, approach the true FPN value. Having an estimate of the FPN, it is then possible to also obtain an assessment of the temporal noise. For full details of the methodology, the reader is referred to the ISO standard. An alternative way to calculate the temporal noise is to calculate the standard deviation of the difference of two images. Since the FPN, by definition, does not change from image to image, it will be canceled out in this case. The variance of the sum or difference of two images is double the variance of the individual images. Therefore, the noise of the difference image should be divided by a factor of $\sqrt{2}$ to obtain the correct result.

It should be mentioned that the calculations used in the ISO standard to estimate the fixed pattern and temporal noise components rely on the fact that there is no correlation between successive images in the set of images needed for the analysis. This assumption may not hold for images captured by modern digital cameras, where temporal noise filtering may have been applied to the images. Such a filter averages several successive images in order to suppress the noise, thus introducing correlations between images. If this type of filter has been applied, the covariances between images must also be taken into consideration in the analysis. Care must therefore be taken when using the currently available ISO 15739 standard (ISO, 2013).

6.9.4 Practical Considerations

To reliably determine the standard deviation of noise in an image, no variations except for the noise should be present. In practice, this may not be the case since nonuniformities, such as shading, may affect the measurement results. If this is the case they should be corrected for before attempting to obtain any noise estimates. For determining the temporal noise, this correction is trivial since such effects will cancel when two images are subtracted. For determining the total noise and FPN, on the other hand, this correction becomes crucial. At least two different methods may be envisaged: high pass filtering or fitting to a plane. The former method assumes that the noise can be separated from the slower varying nonuniformities by a sufficiently sharp highpass filter. Such a filter may be constructed from the function (Baxter and Murray, 2012)

$$H(v) = \frac{2}{1 + \exp(-bv)} - 1 \tag{6.67}$$

Figure 6.33 shows the spatial frequency response of this filter for a few different values of the variable b that controls the cutoff frequency of the filter.

Another way to compensate for nonuniformities is to assume that within the cropped part of the image where the noise is to be measured, the variations are very close to linear. In that case, a plane can be fitted to the image data within the crop. The fitted values are then subtracted from the image data, whereafter the standard deviation can be calculated.

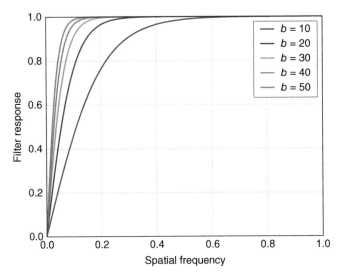

Figure 6.33 Spatial frequency characteristics of a highpass filter proposed to correct for image nonuniformities in noise measurements.

6.10 Color Fringing

Chromatic aberrations were discussed in Chapter 4. There are two types of such aberrations, longitudinal and lateral. Longitudinal chromatic aberrations are basically a change in sharpness as a function of wavelength and can therefore in principle be measured using the E-SFR or S-SFR methods on the different color channels (Cao *et al.*, 2008).

Lateral chromatic aberrations (LCA) give rise to a change in magnification as a function of wavelength and field positions. This is therefore very similar to optical distortion, and therefore, the same test charts can be used. In fact, the same algorithms for finding the dot centers or grid line intersections can also be used. However, instead of calculating the deviation from ideal distances to the image center, an LCA metric measures the relative radial distances between the dot or grid intersections in the different color channels as a function of image height. Figure 6.34 shows the principle. The methodology

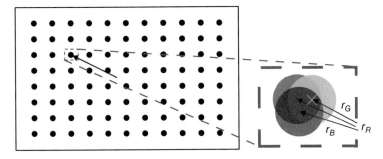

Figure 6.34 Calculating the distances between green and red and green and blue pixels for the LCA metric.

is described in detail in the CPIQ documentation (I3A, 2009b; IEEE, 2017), where in the newest edition it is referred to as lateral chromatic displacement, and also in the ISO 19084 standard (ISO, 2015c). The result can be presented in a variety of ways, for instance in a plot with red-green and blue-green distances as a function of image height, or even a two dimensional map.

6.11   Image Defects

As mentioned in other chapters, image defects usually concern either contaminants or defects in optical elements, or defective pixels. Defects are generally not measured as part of camera benchmarking, since their occurrence can vary significantly between individual instances, and the presence of defects in one camera may therefore not represent the general performance with respect to this parameter. For DSLRs with exchangeable lenses, the situation is even more complicated, since by removing the lens the image sensor will be exposed to dust from the outside environment.

For pixel defects, it can be difficult to separate dead or hot pixels from FPN in an image. In one sense, a pixel fixed at a high signal level will contribute to the tail of the FPN histogram. However, sufficiently bright stuck pixels may be deviating enough from the FPN distribution to stand out from the more normally noisy pixels. Therefore, defect pixels are usually defined as those pixels that have values above some threshold in a completely dark or bright image, respectively. Then, the estimation of pixel defects comes down to counting pixels having values above or below the defined thresholds for bright and dark defective pixels.

6.12   Video Quality Metrics

Contributed by Hugh Denman

There are fewer established standard metrics for video camera performance than for still image performance. This is perhaps principally because developing and standardizing test scenes for video metrics is considerably more difficult than for still image metrics. In any test scene, some ground truth information must be available to serve as a reference for assessment. For still image assessment, static charts can provide this ground truth in the form of known spatial frequency characteristics or particular geometric features, for example. In the case of video, the test scene must include temporally varying features, including various kinds of motion, to fully characterize the video acquisition pipeline. Assessment of performance requires that suitable ground truth information for these features be known. At present, this can be achieved only to a limited degree. These considerations are discussed in more detail in Chapter 7.

This section outlines some approaches for assessing temporal aspects of video performance. These metrics have not yet been internationally standardized, but they are supported in a number of proprietary camera assessment software packages.

6.12.1   Frame Rate and Frame Rate Consistency

The essential temporal characteristic of video is the frame rate. As described in Chapter 3, the desired frame rate may be selected by the user, in which case the assessment must ascertain whether the selected frame rate is sustained, or alternatively the camera may automatically select and adjust the frame rate in response to conditions, in which case the attained frame rate must be assessed in relation to the requirements for the intended use.

The attained frame rate can be accurately measured using a high-speed timer of the type which will be described in Chapter 8. The timer is configured to update at an integer multiple of the claimed capture frame rate; the higher the multiple, the more precise the results. For example, for a camera claiming capture rates of up to 100 fps, the clock should be set to update the display at intervals of 100 μs. The time of capture for each frame, relative to the start of the timer sequence, is then extracted for each frame, and each inter-frame interval computed by subtraction.

The video frame rate metric is then formed by the descriptive statistics of the inter-frame intervals, such as the minimum, maximum, mean, and standard deviation. The inter-frame intervals can also be displayed graphically, such that visual inspection can determine whether the desired performance has been sustained.

For cameras featuring an adaptive frame rate, the metric should be obtained for various scene illumination levels. This is because these cameras are often configured to use a lower frame rate in lower light levels, so as to enable longer exposure times and thus reduce the effect of noise.

6.12.2   Frame Exposure Time and Consistency

Measurement of exposure index for still image capture was discussed earlier in this chapter, and the same metric can be used to measure the exposure index for video. However, as noted in Chapter 3, exposure time affects not only image tone, but also the level of motion blur. The exposure time optimal for each of these may be different, and the appropriate compromise is a question of tuning. Determining the exposure level optimal with regard to motion blur is a question of artistic intent, and it is difficult to establish an appropriate quantitative metric. Visual inspection of the test scene footage, and experience, are required for practical assessment.

In video it is also important to assess the temporal consistency of exposure time, as this is a factor influencing whether the video has a consistent look. This is achieved by measuring the exposure time of each frame, using the same high speed timer device used to measure frame rate. Here the timer is configured to update quickly enough that multiple time indications (individual LEDs) will be lit over the course of the exposure. The exposure time of the frame can then be computed as the product of the number of LEDs visibly lit in the frame and the duration of illumination of each LED.

As in the case of frame rate, the metric then consists of statistics such as the mean, standard deviation, minimum, and maximum of the exposure values over the sequence, and values should be established for a variety of lighting conditions, including illumination levels as well as illuminants.

The auto exposure algorithm may take some time to stabilize after video recording has begun. Assessment of the exposure time consistency should disregard the stabilization time, but this stabilization time is itself a metric of interest. Stabilization can be measured by considering the variation of exposure times within a sliding window of video frames. The specific parameters of window length and condition of variability must be chosen with respect to the frame rate, complexity of the scene, and observed behavior. A reasonable general approach is to use a window size of ten frames, and to consider the auto exposure to have stabilized at the time of the first frame in the first window where no frame has an exposure time deviating by more than 5% from the mean of the exposure times within the window.

Most auto exposure algorithms will adjust the exposure time to adapt to changing lighting conditions as the video frames are captured. The responsiveness of the camera in this regard can be tested by varying the illumination within a video capture, for example by gradually changing from dark to bright, or triggering a brief, bright flash. The stabilization time after an illumination change can then be measured as described above, to provide a responsiveness metric. Research has shown that observers prefer monotonic, gradual transitions with no abrupt changes (Oh *et al.*, 2017), so this preference in responsiveness should be considered when making stabilization measurements.

6.12.3   Auto White Balance Consistency

Quantifying the stabilization time of the auto white balance (AWB) algorithm is a little more challenging than quantifying the auto exposure algorithm, as the active white balance setting is not readily reducible to a single value. Thus, in practice, this aspect of camera performance is often assessed by inspection of the captured video.

It is possible to quantify variation in white balance by examining the color values for a fixed region in the test scene, for example one of the neutral patches in a color checker chart. In outline, a ΔE color difference (described above) is used to compute the frame-to-frame difference for this patch. Stabilization of the AWB algorithm can then be determined in a manner similar to that described for exposure time, on the basis of the RMS of the ΔE values within a sliding window of frames. Very roughly, a ΔE value of 1 corresponds to a just noticeable color difference, and therefore an RMS value of less than 1 can be taken as indicating stable white balance.

Again, the assessment should include videos recorded under a variety of fixed lighting conditions, and also videos recorded under both slowly and suddenly varying illumination. The illuminants chosen should cover a range of color temperatures.

6.12.4   Autofocusing Time and Stability

The simplest way to assess autofocus responsiveness is to mount the camera on a motorized slider track, directed toward an SFR chart. The video is recorded while the camera is displaced through a range of distances from the chart. The time taken for the camera to correctly obtain focus in each position is then representative of autofocus responsiveness.

The time at which the camera obtains focus corresponds to the video frame in which the SFR chart has the maximal measured sharpness, for that distance. A simple MTF50 (the spatial frequency at which an SFR value of 0.5 is attained) measurement of sharpness at the center of the image field suffices for this purpose. It is prudent to compare the maximal MTF50 values found in the video to an MTF50 measurement obtained under ideal conditions (i.e., with the camera statically mounted). This ensures that the maximal values do represent successful autofocusing.

The set of distances evaluated should be logarithmically spaced. The distances should not be evaluated in monotonic order, as autofocus performance at a given distance can depend on whether the camera moved to this distance from a closer or farther distance (i.e., autofocus algorithms can exhibit hysteresis). Therefore, the set of distances should be evaluated in a randomized order.

Evaluation of sharpness from a distance of over two meters or so requires a chart with a large center edge. However, larger charts are typically not produced with sufficiently high resolution to function reliably at close distances. It may therefore be necessary to arrange a test scene containing both a large and a small SFR chart, depending on the range of distances being evaluated. The MTF50 reading from the small chart should be used as long as it is large enough in the field of view to obtain a reliable reading, otherwise the reading from the large chart should be used.

The evaluation of each camera being assessed must follow the same order of camera to chart distances and use the same track slider system. Autofocus systems will adapt to the scene while the camera is in motion, and therefore the measurement of the time taken to obtain focus should commence when the motion from one position to another commences, rather than when the position is obtained. The time of autofocus should be measured as the time of the first frame of a contiguous set of well-focused frames that includes four seconds of video captured at the target distance. This provision prevents the measurement being affected where the camera obtains focus at an intermediate distance in the course of the motion between distances. The camera may obtain and sustain focus from a frame before the target distance is reached, or indeed sustain focus continuously throughout the motion. Thus, the autofocus time can be less than the time taken to move between positions.

The camera track slider should be configured to move the camera between the evaluation distances as quickly as possible, to provide assessment under challenging circumstances. The speed of camera motion must be consistent across the cameras being compared. The motorized track slider must be powerful enough to move the cameras at the same speed regardless of any differences in camera weight.

6.12.5   Video Stabilization Performance

Video stabilization performance may be measured by placing the camera on an electronically controlled motorized platform, with up to six degrees of freedom of motion. A set of platform motions is devised to emulate the camera motions typical of handheld shake or filming while walking. Here we will term the set of motions to be used a *motion profile*.

The camera is directed toward an SFR chart with fiducial markers, and a video recording is made while the chosen motion profile is played back on the platform. The stabilization performance can then be characterized using two measures. Firstly, the temporal stability of the position of the fiducial markers over the video indicates how effective the stabilization has been in bringing the frames into alignment. The mean, standard deviation, and maximum absolute value of the frame-to-frame differences in the position of each fiducial marker, measured as a Euclidean distance, can serve as a metric for this aspect. Secondly, the SFR readings extracted from each frame indicate whether motion blur due to the camera motion has been attenuated or removed by the stabilization system. Here, the mean, standard deviation, and minimum value of the MTF50 values are used as the metric. A texture performance chart, for example using the dead leaves pattern, may be used as an alternative to the SFR chart.

It is important to verify that the camera has correctly focused at the start of the test, by checking that the SFR reading for the first video frames, recorded before the platform has been set in motion, correspond to the best performance of the camera under test. When multiple cameras are being compared, the cameras should be positioned so that the SFR chart fills the same angle of view for each camera, which may entail using different distances for cameras with different focal lengths.

For the most thorough benchmarking, each camera should be assessed for each motion profile available. If supported by the assessment software, a motion profile should be included that emulates handheld shake while the camera is panned. This will enable characterization of any delay or latency introduced to desired, deliberate camera motion by the stabilization system.

6.12.6 Audio-Video Synchronization

Audio-video synchronization is traditionally measured by using a clapperboard, essentially a pair of wooden sticks connected by a hinge. The sticks are brought together rapidly, which makes a clear, short "clap" sound. The video frame in which the sticks made contact with each other is determined by inspection, and the audio/visual delay determined by comparing the time of this frame to the time of the audible "clap" in the sound track.

The deficiency of this method is that its temporal precision is limited to the inter-frame interval (the inverse of the frame rate). This precision is sufficient for most applications, but greater accuracy may be required in certain cases, for example if a camera is being assessed for use as part of a multichannel recording setup. Greater accuracy could be obtained by modifying a timer box to play a sound after a precisely specified interval. The audiovisual delay can then be quantified by comparing the time of the sound to the time displayed on the timer box.

Audio-visual delay should be measured at the start and end of video capture to account for any possible drift over time. Slow rates of drift require long video captures to identify when using a frame-accurate approach, but can be detected in shorter sequences when measuring the synchronization delay to sub-frame-time accuracy.

6.13 Related International Standards

Many of the metrics mentioned in this chapter are defined in international standards. This section lists the standards relevant to camera benchmarking and characterization.

- ISO 3664: Graphic Technology and Photography—Viewing Conditions
- ISO 12231: Photography—Electronic Still Picture Imaging—Vocabulary

- ISO 12232: Photography—Digital Still Cameras—Determination of Exposure Index, ISO Speed Ratings, Standard Output Sensitivity, and Recommended Exposure Index
- ISO 12233: Photography—Electronic Still Picture Imaging—Resolution and Spatial Frequency Responses
- ISO 12234: Electronic Still Picture Imaging—Removable Memory—Part 1: Basic Removable-Memory Model
- ISO 14524: Photography—Electronic Still-Picture Cameras—Methods for Measuring Opto-Electronic Conversion Functions (OECFs)
- ISO 15739: Photography—Electronic Still Picture Imaging—Noise Measurements
- ISO 15781: Photography—Digital Still Cameras—Measuring Shooting Time Lag, Shutter Release Time Lag, Shooting Rate, and Start-up Time
- ISO 17321: Graphic Technology and Photography—Colour Characterization of Digital Still Cameras (DSCs)
- ISO 20462: Photography—Psychophysical Experimental Methods for Estimating Image Quality
- ISO 22028: Photography and Graphic Technology—Extended Colour Encodings for Digital Image Storage, Manipulation, and Interchange
- ISO 9039: Optics and Photonics—Quality Evaluation of Optical Systems—Determination of Distortion
- ISO 9358: Optics and Optical Instruments—Veiling Glare of Image Forming Systems—Definitions and Methods of Measurement

Apart from these standards, the following new standards have been recently published or will be published in the near future:

- ISO 17850: Photography—Digital Cameras—Geometric Distortion (GD) Measurements
- ISO 17957: Photography—Digital Cameras—Shading Measurements
- ISO 18844: Photography—Digital Cameras—Flare Measurement Techniques for Digital Camera Systems
- ISO 19084: Photography—Digital Cameras—Chromatic Displacement Measurements
- ISO 19093: Photography—Digital Cameras—Measuring Low Light Performance
- ISO 19567: Photography—Digital Cameras—Texture Reproduction Measurements
- ISO 19247: Photography—Digital Cameras—Guidelines for Reliable Camera Testing

The CPIQ initiative has released a set of documents describing measurement of some key image quality attributes as well as a white paper covering general image quality aspects. These documents are:

- Camera Phone Image Quality—Phase 1: Fundamentals and Review of Considered Test Methods
- Camera Phone Image Quality—Phase 2: Subjective Evaluation Methodology
- Camera Phone Image Quality—Phase 2: Acutance—Spatial Frequency Response
- Camera Phone Image Quality—Phase 2: Color Uniformity
- Camera Phone Image Quality—Phase 2: Initial Work on Texture Metric
- Camera Phone Image Quality—Phase 2: Lateral Chromatic Aberration

- Camera Phone Image Quality—Phase 2: Lens Geometric Distortion (LGD)
- IEEE 1858–2016 Standard for Camera Phone Image Quality

Other standards related to the topics described in this and previous chapters include:

- SMIA 1.0 Part 5: Camera Characterisation Specification, Rev A
- CIPA Standard DC-011-2015 Measurement and Description Method for Image Stabilization Performance of Digital Cameras (Optical System)
- EMVA Standard 1288: Standard for Characterization of Image Sensors and Cameras
- ITU-R BT.500-13 (2012): Methodology for the subjective assessment of the quality of television pictures
- ITU-R BT.1359 (1998): Relative timing of sound and vision for broadcasting
- ITU-R BT.1683 (2004): Objective perceptual video quality measurement techniques for standard definition digital broadcast television in the presence of a full reference
- ITU-R BT.1788 (2007): Subjective assessment of multimedia video quality (SAMVIQ)
- ITU-R BT.1866 (2010): Objective perceptual video quality measurement techniques for broadcasting applications using low definition television in the presence of a full reference signal
- ITU-R BT.1867 (2010): Objective perceptual visual quality measurement techniques for broadcasting applications using low definition television in the presence of a reduced bandwidth reference
- ITU-R BT.1885 (2011): Objective perceptual video quality measurement techniques for standard definition digital broadcast television in the presence of a reduced bandwidth reference
- ITU-T J.100 (1990): Tolerances for transmission time differences between the vision and sound components of a television signal
- ITU-T J.143 (2000): User requirements for objective perceptual video quality measurements in digital cable television
- ITU-T J.144 (2004): Objective perceptual video quality measurement techniques for digital cable television in the presence of a full reference
- ITU-T J.148 (2003): Requirements for an objective perceptual multimedia quality model
- ITU-T J.149 (2004): Method for specifying accuracy and cross-calibration of Video Quality Metrics (VQM)
- ITU-T J.244 (2008): Full reference and reduced reference calibration methods for video transmission systems with constant misalignment of spatial and temporal domains with constant gain and offset
- ITU-T J.246 (2008): Perceptual visual quality measurement techniques for multimedia services over digital cable television networks in the presence of a reduced bandwidth reference
- ITU-T J.247 (2008): Objective perceptual multimedia video quality measurement in the presence of a full reference
- ITU-T J.249 (2010): Perceptual video quality measurement techniques for digital cable television in the presence of a reduced reference
- ITU-T J.340 (2010): Reference algorithm for computing peak signal to noise ratio (PSNR) of a processed video sequence with constant spatial shifts and a constant delay
- ITU-T J.341 (2016): Objective perceptual multimedia video quality measurement of HDTV for digital cable television in the presence of a full reference

- ITU-T J.342 (2011): Objective multimedia video quality measurement of HDTV for digital cable television in the presence of a reduced reference signal
- ITU-T J.343 (2014): Hybrid perceptual bitstream models for objective video quality measurements
- ITU-T P.910 (2008): Subjective video quality assessment methods for multimedia applications
- ITU-T P.912 (2008): Subjective video quality assessment methods for recognition tasks
- ITU-T P.913 (2014): Methods for the subjective assessment of video quality, audio quality and audiovisual quality of Internet video and distribution quality television in any environment

Summary of this Chapter

- Objective measurements are independent of human judgment.
- It is possible to correlate objective metrics with results from subjective investigations.
- Sharpness and resolution are measured by determining the Modulation Transfer Function (MTF).
- For a camera, the MTF is at present best determined using the slanted edge technique or a sinusoidal Siemens star.
- Texture blur can be measured using a *dead leaves target*.
- The noise in an image is determined through the signal to noise ratio (SNR). To obtain an accurate measure of the SNR, aspects such as color, spatial frequency dependence, and linearity must be taken into account.
- The dynamic range of a camera is measured by calculating the ratio of the largest possible signal and the smallest possible signal, referred to the scene.
- The assessment of color depends on type of light source, the observer, and the scene.
- The preferred color space for calculating color differences is CIELAB.
- To calculate color differences, either the ΔE_{94} or CIEDE2000 difference metrics are preferred.
- Image distortion is best determined by calculating the optical distortion, not the TV distortion.
- Shading can show up both in color and luminance. For a reasonably perceptually correlated shading metric, CIELAB is an appropriate color space.
- Stray light consists of ghosting effects as well as flare. The latter results in an overall contrast reduction in the image. This can be measured using the veiling glare index (VGI).
- Image quality metrics specific for video include frame rate, frame exposure time, auto white balance consistency, autofocusing stability, video stabilization performance, and audio-video synchronization.

References

Adams, A. (1995a) *The Camera*, Bulfinch Press, New York, USA.
Adams, A. (1995b) *The Negative*, Bulfinch Press, New York, USA.

Adams, A. (1995c) *The Print*, Bulfinch Press, New York, USA.

ANSI (2015) Photography – Digital Still Cameras – Guidelines for Reporting Pixel-Related Specifications. ANSI.

Arfken, G. (1985) *Mathematical Methods for Physicists*, Academic Press, San Diego, CA, USA, 3rd edn.

Artmann, U. (2015) Image quality assessment using the dead leaves target: Experience with the latest approach and further investigations. *Proc. SPIE*, **9404**, 94040J.

Artmann, U. (2016) Linearization and normalization in spatial frequency response measurements. *Electronic Imaging: Image Quality and System Performance*, **2016**, 1–6.

Artmann, U. and Wueller, D. (2009) Interaction of image noise, spatial resolution, and low contrast fine detail preservation in digital image processing. *Proc. SPIE*, **7250**, 72500I.

Artmann, U. and Wueller, D. (2012) Improving texture loss measurement: Spatial frequency response based on a colored target. *Proc. SPIE*, **8293**, 829305.

Baxter, D., Phillips, J.B., and Denman, H. (2014) The subjective importance of noise spectral content. *Proc. SPIE*, **9016**, 901603.

Baxter, D.J. and Murray, A. (2012) Calibration and adaptation of ISO visual noise for I3A's camera phone image quality initiative. *Proc. SPIE*, **8293**, 829303.

Birch, G.C. and Griffin, J.C. (2015) Sinusoidal Siemens star spatial frequency response measurement errors due to misidentified target centers. *Opt. Eng.*, **54**, 074104.

Brauers, J., Seilers, C., and Aach, T. (2010) Direct PSF estimation using a random noise target. *Proc. SPIE*, **7537**, 75370B.

Burns, P.D. (2000) Slanted-edge MTF for digital camera and scanner analysis. *Proc. IS&T PICS Conference*, pp. 135–138.

Burns, P.D. (2005) Tone-transfer (OECF) characteristics and spatial frequency response measurements for digital cameras and scanners. *Proc. SPIE*, **5668**, 123–128.

Burns, P.D. and Williams, D. (2002) Refined slanted-edge measurement for practical camera and scanner testing. *Proc. IS&T PICS Conference*, pp. 191–195.

Buzzi, J. and Guichard, F. (2005) Noise in imaging chains: Correlations and predictions. *Proc. ICIP*, **1** (409-12).

Cadou, J. (1985) *Comparison of Two MTF Measurement Methods: Sine-Wave vs Edge Gradient Analysis*, Master's thesis, Rochester Institute of Technology.

Cao, F., Guichard, F., and Hornung, H. (2009) Measuring texture sharpness of a digital camera. *Proc. SPIE*, **7250**, 72500H.

Cao, F., Guichard, F., and Hornung, H. (2010) Dead leaves model for measuring texture quality on a digital camera. *Proc. SPIE*, **7537**, 75370E.

Cao, F., Guichard, F., Hornung, H., and Sibade, C. (2008) Characterization and measurement of color fringing. *Proc. SPIE*, **6817**, 68170G.

CIE (2004) CIE Publication No. 15.2, Colorimetry, *Tech. Rep.*, Commission Internationale de l'Éclairage, Vienna, Austria.

CIE (2013) CIE 204:2013 Methods for Re-defining CIE D Illuminants, *Tech. Rep.*, Commission Internationale de l'Éclairage, Vienna, Austria.

Coltman, J.W. (1954) The specification of imaging properties by response to a sine wave input. *J. Opt. Soc. Am.*, **44**, 468–471.

Dillon, P.L.P., Lewis, D.M., and Kaspar, F.G. (1978) Color imaging system using a single CCD area array. *IEEE Trans. Electron. Devices*, **25**, 102–107.

Easton, R.L. (2010) *Fourier Methods in Imaging*, John Wiley & Sons Ltd, Chichester, UK.

Fairchild, M.D. (2013) *Color Appearance Models*, John Wiley & Sons Ltd, Chichester, UK, 3rd edn.

FFTW (2015) FFTW website. http://www.fftw.org, (accessed 29 May 2017).

Fischer, T.A. and Holm, J. (1994) Electronic still picture camera spatial frequency response measurement. *Proc. IS&T PICS Conference*, pp. 626–630.

Hecht, E. (1987) *Optics*, Addison-Wesley, Reading, MA, USA, 2nd edn.

Hunt, R.W.G. (2004) *The Reproduction of Colour*, John Wiley & Sons Ltd, Chichester, UK, 6th edn.

Hunt, R.W.G. and Pointer, M.R. (2011) *Measuring Colour*, John Wiley & Sons Ltd, Chichester, UK, 4th edn.

I3A (2009a) Camera Phone Image Quality – Phase 2 – Color Uniformity. IEEE.

I3A (2009b) Camera Phone Image Quality – Phase 2 – Lateral Chromatic Aberration. IEEE.

I3A (2009c) Camera Phone Image Quality – Phase 2 – Lens Geometric Distortion (LGD). IEEE.

IEC (1999) IEC 61966-2-1:1999 Colour Measurement and Management in Multimedia Systems and Equipment – Part 2-1: Default RGB Colour Space – sRGB.

IEEE (2017) IEEE 1858-2016, IEEE Standard for Camera Phone Image Quality. IEEE.

ISO (1994) ISO 9358:1994 Optics and Optical Instruments – Veiling Glare of Image Forming Systems – Definitions and Methods of Measurement. ISO.

ISO (2006) ISO 12232:2006 Photography – Digital Still Cameras – Determination of Exposure Index, ISO Speed Ratings, Standard Output Sensitivity, and Recommended Exposure Index. ISO.

ISO (2008) ISO 9039:2008 Optics and Photonics – Quality Evaluation of Optical Systems – Determination of Distortion. ISO.

ISO (2009) ISO 14524:2009 Photography – Electronic Still Picture Cameras – Methods For Measuring Opto-Electronic Conversion Functions (OECFs). ISO.

ISO (2009) ISO 3664:2009 Graphic Technology and Photography – Viewing Conditions. ISO.

ISO (2012a) ISO 12231:2012 Photography – Electronic Still Picture Imaging – Vocabulary. ISO.

ISO (2012b) ISO 12234:2012 Electronic Still Picture Imaging – Removable Memory – Part 1: Basic Removable-Memory Model. ISO.

ISO (2012c) ISO 17321:2012 Graphic Technology and Photography – Colour Characterization of Digital Still Cameras (DSCs). ISO.

ISO (2012d) ISO 20462:2012 Photography – Psychophysical Experimental Methods for Estimating Image Quality. ISO.

ISO(2013) ISO 15739:2013 Photography – Electronic Still Picture Imaging – Noise Measurements. ISO.

ISO (2013) ISO 22028:2013 Photography and Graphic Technology – Extended Colour Encodings for Digital Image Storage, Manipulation and Interchange. ISO.

ISO (2014) ISO 12233:2014 Photography – Electronic Still Picture Imaging – Resolution and Spatial Frequency Responses. ISO.

ISO (2015a) ISO 15781:2015 Photography – Digital Still Cameras – Measuring Shooting Time Lag, Shutter Release Time Lag, Shooting Rate, and Start-up Time. ISO.

ISO (2015b) ISO 17957:2015 Photography – Digital cameras – Shading measurements. ISO.

ISO (2015c) ISO 19084:2015 Photography – Digital cameras – Chromatic displacement measurements. ISO.

Kirk, L., Herzer, P., Artmann, U., and Kunz, D. (2014) Description of texture loss using the dead leaves target: Current issues and a new intrinsic approach. *Proc. SPIE*, **9023**, 90230C.

Kohm, K. (2004) Modulation transfer function measurement method and results for the OrbView-3 high resolution imaging satellite. *Proc. XXth ISPRS Congress, Istanbul, Turkey*, pp. 7–12.

Loebich, C., Wueller, D., Klingen, B., and Jaeger, A. (2007) Digital camera resolution measurement using sinusoidal Siemens stars. *Proc. SPIE*, **6502**, 65020N.

Masaoka, K., Yamashita, T., Nishida, Y., and Sugawara, M. (2014) Modified slanted-edge method and multidirectional modulation transfer function estimation. *Opt. Express*, **22**, 6040–6046.

McElvain, J., Campbell, S.P., Miller, J., and Jin, E.W. (2010) Texture-based measurement of spatial frequency response using the dead leaves target: Extensions, and application to real camera systems. *Proc. SPIE*, **7537**, 75370D.

Oh, S., Passmore, C., Gold, B., Skilling, T., Pieper, S., Kim, T., and Belska, M. (2017) A framework for auto-exposure subjective comparison. *Electronic Imaging: Image Quality and System Performance*, **2017**, 202–208.

Ojanen, H.J. and Tervonen, A. (2009) Method for characterizing a digital imaging system. URL http://www.google.ca/patents/US7499600, (accessed 29 May 2017), US Patent 7,499,600.

Parulski, K.A., D'Luna, L.J., Benamati, B.L., and Shelley, P.R. (1992) High-performance digital color video camera. *J. Electron. Imaging*, **1**, 35–45.

Phillips, J.B. and Christoffel, D. (2010) Validating a texture metric for camera phone images using a texture-based softcopy attribute ruler. *Proc. SPIE*, **7529**, 752904.

Reichenbach, S.E., Park, S.K., and Narayanswamy, R. (1991) Characterizing digital image acquisition devices. *Opt. Eng.*, **30** (2), 170–177.

Reif, F. (1985) *Fundamentals of Statistical and Thermal Physics*, McGraw-Hill, Singapore, intl. edn.

Reinhard, E., Khan, E.A., Akyuz, A.O., and Johnson, G.M. (2008) *Color Imaging – Fundamentals and Applications*, A. K. Peters, Wellesley, MA, USA.

Saxby, G. (2002) *The Science of Imaging – An Introduction*, IOP Publishing Ltd, Bristol, UK.

Sonka, M., Hlavac, V., and Boyle, R. (1999) *Image Processing, Analysis, and Machine Vision*, Brooks/Cole Publishing Company, Pacific Grove, CA, USA.

Vollmerhausen, R.H. and Driggers, R.G. (2000) *Analysis of Sampled Imaging Systems*, SPIE Press, Bellingham, WA, USA.

Williams, D. (1998) Benchmarking of the ISO 12233 slanted-edge spatial frequency response plug-in. *Proc. IS&T PICS Conference*, pp. 133–136.

Williams, D. (2003) Debunking of specmanship: Progress on ISO/TC42 standards for digital capture imaging performance. *Proc. IS&T PICS Conference*, pp. 77–81.

Williams, D. and Burns, P.D. (2001) Diagnostics for digital capture using MTF. *Proc. IS&T PICS Conference*, pp. 227–232.

Williams, D. and Burns, P.D. (2004) Low-frequency MTF estimation for digital imaging devices using slanted edge analysis. *Proc. SPIE*, **5294**, 93–101.

Williams, D., Wueller, D., Matherson, K., Yoshida, H., and Hubel, P. (2008) A pilot study of digital camera resolution metrology protocols proposed under ISO 12233, edition 2. *Proc. SPIE*, **6808**, 680804.

Wyszecki, G. and Stiles, W.S. (1982) *Color Science: Concepts and Methods, Quantitative Data and Formulae*, John Wiley & Sons Ltd, Chichester, UK.

Yotam, E., Ephi, P., and Ami, Y. (2007) MTF for Bayer pattern color detector. *Proc. SPIE*, **6567**, 65671M.

7

Perceptually Correlated Image Quality Metrics

In Chapter 6, a set of objective image quality metrics was presented. In order to build up a camera image quality benchmarking framework, the results obtained using these metrics need to be correlated with subjective perceptions of the attributes that are meant to be measured. Consequently, the aim of this chapter is to provide a link between the objective metrics in Chapter 6 and results from subjective studies performed using the methodologies described in Chapter 5. To facilitate this task, we need to construct a model of the human visual system (HVS) that is complex enough to provide a reasonably accurate description, while at the same time not involving too many parameters or adjustment possibilities, making the model too difficult to use.

Apart from having a useful model of the HVS, we also require a methodology for relating the results of subjective testing to the data obtained from objective measurements that may (or may not, depending on the metric) have been modified using HVS modeling according to the discussion above. After having read this chapter, it should be possible for the reader to perform her/his own experiments giving results that are well-correlated with the perceived experience of the attributes being tested.

7.1 Aspects of Human Vision

As already raised in the introduction, a simple model of the human visual system will make it easier to provide a link between the subjective experience of a particular image quality attribute and the results from objective measurements of that attribute. To obtain such a model, it is necessary to have a basic understanding of the processes involved, at least on a phenomenological level. The description below is necessarily quite limited, and the interested reader is encouraged to study the available literature, which includes books by, for example, Wandell (1995), Wyszecki and Stiles (1982), and Reinhard *et al.* (2008), as well as references therein. The online version of David Hubel's work *Eye, Brain, and Vision* (Hubel, 1995), should also be of interest.

7.1.1 Physiological Processes

In order to produce a typical visual sensation, light from objects in the environment must enter the eye and impinge upon the *retina* at the back of the eye. The retina consists of light sensitive cells, called *rods* and *cones*. Through chemical processes, electrical signals will be generated when these cells are stimulated by light. The rods have a much

Camera Image Quality Benchmarking, First Edition. Jonathan B. Phillips and Henrik Eliasson.
© 2018 John Wiley & Sons Ltd. Published 2018 by John Wiley & Sons Ltd.
Companion website: www.wiley.com/go/benchak

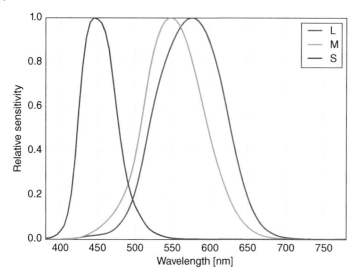

Figure 7.1 Cone spectral sensitivities calculated from the CIE standard observer and the Hunt–Pointer–Estevez transformation in Eq. (7.1). The red, green, and blue curves describe the sensitivities of the L, M, and S cones, respectively.

higher sensitivity than the cones and become saturated in normal, daylight conditions. At such high light levels, the cones will therefore be the active photoelements. This is known as *photopic vision*. In low-light conditions, when the rods are the active elements, *scotopic vision* is in operation. In the intermediate, *mesopic*, region, both types of cells are contributing to vision.

There are in principle three types of cones, *long* (L), *medium* (M), and *short* (S). They are given these names according to the wavelength region in which they are most sensitive. Figure 7.1 shows typical spectral sensitivities of the human cones. The standard observer curves shown in Figure 6.7 were derived before the cone sensitivities were known and had therefore to be determined through indirect experiments. Due to other considerations, such as having the $\bar{y}(\lambda)$ function match the luminance sensitivity $V(\lambda)$, the end result is therefore different from the curves shown in Figure 7.1. However, there exist linear transformations between the two representations. Due to nonlinearities, these transformations are not exact and variants exist. One of the most commonly used transformations is based on the Hunt–Pointer–Estevez approximation (Fairchild, 2013):

$$\left(L \ M \ S \right) = \left(X \ Y \ Z \right) \begin{pmatrix} 0.38971 & -0.22981 & 0 \\ 0.68898 & 1.18340 & 0 \\ -0.07868 & 0.04641 & 1 \end{pmatrix} \tag{7.1}$$

These transformations become important in color appearance models and chromatic adaptation transformations, to be discussed further in Section 7.5.

The trichromatic representation was challenged early on by findings that colorblindness was manifested in the disability to distinguish red and green or yellow and blue colors. Also, when describing mixtures of colors, combinations such as red-blue, green-yellow, and so on, are found in abundance, while other combinations, such as red-green and yellow-blue are not detected. This led to the description of color vision in terms of

Figure 7.2 The top images have been lowpass filtered in the luminance channel in increasing amounts from left to right. In the bottom images, the same amount of blurring was instead applied to the chrominance channels. Notice the distinct difference in appearance between the image sets.

opponent colors. Today, the trichromatic and opponent descriptions have been joined in a stage theory, where the first stage of trichromatic signals is transformed to the second opponent stage before being transferred to the brain (Fairchild, 2013).

By this division into luminance and chrominance information, the amount of visual data transferred to the brain through the optic nerve can be reduced substantially. Consider Figure 7.2. The top images show a successive increase in blurring of the luminance channel, and the bottom images a corresponding amount of blur in the chrominance channels. Evidently, the spatial information in the chromatic channels can be reduced to a great extent without losing much information. This property of the HVS is made use of in image and video compression standards, such as JPEG and MPEG, where the chrominance channels are subsampled in order to reduce data volumes.

The sensitivity to detail in luminance and chrominance is manifested in the *contrast sensitivity functions* (CSFs) of the HVS. Figure 7.3 shows a typical set of such functions

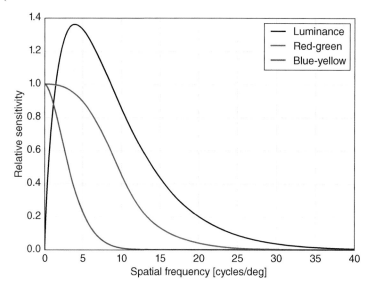

Figure 7.3 Human contrast sensitivity functions plotted. *Source*: Data from Johnson and Fairchild 2003.

(Johnson and Fairchild, 2003). It should be noted that these curves may change considerably depending on a wide selection of parameters, depending on, for example, age and adaptation to light, spatial frequency, contrast, etc. (Barten, 1999; Fairchild, 2013; Johnson, 2003; Fujii *et al.*, 2013; Wandell, 1995). Temporal aspects are also important, and will be discussed in more detail in section 7.8.

As can be seen from the figure, the luminance CSF behaves like a bandpass filter, with a peak in sensitivity around 4 cycles per degree. The two chrominance channels, on the other hand, show a lowpass behavior, and attenuate higher spatial frequencies considerably more than the luminance channel. This corroborates the fact that detail information is mainly carried in the achromatic channel.

The luminance CSF can also be visualized using a *Campbell–Robson* chart (Campbell and Robson, 1968), shown in Figure 7.4. This chart consists of a sinusoidal pattern, increasing in spatial frequency from left to right and increasing in contrast from top to bottom. The contrast is kept constant in the horizontal direction across the chart. However, to the observer it looks as if the position where the contrast seems to approach zero is changing with spatial frequency, and the limit where the bars seem to disappear forms an outline of a curve with a peak somewhere in the middle of the chart in the horizontal direction. This outline actually corresponds to the viewer's own contrast sensitivity function. Notice that the peak position changes as you move closer and farther away from the chart. This links back to the dependence of the CSF on spatial frequency expressed in units of cycles per degree. Since objects of a certain size will occupy a larger or smaller visual angle depending on the distance to the observer, a particular spatial frequency at the position of the object, expressed in cycles per degree, will slide up and down the CSF curve as the distance is changing. Implications of this fact, together with other viewing conditions, will be revisited in Section 7.3. Examples of Campbell–Robson charts for the chrominance channels are shown in Figure 7.5.

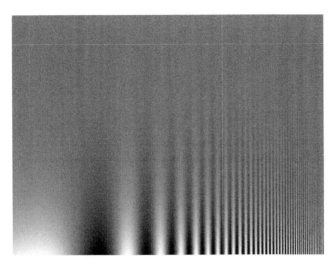

Figure 7.4 A Campbell–Robson chart.

Figure 7.5 Campbell–Robson charts for the chrominance channels.

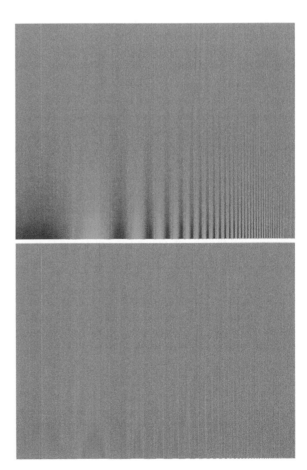

In Chapter 5, the Weber, Fechner, and Stevens laws were presented. These laws are the manifestation of the nonlinear nature of human perception. This was also touched upon in Chapter 6, when discussing the CIELAB and CIELUV color spaces. In a reasonably accurate model of human vision, these nonlinearities need to be addressed as well.

7.2 HVS Modeling

From the above discussion, it is clear that a simple model of the HVS must at least take into account the opponent representation, the contrast sensitivity functions, and the nonlinear nature of perception. Such processing is found in image quality metrics such as the S-CIELAB method (Zhang and Wandell, 1997; Zhang *et al.*, 1997; Zhang and Wandell, 1998; Johnson and Fairchild, 2003) as well as the iCAM framework (Fairchild and Johnson, 2004), to be discussed in more detail later in this chapter.

7.3 Viewing Conditions

The CSF is a function of spatial frequency expressed in units of cycles per degree, since the distance between observer and test object, as well as the size of the test chart, have an impact on, for example, the perceived sharpness. Therefore, a visual metric needs to take into account the conditions under which the scene is viewed. In order to be manageable, it is necessary to define a limited set of viewing conditions[1] that span a large enough part of the space of all possible viewing situations. This set of conditions must also include the medium on which the image is observed, for example, prints (hardcopy) or displays (softcopy). Table 7.1 describes one suggestion of suitable conditions; other conditions are described in the ISO 15739 and CPIQ standards (ISO, 2013; IEEE, 2017). These conditions will have to be adapted for specific use cases, and certainly need be updated over time as technology evolves.

The conversion from spatial frequency in angular units (e.g., cycles per degree) to spatial units (e.g., cycles per mm) can be made straightforwardly as follows. The angular position, θ, of an object a vertical distance h from the horizontal line between the observer and the object, viewed at a distance D, is given by

$$\tan\theta = \frac{h}{D}$$

This yields a nonlinear relation between the angle and the distance h. By making the simplification that the object is located on the surface of a sphere with radius D, and assuming $D \gg h$, the relation becomes $h \approx D\theta$, with θ expressed in radians. Then, any variation in h will be proportional to variations in θ, that is,

$$\Delta h \approx D\Delta\theta$$

---

1 Usually, when specifying viewing conditions, factors such as illumination, background luminance, amount of glare, and so on are included. Therefore, only including the distance between image and observer and image size in a definition of viewing conditions is in a strict sense too limited. However, for lack of a better term, we have chosen this limited definition in the discussion of the spatial metrics in this chapter. If not explicitly noted, therefore, any reference to viewing conditions henceforth only includes the spatial aspects as stated in this section.

Table 7.1 Example viewing conditions for visual image quality metrics. The last item is an example of a viewing condition likely to become relevant for a broad consumer range in the near future. This last item demonstrates the need to update the standard viewing conditions regularly.

Condition	Viewing distance (mm)
8" × 10" print	400
1:1 monitor	600
4" mobile phone screen	200
40" 1080p HDTV	2000
60" UHD TV	2000

From the fact that $\Delta\theta$ and Δh are the periods in angular and spatial units, respectively, the relation between spatial frequencies becomes

$$v_{cy/deg} = \frac{\pi D}{180} v_{cy/mm} = \frac{\pi D}{180p} v_{cy/pixel} \tag{7.2}$$

where p is equal to the pixel pitch. Keep in mind the units in these conversions: D must be given in millimeters to make the correct conversion from cycles per millimeters, and p and D should be specified in the same units in the conversion from cycles per pixel.

It may also be desirable to take into account the medium used for viewing the image, that is, the print or display. As described by Baxter *et al.* (2012), this may be done by incorporating simple models of the MTF of the chosen media. The display MTF is here given by

$$m_{disp}(v) = |\text{sinc}(v)| \tag{7.3}$$

with the spatial frequency, v, expressed in units of cycles per pixel. For prints, a representative MTF (Koopipat *et al.*, 2002; Bonnier and Lindner, 2010) was chosen as

$$m_{print}(v) = \exp\left(-\frac{v}{k}\right) \tag{7.4}$$

where in this case the spatial frequency is expressed in units of cycles per millimeter.

Since the CSF is expressed as a function of spatial frequency in units of cycles per degree, the units of spatial frequency for the display and print MTFs must be converted when used together with the CSF. To do this, Eqs. (7.3) and (7.4) may be expressed in the following forms:

$$m_{disp}(v) = |\text{sinc}(k_{disp}v)| \tag{7.5}$$

and

$$m_{print}(v) = \exp\left(-\frac{v}{k_{print}}\right) \tag{7.6}$$

where now the spatial frequency is expressed in units of cycles per degree and the two constants k_{disp} and k_{print} are calculated using Eq. (7.2) for the specific viewing condition at hand.

7.4 Spatial Image Quality Metrics

The HVS modeling described above has been used as a basis for the CPIQ *spatial metrics* (Baxter *et al.*, 2012; IEEE, 2017). Therefore, we will use these metrics, proposals for which have been described by Baxter *et al.* (2012) and Baxter and Murray (2012), as model examples in the following. For a complete description of the metrics, the reader is referred to the CPIQ standard (IEEE, 2017). The common factor for these metrics is that they are all dependent on spatial frequency, and therefore require well-defined viewing conditions according to the previous discussion. That being the case, it makes sense to use a common framework that incorporates contrast sensitivity functions, MTFs for printer and display, and the opponent color space in which to apply the CSFs when deriving visually correlated metrics for these attributes.

The CSFs proposed by Baxter *et al.* (2012) are shown in Figure 7.3 and have been adapted from the work of Johnson and Fairchild (2003). They can be described by the formula

$$\text{CSF}(v) = a_1 v^{c_1} \exp(-b_1 v^{c_2}) + a_2 \exp(-b_2 v^{c_3}) \tag{7.7}$$

where the spatial frequency v is expressed in cycles per degree. The values of the coefficients are summarized in Table 7.2.

As has already been noted earlier in this chapter, the luminance CSF shows a bandpass behavior with zero response at the lowest spatial frequency. In the visual noise model described below, the CSF is used as a spatial filter. For the luminance channel, this means that there will be no signal at zero spatial frequency and large uniform areas will therefore turn black. This is certainly not what is seen by a human observer. In some models, such as the ISO 15739 visual noise model (ISO, 2013), this is handled by normalizing the luminance CSF such that the zero frequency response is set to 1. This has the side effect of amplifying low and intermediate frequency signals, which results in clipping problems (Baxter and Murray, 2012) and will be revisited in more detail in Section 7.4.3. With this discussion in mind, it should be noted that the luminance CSF as expressed in Eq. (7.7) using the values in Table 7.2 will amplify the contrast around the peak of 4 cycles per degree. To avoid this, the normalization constant a_1 may be given a value of $4^{-0.80} \exp(0.20 \times 4) \approx 0.73$. However, this is less of an issue for the sharpness and texture acutance metrics, described later in this chapter, since

Table 7.2 Coefficients defining the luminance and chrominance CPIQ CSFs, *Source*: adapted from Johnson and Fairchild (2003).

Coefficient	CSF_A	CSF_{C1}	CSF_{C2}
a_1	1.0	0.54	1.0
a_2	0	0.46	0
b_1	0.20	0.00040	0.10
b_2	0	0.0037	0
c_1	0.80	0	0
c_2	1.0	3.4	1.6
c_3	0	2.2	0

the normalization constant will be canceled in the acutance calculation. For the visual noise metric, it will still be important.

Since the CSFs are defined for luminance and chrominance channels, the image data must be transformed to a color opponent space. This transformation may be done in a variety of ways (Zhang and Wandell, 1997; Johnson and Fairchild, 2004; ISO, 2013). The definition adopted by CPIQ (IEEE, 2017; Baxter *et al.*, 2012; Baxter and Murray, 2012) is the ISO 15739 AC_1C_2 color space.

7.4.1   Sharpness

7.4.1.1   Edge Acutance

The goal with a perceptually correlated metric is to produce a number that should reflect the perceived sensation of the attribute tested. In the case of sharpness, the quantity being measured is the MTF (SFR). This yields a multitude of values for a range of spatial frequencies, which are not easy to correlate with the visual sensation of sharpness. A well-established method to produce a number reflecting the visual impression of sharpness is the measure of *acutance* (Kriss, 1998). This may be expressed as (Baxter *et al.*, 2012)

$$Q_{edge} = \frac{\int_0^{v_c} S(v)C(v)M(v)dv}{\int_0^{\infty} C(v)dv} \tag{7.8}$$

that is, through the integral of the measured MTF, $S(v)$, weighted by the CSF, $C(v)$, and the MTF of the display or print, $M(v)$. Since the impression of sharpness is mainly carried in the luminance channel (see Figure 7.2), the CSF used is the luminance CSF, which from Eq. (7.7) and Table 7.2 becomes

$$C(v) = v^{0.8} \exp(-0.2v) \tag{7.9}$$

The upper limit of the integral, v_c, is a cutoff spatial frequency. It should be set to the lowest of either the half sampling (Nyquist) frequency of the camera or the printer/display. Observe that in some descriptions of the acutance, the upper limit of integration in the denominator is also limited to v_c, and not infinity. Consider a measured MTF equal to 1 for spatial frequencies up to v_c, and zero elsewhere (which is somewhat contrived, but illustrates the problem). If, neglecting the print/display MTF, the integral in the denominator is integrated only up to the Nyquist frequency, the acutance will evaluate to 1, independent of which cutoff frequency was used. The acutance value will then not reflect the lower sharpness for systems with lower resolution (and therefore lower Nyquist frequency). Therefore, the correct expression for the acutance, at least in the view of the authors, is to integrate the expression in the denominator to infinity.

Since we assume the SFR is typically measured using the slanted edge method described in Chapter 6, we will refer to this type of acutance as *edge acutance* to distinguish it from acutance values obtained from texture blur measurements.

It being a well known fact that the luminance channel carries most of the sharpness information, the acutance should be evaluated for the luminance channel only.

As an example using the above methodology, let us consider a case with a camera having an f-number of 2.8 and 8 megapixels resolution in an optical format of 1/3 inches. This gives a pixel pitch of 1.4 μm. We assume that the on-axis MTF of the lens is diffraction-limited. For the case of 100% magnification viewing on

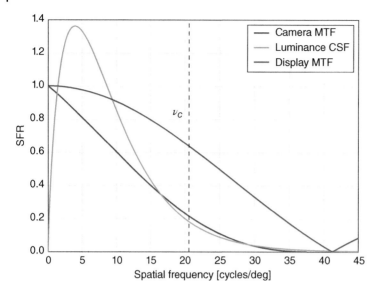

Figure 7.6 MTF and CSF curves used to calculate the acutance for a 100% magnification viewing on a computer screen, as discussed in the text.

a computer screen at a distance of 600 mm, the half sampling frequency is determined by the display pixel pitch (assumed to be 0.254 mm) and is therefore equal to $\pi \times 600/(180 \times 2 \times 0.254) = 20.6$ cycles per degree. For this particular viewing condition, the conversion from spatial frequency in cycles per mm to cycles per degree is $v_{cy/deg} = \pi \times 600/180 v_{cy/mm} = 10.47 v_{cy/mm}$. In Figure 7.6, the SFR of the combined lens and pixel MTF together with the CSF and printer MTF are plotted. With this data, the acutance comes out to be 0.61.

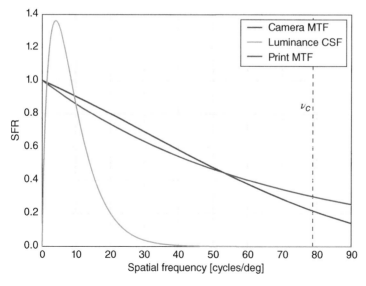

Figure 7.7 MTF and CSF curves used to calculate the acutance for viewing on a 8″ × 10″ print, as discussed in the text.

For comparison, we can take the same camera data as above, but for a different viewing condition. If we choose the 8″ × 10″ print viewing condition with a viewing distance of 75 cm, the curves will look as in Figure 7.7. For our 8 megapixel camera, the number of pixels in the vertical direction is 2448. Over 8″ of height, this gives a half sampling frequency of 6.02 cycles per mm. A print resolution of 600 ppi corresponds to a sampling frequency of 23.6 cycles per mm, and consequently $v_c = 11.8$ cycles per mm. This is larger than the camera half sampling frequency of 6.02 cycles per mm, and consequently the cutoff frequency in units of cycles per degree is $v_c = \pi \times 750/180 \times 6.02 = 78.8$ cycles per degree. In this case, the calculation of the acutance yields a value of 0.80, a little higher than the 100% display case.

7.4.1.2 Mapping Acutance to JND Values

The acutance value provides the link between the measured MTF/SFR values and the perception of sharpness. In the CPIQ formalism, however, one more step is needed to obtain a metric that expresses the perceived sharpness in units of JNDs (IEEE, 2017). As already mentioned in Chapter 5, the ISO 20462-3 standard describes a method for performing subjective studies aiming at assessing overall image quality. In this method, a quality ruler is used, which consists of a series of images degraded in a controlled way and calibrated against JND values. The degradation in this case happens to be sharpness. The ruler images were made by applying spatial filtering with specific MTF curves given by the expression (Jin *et al.*, 2009)

$$m(v) = \begin{cases} \frac{2}{\pi}\left(\cos^{-1}(kv) - kv\sqrt{1-(kv)^2}\right) & kv \le 1 \\ 0 & kv > 1 \end{cases} \tag{7.10}$$

The parameter k, describing the degradation in sharpness for the ruler images, takes on values between 0.01 and 0.26. The spatial frequency, v, is here expressed in units of cycles per degree at the retina of the observer. From the calibration procedure described in ISO 20462-3, the following relation between JNDs of image quality and the parameter k has been established:

$$\text{JND}(k) = \frac{17\,249 + 20\,3792k - 114\,950k^2 - 3\,571\,075k^3}{578 - 1304k + 357\,372k^2} \tag{7.11}$$

With this relation and Eq. (7.10), it is now possible to find the relationship between JND values of quality and acutance values calculated according to the method described in the previous section (Baxter *et al.*, 2012). The result, in the form of a plot, is shown in Figure 7.8. Three observations can be made when studying this plot. First, the relationship between JND values of quality and acutance are very close to linear over a large range of values. Second, for acutance values above approximately 0.85, the curve levels out, signaling saturation. Consequently, acutance values above 0.85 will not produce images with higher perceived sharpness. Third, a JND value of zero does not correspond to an acutance value of zero. The SQS JND scale, as described and implemented in the ISO 20462 standard, only blurs images up to a certain level. A blurred image, corresponding to an SQS JND value of zero for a 10 cm wide image viewed at a distance of 40 cm is shown in Figure 7.9. The image quality is undoubtedly very poor, but it would still be possible to blur the image more, even to a point where only the average value of the image survives, showing just a uniform colored patch. Thus, it is important to remember that the SQS is still a relative scale, not by itself associated with absolute image

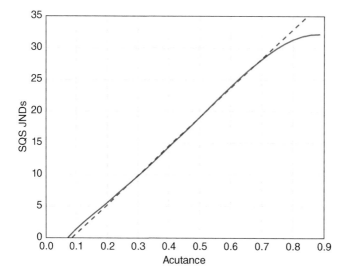

Figure 7.8 Plot relating quality JND to edge acutance values. Dashed blue curve is a straight line with equation $y = 46x - 4.0$.

Figure 7.9 Image approximately corresponding to an SQS JND value of 0. The image should be viewed at a distance of 40 cm.

quality values, in analogy with, for example, the Kelvin temperature scale. However, since it was decided that the zero point of the scale should be anchored at a quality level corresponding to what is shown in Figure 7.9, then if the procedures described in the ISO standard are followed, an absolute scale is in fact obtained, since it will be possible to compare results from different measurements using the scale. Going back to the example calculations of edge acutance, we found acutance values of 0.61 and 0.80, respectively in both cases. From Figure 7.8, these values yield JND values of 24 and 31, respectively.

Therefore, we predict that the perceived sharpness in the 8″×10″ print viewing condition is about 7 JND units higher than for the 100% computer monitor condition.

7.4.1.3 Other Perceptual Sharpness Metrics

An often encountered metric for perceptual sharpness is the *subjective quality factor* (SQF), developed by Granger and Cupery (1973). Here also, the measured MTF is weighted with the CSF and integrated to obtain one single number of sharpness. However, apart from the fact that the MTF of the print or display is missing, a compressive action is included by integrating with respect to the logarithm of the spatial frequency, that is:

$$SQF = \frac{\int_0^\infty S(v)C(v)d\log v}{\int_0^\infty C(v)d\log v} \tag{7.12}$$

where, as before, $S(v)$ is the measured spatial frequency response, and $C(v)$ the CSF. At the time when this metric was developed, computing power was severely limited, which motivated a simplification by exchanging the CSF for the function

$$C(v) = \begin{cases} 1, & 3 \text{ cy/deg} < v < 12 \text{ cy/deg} \\ 0, & \text{otherwise} \end{cases} \tag{7.13}$$

This instead gave the equation

$$SQF = 1.386 \int_{3 \text{ cy/deg}}^{12 \text{ cy/deg}} S(v)d\log v = 1.386 \int_{3 \text{ cy/deg}}^{12 \text{ cy/deg}} S(v)\frac{dv}{v} \tag{7.14}$$

Since this expression will surely produce different results compared with the original formula, it is important to be aware of the existence of both.

Another metric is the *square root integral* (SQRI) (Barten, 1989, 1999). This is calculated using the formula

$$SQRI = \frac{1}{\log 2} \int_0^\infty \sqrt{S(v)C(v)} \frac{dv}{v} \tag{7.15}$$

In this expression, two compressive factors are used to model the human visual response: the square root and the logarithmic integration. When tested with a comparatively large set of SFR curves, this metric has been shown to produce results that are well-correlated with perception (Barten, 1999).

7.4.2 Texture Blur

The texture blur metric based on the dead leaves approach, previously discussed in Section 6.8, produces a curve similar to the MTF. Therefore, it is possible to calculate an acutance value in exactly the same way as for the edge acutance. However, since texture acutance is different in appearance to sharpness, correlating texture acutance with JND values using the simple approach for edge acutance is not applicable. Instead, a full subjective experiment using, for example, the ISO 20462-3 subjective quality ruler must be carried out to find the relation between texture acutance and JNDs of overall image quality. Two such studies have been performed (Phillips *et al.*, 2009; Phillips and Christoffel, 2010) in which several test images were processed using an adaptive noise cleaning filter, yielding varying levels of texture loss. By using the methodology of ISO

20462-3, previously described in detail in Chapter 5, the result was a linear relationship between texture acutance and SQS JND values according to the equation (Phillips and Christoffel, 2010)

$$\text{Texture JND} = 25.4Q_{\text{texture}} + 4.16 \tag{7.16}$$

It should be observed that the range of values over which the fit of acutance values to JNDs was made was limited to acutance values between 0.5 and 1.0. Furthermore, the acutance values in this study were calculated without using a display or print MTF, and only for one specific use case. Therefore, additional correlation studies are needed to establish a firm relationship between JNDs and texture acutance values. The CPIQ standard (IEEE, 2017) includes an updated JND mapping.

7.4.3 Visual Noise

Figure 6.31 in Chapter 6 demonstrates that noise of different spatial content can be perceived in quite different ways. A simple noise metric where only the variance is calculated is clearly not enough in these cases, since all three examples in the figure will have the same calculated variance. From this example only, it should be obvious that the spatial content of noise must be taken into account when constructing a visually correlated noise metric. The ISO 15739 standard does describe such a method. Briefly, it works in the following way. First, an image of a test chart containing uniform test patches of varying reflectance is captured by the camera. Crops of the resulting image containing only the patch data are extracted and the RGB values are transformed via the XYZ space to an opponent color representation. In this opponent space, the luminance and chrominance channels are filtered using CSFs. The image data is thereafter transformed back to XYZ and then to the CIELUV space, in which variances are calculated for the luminance and chrominance channels. These values are then weighted together to obtain an ISO visual noise metric (ISO, 2013).

In the noise metric proposed by Baxter and Murray (2012) and implemented in CPIQ (IEEE, 2017) (with some alterations), the methodology is based on the ISO method, but with some modifications and additional steps in order to transform the final visual noise value into JNDs of image quality.

The image of the noise test chart is first transformed into an opponent color space, as in the ISO standard. So far, the procedure is identical to that used in the ISO standard. In the subsequent steps, however, the methods start to deviate, first in the choice of CSF, then by including filtering with the display/print MTFs as described earlier, and also by the use of a highpass filter, which serves to remove nonuniformities due to shading. The filter used was described in Eq. (6.67). The choice of a different CSF warrants a deeper discussion. In the ISO implementation, the CSF is normalized to a value of one at zero spatial frequency. This is done in order to preserve the mean value in the image, which would otherwise have been set to zero if the CSF had a value of zero at zero spatial frequency. This choice has the unwanted side effect of amplifying low to intermediate spatial frequencies to such high levels that image values due to noise in many situations will become negative and therefore clipped, leading to wrong estimations of the visual noise at the other end of the processing chain. In order to avoid this, the CSF described in Eq. (7.7) and Table 7.2 is used together with a subtraction of the mean level prior to filtering and adding it back after the filtering step. This methodology has previously been described by Johnson and Fairchild (2003).

The image data is subsequently converted back to the CIEXYZ representation. In the ISO standard, the data is then transformed to the CIELUV color space followed by a weighted sum of the standard deviations of the L^*, u^*, and v^* channels. In the CPIQ case, the CIELAB space was chosen instead of CIELUV. The reason for this is mainly that CIELAB is a *de facto* standard and is usually recommended to be used instead of CIELUV (Fairchild, 2013).[2] For the next step, a similar methodology as the edge and texture acutance metrics is used, where an objective metric of quality is first calculated in an intermediate step, followed by a transformation into JND values. In the sharpness and texture blur cases, the objective metric was the acutance, described by Eq. (7.8). In the visual noise case, the objective metric, Ω, is calculated as follows (Keelan *et al.*, 2011):

$$\Omega = \log_{10}[1 + w_1\text{Var}(L^*) + w_2\text{Var}(a^*) + w_3\text{Var}(b^*)$$
$$+ w_4\text{Covar}(L^*, a^*) + w_5\text{Covar}(L^*, b^*) + w_6\text{Covar}(a^*, b^*)] \tag{7.17}$$

The logarithm of this weighted sum is calculated, since noise perception has been found to be approximately logarithmic above the threshold of detection (Baxter and Murray, 2012; Bartleson, 1985; Keelan *et al.*, 2011).

The final transformation to JND values may be made using the integrated hyperbolic increment function (IHIF), described by Keelan (2002):

$$\Delta Q(\Omega) = \begin{cases} \frac{\Omega - \Omega_r}{\Delta\Omega_\infty} - \frac{R_r}{\Delta\Omega_\infty^2}\ln\left(1 + \frac{\Delta\Omega_\infty(\Omega - \Omega_r)}{R_r}\right), & \Omega > \Omega_r \\ 0, \Omega \leq \Omega_r \end{cases} \tag{7.18}$$

where $\Delta Q(\Omega)$ is the JND value corresponding to objective quality value Ω, and Ω_r, R_r, and $\Delta\Omega_\infty$ are constants with values 0.4627, 0.1418, and 0.02312, respectively.

What remains to make the visual noise metric complete is to determine the weights w_1–w_6. From a previous study (Keelan *et al.*, 2011), a set of noisy colored patches and their associated values of overall quality JNDs have been made available. The JND values were in this case found through a subjective study using the ISO 20462 quality ruler. Using these patches, and calculating the variances and covariances of the CIELAB coordinates, it is possible to do a least squares fit to the data. By inverting the IHIF, which unfortunately can only be done numerically, the objective values corresponding to the JND values of the noisy patches can be found and used to make a linear least squares fit to the data in Eq. (7.17). Figure 7.10 shows the IHIF as applied to this data set.

Other visual noise metrics also exist. One is the vSNR metric (Farrell *et al.*, 2010), which uses the formalism of S-CIELAB, to be explained further in Section 7.5. This formalism also includes CSFs as well as a color opponent representation, and is therefore to a large part quite similar to the ISO and CPIQ visual noise metrics. However, the final result of the metric is a signal to noise value, obtained by calculating the average color difference, $\overline{\Delta E}$, of a uniform patch processed by the S-CIELAB algorithm and then evaluating

$$\text{vSNR} = \frac{1}{\overline{\Delta E}}$$

In order to characterize the imaging performance of image sensors, it is useful to perform a noise analysis that is reasonably representative of how an image from the sensor

---

2 However, CIELUV is better suited for handling additive color mixtures, which makes it useful for display applications.

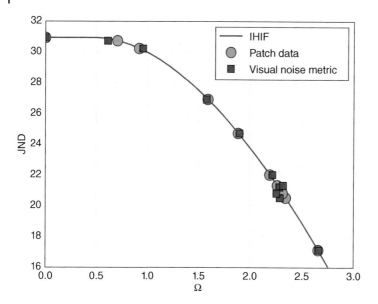

Figure 7.10 The IHIF fitted to the Baxter and Murray (2012) visual noise metric data (red squares) and the JND values of the noisy patches that were used to calibrate the proposed CPIQ metric (green circles). *Source*: Adapted from Keelan, Jin, Prokushkin, Development of a perceptually calibrated objective metric of noise, *Proc. SPIE* **7867**, 2011.

would be perceived. However, since the noise present in a raw image is quite different from that in the final image, it is not useful to utilize a full visual metric such as the one described in this chapter. Instead, a simplified method may be applied in which at least the influence of the color matrix (discussed in Chapter 6) is taken into account. One attempt at such a characterization which has become widespread is what is currently known as the SNR 10 metric (Alakarhu, 2007). This metric calculates a luminance noise value from a color corrected raw image sensor signal. The noise in the red, green, and blue channels after color correction is calculated using Eq. (6.63), with a diagonal covariance matrix **C**. This is reasonable, since the raw color noise can be considered to be uncorrelated, and consequently the covariance terms of the matrix **C** are 0. However, in the resulting covariance matrix, **C′**, the off-diagonal elements will *not* be zero. When calculating the luminance noise, another transformation matrix is used, transforming the color corrected RGB signal into a color space with one luminance channel and two chrominance channels. Since only the luminance information is considered in this case, the chrominance signals are disregarded. If we let the weights for red, green, and blue signals be α_L, β_L, and γ_L for the luminance signal, and with the subscripts $C1$ and $C2$ for the chrominance channels, the transformation matrix becomes

$$\mathbf{M}_L = \begin{pmatrix} \alpha_L & \alpha_{C1} & \alpha_{C2} \\ \beta_L & \beta_{C1} & \beta_{C2} \\ \gamma_L & \gamma_{C1} & \gamma_{C2} \end{pmatrix} \qquad (7.19)$$

Then, the resulting covariance matrix is calculated as

$$\mathbf{C}'' = \mathbf{M}_L^T \mathbf{C}' \mathbf{M}_L \qquad (7.20)$$

The top left element in the resulting matrix $\mathbf{C}''$ is the noise variance in the luminance channel, and is evaluated to be

$$
\begin{aligned}
\mathrm{Var}(L) = {} & \alpha_L^2 \mathrm{Var}(R') + \beta_L^2 \mathrm{Var}(G') + \gamma_L^2 \mathrm{Var}(B') \\
& + 2\alpha_L \beta_L \mathrm{Covar}(R', G') + 2\alpha_L \gamma_L \mathrm{Covar}(R', B') \\
& + 2\beta_L \gamma_L \mathrm{Covar}(G', B')
\end{aligned}
\tag{7.21}
$$

In the SNR 10 metric as presented by Alakarhu (2007), the covariance terms are neglected. Therefore, this metric does not represent the luminance noise in the strictest sense. Furthermore, a subjective correlation was never established for the metric at the time of its introduction. The literature on this type of metric is not abundant, which makes it more difficult to provide a metric better correlated with actual subjective experiments. However, one study has been performed that does provide data on this type of correlation (Kelly and Keelan, 2005). The results from this study were used in the ISO 12232 standard for calculating the noise-based ISO speed. Although the luminance noise usually dominates the total noise perception, the contributions from the chrominance noise should not be neglected altogether, as is done in the SNR 10 metric. For this reason, the chrominance noise was also included in the cited study. Here, the transformation matrix from color corrected RGB values to the opponent space containing luminance and chrominance channels is

$$
\left(\, Y \quad R - Y \quad B - Y \,\right) = \left(\, R' \quad G' \quad B' \,\right) \mathbf{M}_{\mathrm{ISO}}
\tag{7.22}
$$

where

$$
\mathbf{M}_{\mathrm{ISO}} = \begin{pmatrix} 0.2126 & 0.7874 & -0.2126 \\ 0.7152 & -0.7152 & -0.7152 \\ 0.07219 & -0.07219 & 0.9278 \end{pmatrix}
\tag{7.23}
$$

Then, the noise metric is calculated as

$$
\sigma_{\mathrm{ISO}} = \sqrt{\mathrm{Var}(Y) + 0.279\mathrm{Var}(R - Y) + 0.088\mathrm{Var}(B - Y)}
\tag{7.24}
$$

where the weights were determined by a psychophysical experiment and the variances are extracted from the diagonal of the covariance matrix resulting from evaluating the expression

$$
\mathbf{C}_{\mathrm{ISO}} = \mathbf{M}_{\mathrm{ISO}}^T \mathbf{C}' \mathbf{M}_{\mathrm{ISO}}
\tag{7.25}
$$

From this result, but also from the discussion earlier about the CPIQ visual noise metric, it should be obvious that contributions from chrominance noise must be incorporated in a noise metric. In a later investigation, the SNR 10 metric was actually found to yield worse perceptual correlation compared with the metrics used in the ISO 12232 standard (Koskinen *et al.*, 2011).

A subjective study on the relationship between color and noise using the softcopy image quality ruler has been made by Keelan *et al.* (2012). The conclusions from this study showed a shift in preferable color saturation, measured using Eq. (7.38), from around 80% at a luminance SNR of 10 (which according to ISO 15739 is defined as just acceptable) to about 110% at the highest SNR values. This illustrates the tradeoff between the noise suppression due to a decreased color saturation and the lower perceived quality due to the resulting less colorful image.

7.5 Color

In Chapter 6, some important color concepts were introduced. The description ended with a discussion about the CIELAB color space and color difference metrics. With this knowledge, it is possible to perform many color measurement tasks. However, the appearance of color is certainly quite complex, and the number of parameters needed to obtain a color model that is capable of predicting color appearance phenomena becomes very large very quickly. It is therefore usually not practical, or even possible, to employ a full color appearance model when doing a camera characterization. However, it is still important to understand the shortcomings of the models and methods used, not least to enable further development.

7.5.1 Chromatic Adaptation Transformations

An important topic when dealing with perceptually correlated metrics, of which color metrics are a very important part, is the concept of adaptation. This was touched upon already in the discussion on the contrast sensitivity function, which can show very different behavior depending on various adaptation mechanisms, for example, with respect to light level, spatial frequency, contrast, and so on. In the realm of color appearance, the concept of *chromatic adaptation* is very important. Any color model that has the ambition to be reasonably well correlated with human perception needs to have some kind of chromatic adaptation mechanism built into it. The CIELAB model, discussed in detail in Chapter 6, does have such a mechanism, where the CIEXYZ values are normalized with the XYZ values of the illuminant. This type of *chromatic adaptation transformation* (CAT) is in certain respects not correct, and is therefore usually referred to as a *wrong von Kries* transformation (Fairchild, 2013).

When discussing chromatic adaptation transformations, it is impossible not to mention the von Kries transformation. The idea of the transformation originates from Johannes von Kries (1853–1928), who postulated that the gain of each of the three cones in the retina could be controlled individually in order to adapt to variations in lighting. As it turns out, the von Kries transformation has proved to work quite well in many situations. In practice, this transformation may be carried out in the following way. If we have the source XYZ values obtained for some illuminant with chromaticities x_s and y_s, and we want to transform these values as if they were obtained under the illuminant with chromaticities x_d and y_d, as a first step, the source XYZ values are transformed into the LMS cone space, using Eq. (7.1). Secondly, the source and destination illuminant chromaticities are converted to XYZ values with a Y value of 1, and then transformed to the LMS cone space. In the third step, the source LMS values are divided by the LMS value of the source illuminant and then multiplied by the destination LMS values. In the final step, the modified LMS values are transformed back to XYZ using the inverse matrix of Eq. (7.1). Mathematically,

$$
\left(X \ Y \ Z \right)_{dest} =
$$
$$
\mathbf{M}^{-1}
\begin{pmatrix}
L_d/L_s & 0 & 0 \\
0 & M_d/M_s & 0 \\
0 & 0 & S_d/S_s
\end{pmatrix}
\mathbf{M}\left(X \ Y \ Z \right)_{src}
\qquad (7.26)
$$

where $\mathbf{M}$ is the Hunt–Pointer–Estevez transformation matrix given by Eq. (7.1), and

$$(L_s \; M_s \; S_s) = \mathbf{M} (X_s \; Y_s \; Z_s)$$

$$X_s = \frac{x_s}{y_s}$$

$$Y_s = 1$$

$$Z_s = \frac{1 - x_s - y_s}{y_s}$$

and similarly for the destination values (L_d, M_d, S_d), (X_d, Y_d, Z_d), and (x_d, y_d).

Even though the von Kries transformation has been proven to work successfully, other transformations giving better results have been developed. Out of these, the *Bradford transformation* is very frequently encountered (Lam, 1985). The full model includes a nonlinear component for the short wavelength cone response, but this is frequently omitted. In this case, the procedure is just the same as for the von Kries transformation, except that the transformation matrix is replaced by

$$\mathbf{M} = \begin{pmatrix} 0.8951 & -0.7502 & 0.0389 \\ 0.2664 & 1.7135 & -0.0685 \\ -0.1614 & 0.0367 & 1.0296 \end{pmatrix} \tag{7.27}$$

In light of the discussion above, camera white balance can also be better understood. If the spectral sensitivities of the camera color filter array correspond to the cone sensitivities of the human eye, white balancing can be said to correspond to a von Kries transformation as described above.

7.5.2 Color Appearance Models

The CIELAB color difference model can be considered a simple *color appearance model* (CAM) in that it takes into consideration adaptation to the white point of the illuminant as well as nonlinearities of the HVS. However, such a simple model cannot take into account even the most common color appearance phenomena, such as simultaneous contrast (discussed in Chapter 1), increase of perceived brightness with color saturation (Helmholtz–Kohlrausch effect), increase in perceived colorfulness with luminance (Hunt effect), increase in contrast with luminance (Stevens effect), or the impact of the surrounding luminance (e.g., Bartleson–Brennan effect) (Fairchild, 2013). For such effects to be considered, more comprehensive color appearance models need to be utilized. Out of the different models available, CIE is recommending the CIECAM02 model, described briefly in the following paragraphs. For a more thorough description, the reader is encouraged to consult the texts by, for example, Reinhard *et al.* (2008), Hunt and Pointer (2011), and Fairchild (2013).

The goal of a color appearance model can be expressed as being able to predict the appearance of color by describing a set of key attributes. CIE has specified these attributes as follows (Hunt and Pointer, 2011):

- *Hue*—describing the appearance of a region as being similar to one of the colors red, yellow, green, and blue, or mixtures thereof.
- *Brightness*—describing the extent to which a region appears more or less light.
- *Colorfulness*—describing how much of its hue a region exhibits.

- *Lightness*—describing the brightness of a region relative to another region with similar illumination that appears white.
- *Chroma*—describing the colorfulness of a region relative to another region with similar illumination that appears white.
- *Saturation*—the relative colorfulness of a region with respect to its brightness.

The last three items in the list describe relative attributes. Since, as is well known, the human visual system is highly adaptive to a wide range of viewing conditions, these attributes are very important descriptors of how color is perceived. While a full color appearance model should include all attributes, other applications, such as for color rendering, may only need descriptions of the relative attributes.

In addition, a definition of viewing conditions[3] is needed in order to obtain a reliable assessment of color appearance. To specify the viewing conditions, the following definitions are made:

- *Stimulus*—the object for which the appearance of color should be assessed. This is typically a uniform patch subtending approximately 2° of the viewing field.
- *Proximal field*—The environment immediately surrounding the stimulus. This is not used in most color appearance models.
- *Background*—The environment just outside the stimulus, or proximal field if present, covering around 10° outside the stimulus.
- *Surround*—The area outside the background.

One typical use for a color appearance model is for color management applications, where colors should match over a large range of devices and other media. Then it is necessary to predict tristimulus values corresponding to the correlates for lightness, brightness, hue, colorfulness, chroma, and saturation. In that respect, for a color appearance model to be truly useful for such purposes, it must be invertible. In the case of CIECAM02, this is readily achieved. In the case of camera benchmarking, the purpose would be to obtain a description of color signals produced by a particular imaging device. Consequently, the forward transform is of most interest within the scope of this book. For a thorough rundown of the calculations for transforming CIEXYZ values to CIECAM02, the reader is referred to Reinhard *et al.* (2008).

The main purpose of the CIECAM02 model is to predict how a certain color stimulus under certain conditions will appear to an observer. As such, it has been proven to work well. At the same time, it is a complex model, requiring the specification of a large number of parameters, like surround factors, the absolute luminance of the adapting white, the degree of chromatic adaptation, and so on. While such parameters can be standardized for certain viewing conditions, the implementation of the model is nevertheless still considerably more complicated compared to, for example, calculating color differences using CIELAB. Furthermore, it only applies to the appearance of colored patches, and does not include spatial correlations between proximal features in an image, e.g., simultaneouos contrast. To take such effects into account, other methods have been developed that enhance the utility of the CIE color appearance models, as described in further detail in the following section.

---

3 Note that this definition of viewing conditions is different from what was discussed previously in this chapter.

7.5.3 Color and Spatial Content—Image Appearance Models

S-CIELAB is, as the name implies, a modification of the CIELAB color space (Zhang and Wandell, 1997). It attempts to take into consideration the spatial aspects of human color vision. An example application of such a system is predicting degradation in color appearance for half-tone prints. If one were just to apply, for example, the CIEDE2000 color difference formula pixel by pixel between an original continuous tone print and a half tone print, the result would not correlate well with the perceived difference in color. In particular, the fact that the perceived color difference will change with viewing distance is not captured at all with such a metric.

In the S-CIELAB metric, images are filtered with contrast sensitivity functions, similar to the ones presented in Section 7.4, before they are compared. The workflow is as follows. The images are first converted to CIEXYZ, followed by a conversion into an opponent color space. In this space, the luminance and chrominance channels are separately filtered by their corresponding CSFs, whereafter the image is converted back to CIEXYZ and then CIELAB. In CIELAB, a pixel by pixel color difference is then calculated. As described in Section 7.4, the CSFs are specified as functions of angular frequency, and this makes S-CIELAB become dependent on the viewing conditions in terms of viewing distance and physical image size. In the original version of S-CIELAB, the contrast sensitivity functions were approximated using convolution kernels consisting of weighted sums of Gaussian functions:

$$f(x, y) = k \sum_i \omega_i F_i(x, y) \tag{7.28}$$

where

$$F_i(x, y) = k_i \exp\left(-\frac{x^2 + y^2}{\sigma_i^2}\right) \tag{7.29}$$

The parameters are summarized in Table 7.3. The transformation from CIEXYZ to the color opponent representation AC_1C_2 is accomplished using

$$\begin{pmatrix} A & C_1 & C_2 \end{pmatrix} = \begin{pmatrix} X & Y & Z \end{pmatrix} \begin{pmatrix} 0.279 & -0.449 & 0.0860 \\ 0.720 & 0.290 & -0.590 \\ -0.107 & -0.0770 & 0.501 \end{pmatrix} \tag{7.30}$$

This model was modified by Johnson and Fairchild (2003) to do the spatial filtering in the Fourier domain instead, using the CSFs described in Section 7.4 and calculating color

Table 7.3 Parameters for the original S-CIELAB CSFs. *Source:* data from Zhang and Wandell (1997).

	Luminance	Red-green	Blue-yellow
ω_1	0.921	0.531	0.488
ω_2	0.105	0.330	0.371
ω_3	−0.108	0	0
σ_1	0.0283	0.0392	0.0536
σ_2	0.133	0.494	0.386
σ_3	4.336	0	0

differences using the CIEDE2000 metric. It should then be noticed that the structure of calculations is similar to the visual noise metric described in Section 7.4.3. In fact, as mentioned previously in this chapter, an alternative visual noise metric based on S-CIELAB has been presented, called vSNR (visual SNR) (Farrell *et al.*, 2010).

The spatial processing described here may be used to extend color appearance models to be more suitable for images. One such model is the iCAM framework (Fairchild and Johnson, 2002, 2004; Reinhard *et al.*, 2008), which is outlined below. The model includes a spatial preprocessing step similar to S-CIELAB, but with the color opponent space given by the transformation matrix

$$(Y' \ C_1 \ C_2) = (X \ Y \ Z) \begin{pmatrix} 0.0556 & 0.9510 & 0.0386 \\ 0.9981 & -0.9038 & 1.0822 \\ -0.0254 & 0 & -1.0276 \end{pmatrix} \tag{7.31}$$

The input CIEXYZ values should be specified for a D65 white point. The spatial filters can be described by Eq. (7.7), but with the parameters shown in Table 7.4. After pre-processing, chromatic adaptation takes place. The CAT used is the same as found in CIECAM02. However, instead of adapting to a global white point, iCAM uses a local adaptation model, where a lowpass filtered version of the image is used as the adapting white, and the CAT is applied on a per-pixel basis. The target white point for the transformation is D65. Thus, the chromatic adaptation transformation may be explicitly expressed as follows:

$$R_c(x, y) = \left(\frac{DR_{D65}}{R_w(x, y)} + D - 1 \right) R(x, y)$$

$$G_c(x, y) = \left(\frac{DG_{D65}}{G_w(x, y)} + D - 1 \right) G(x, y)$$

$$B_c(x, y) = \left(\frac{DB_{D65}}{B_w(x, y)} + D - 1 \right) B(x, y)$$

In a next step, the adapted color signals are transformed into LMS cone space. A nonlinear compression using a power function follows, whereafter transformation into XYZ space is performed. From there, a transformation into the IPT color opponent space

Table 7.4 Coefficients defining the luminance and chrominance CSFs for the iCAM model. *Source*: data from Reinhard *et al.* (2008).

Coefficient	$CSF_{y'}$	CSF_{C1}	CSF_{C2}
a_1	0.63	91.228	5.623
a_2	0	74.907	41.9363
b_1	0.085	0.0003	0.00001
b_2	0	0.0038	0.083
c_1	0.616	0	0
c_2	1	2.803	3.4066
c_3	0	2.601	1.3684

(Ebner and Fairchild, 1998) is made according to

$$
(L\ M\ S) = (X\ Y\ Z) \begin{pmatrix} 0.4002 & -0.2280 & 0 \\ 0.7075 & 1.150 & 0 \\ -0.08070 & 0.06120 & 0.9184 \end{pmatrix} \tag{7.32}
$$

where, again, the XYZ values are specified for a D65 white point. Next,

$$
L'(x,y) = \begin{cases} L(x,y)^{0.43F_L(x,y)}, & L(x,y) \geq 0 \\ -|L(x,y)|^{0.43F_L(x,y)}, & L(x,y) < 0 \end{cases}
$$

$$
M'(x,y) = \begin{cases} M(x,y)^{0.43F_L(x,y)}, & M(x,y) \geq 0 \\ -|M(x,y)|^{0.43F_L(x,y)}, & M(x,y) < 0 \end{cases}
$$

$$
S'(x,y) = \begin{cases} S(x,y)^{0.43F_L(x,y)}, & S(x,y) \geq 0 \\ -|S(x,y)|^{0.43F_L(x,y)}, & S(x,y) < 0 \end{cases}
$$

Here, $F_L(x,y)$ is the brightness parameter from CIECAM02, but calculated per pixel in the image. The IPT values are now obtained as

$$
(I\ P\ T) = (L'\ M'\ S') \begin{pmatrix} 0.4000 & 4.455 & 0.8056 \\ 0.4000 & -4.851 & 0.3572 \\ 0.2000 & 0.3960 & -1.1628 \end{pmatrix} \tag{7.33}
$$

From the IPT values, correlates for, for example, lightness, J, chroma, C, and hue angle, h, can be found using the identities

$$
J = I \tag{7.34}
$$

$$
C = \sqrt{P^2 + T^2} \tag{7.35}
$$

$$
h = \tan^{-1}\left(\frac{P}{T}\right) \tag{7.36}
$$

The iCAM framework is so far not a finished image appearance model, and should be regarded as still being in the research phase. Therefore, the above description is necessarily incomplete and, thus, far from being a "cookbook." For the purposes of camera and image characterization and benchmarking, it may therefore still be too rudimentary, and more research is needed to establish a full model. Nevertheless, portions of the color and image appearance models may be used for the development of custom metrics addressing some particular aspect of image quality. We have seen, for instance, how similar ideas to that of the S-CIELAB metric have been used in the spatial metrics of ISO and CPIQ.

7.5.4 Image Quality Benchmarking and Color

Establishing a quality metric for color is not trivial. Apart from the aspects already discussed, color is also a preferential attribute, that is, different persons will judge what is good color reproduction differently. Therefore, the straightforward approach of calculating ΔE values for a set of colored patches (e.g., the X-Rite ColorChecker) using CIELAB values derived from the standard observer as reference, may not provide an adequate assessment of color reproduction. This is further complicated by the fact that the "true" colors are in most cases deemed as too dull and not representative of a scene as a person typically remembers it, as has already been pointed out in Chapter 3.

Yet another complication is *metamerism*. There are two basic types: illuminant and observer metamerism. Illuminant metamerism illustrates the phenomenon that two colors that appear to match under one light source may not match under another. Observer metamerism is the fact that two colors that appear to match for one observer may not match for another. This is of course true also for color cameras, and "observer metamerism" between a human observer and the camera is for this reason a factor that has to be taken into account. This means that it is not sufficient to establish the color reproduction of the camera for one light source only, but several different ones have to be used.

Nevertheless, for a camera manufacturer who is, at least in principle, free to choose how colors should be reproduced, from experience as well as preference, it is a comparatively simple task to establish reference colors in, for example, the CIELAB color space for a chosen set of illuminants and then optimize the colors reproduced by their products with respect to these references. For benchmarking, on the other hand, as already pointed out, this will be difficult since there is no clear reference available. Due to the preferential nature of color, the margins of acceptability must be made quite wide.

For all these reasons, an initial solution to the problem for benchmarking may be to use a simple metric that quantifies the amount of *color saturation* that images from a certain camera exhibit under certain conditions. This metric is particularly interesting if one considers the fact that there is a strong relationship between color saturation and signal to noise ratio in an image (Keelan *et al.*, 2012). Because of this, many camera ISPs reduce the color saturation as the ambient light level is decreased.

A metric for color saturation may be constructed as follows (Imatest, 2015; IEEE, 2017). With the camera under test, capture an image of a color test target, for example, the X-Rite ColorChecker. Obtain CIELAB values for the colored patches of the image. Calculate the chroma for each patch i as

$$c_i = \sqrt{a_i^{*2} + b_i^{*2}} \tag{7.37}$$

Calculate also the chroma values for the patches using the standard CIE observer, c_i^{ref}. The saturation metric is then given by the ratio of the means of the chroma values:

$$S = 100 \frac{\sum_{i=1}^{N} c_i}{\sum_{i=1}^{N} c_i^{\text{ref}}} \tag{7.38}$$

A correlation with JND values has been made by Keelan *et al.* (2012) specific to the color patches of the X-Rite ColorChecker Classic target. The relation between the just described saturation values, S, and the relative perceptual quality in JNDs, ΔQ, was here found to be

$$\Delta Q = 27.44 \left[\exp\left(-(0.00786|S - 110.6|)^{1.414}\right) - 1 \right] \tag{7.39}$$

This is a peaked curve with maximum value at $S = 110.6$, illustrating the preferential nature of color saturation. Note that CPIQ implements a different formula for this mapping due to use of a different target.

Another aspect of camera color reproduction is white balancing. It might seem simple to construct an objective (and visually correlated) metric to quantify this attribute. However, considering the preferential nature of color appearance, there are many situations where a "true" white balance (i.e., where all surfaces with spectrally neutral reflectance are rendered gray) is not desired. For instance, the golden colors of a sunset scene would look quite unnatural if the white balancing were to completely remove the color cast in objects with spectrally neutral reflectance. The same is true for indoor incandescent lighting. Most cameras therefore put restrictions on the auto white balancing algorithms at both low and high CCT values. In this way, spectrally neutral objects present in, for example, an indoor scene with incandescent lighting will to some degree be rendered as yellow-orange rather than gray in order to preserve the atmosphere of the scene. Similarly, the same objects might be rendered with a bluish cast in outdoor shadow scenes. Therefore, care should be taken in white balance measurements and much consideration must be made with regard to the light source used in the test setup. Keeping these complexities in mind, a general white balance metric can be designed that measures the CIELAB values of spectrally neutral gray patches in a test chart, for example, an X-Rite ColorChecker, and compares against some aim value. Since the lightness should remain constant, it is possible to use a simpler ΔE metric, where the lightness difference is omitted:

$$\Delta C_{wb} = \sqrt{(a^* - a_0^*)^2 + (b^* - b_0^*)^2} \tag{7.40}$$

In the simplest case, where completely neutral gray is desired, the reference values would simply be given the values $a_0^* = 0$ and $b_0^* = 0$. To extend this metric for other situations, it should be possible to perform a subjective study aiming at finding the optimal values of a_0^* and b_0^* for a range of commonly used illuminants.

7.6 Other Metrics

The set of metrics described in this chapter is not complete for obtaining a full benchmark. Visual correlation for optical distortion, chromatic aberrations, shading, etc. are missing since too few subjective experiments have been performed with regard to these attributes. It is, however, comparatively straightforward to set up and perform an experiment to obtain perceptual correlation for the remaining attributes. By using the softcopy ruler described in Chapter 5, it is relatively easy to construct custom objective metrics that produce JNDs of quality.

One such experiment investigating the subjective impression of exposure quality has been made by He *et al.* (2016). In this case, several images of a variety of different scenes were captured, all with varying exposure. The objective metric in this case was the normalized measured value in the green channel of one of the gray patches of the 24 patch X-Rite ColorChecker chart. Through subjective experiments using the softcopy image quality ruler (see Chapter 5), a relationship between the objective metric and subjective image quality in JNDs could be established as

$$Q = 250 \left[1 - \exp\left(-0.218|G_n - 0.537|^{1.74}\right)\right] \tag{7.41}$$

where Q is the quality loss in JNDs, and G_n the normalized green channel pixel value. Observe that this function yields a U-shaped curve with its minimum at $G_n = 0.537$. As expected for an exposure metric, there is an optimum exposure yielding the best image quality that is neither too dark or too bright.

7.7 Combination of Metrics

In order to construct a comprehensive general image quality metric out of a set of individual measurements on simple attributes, a combination scheme of some kind needs to be employed. The simplest such scheme might be to calculate the average of the JND degradations of the individual attributes measured using the methods described above. However, in so doing, one will most certainly discover relatively quickly that a set of "average" results from a measurement on one particular camera may yield the same end result as measurements on another camera that gives excellent results on all metrics apart from one or two that on the other hand are extremely poor. On visual inspection of images produced by the two cameras, it will be immediately apparent that the perceptual quality of the images produced by the latter camera is considerably worse compared with the former. For instance, consider the case with the former camera producing images with average sharpness, signal to noise ratio, color, etc., while the latter camera gives excellent results for all metrics except color, where a greenish cast is present in all images. For an observer, such an error will dominate the impression of image quality to the extent that any other smaller errors in the other metrics will not be important. In order to capture this important fact in a combination metric, one may employ a Minkowski metric (Keelan, 2002; Wang and Bovik, 2006), as described in Section 5.6.5.

Using this formalism, it is important that the individual metrics produce values with the same units, for example, JND, otherwise combination metrics like the above will not provide relevant results.

7.8 Full-Reference Digital Video Quality Metrics

Contributed by Hugh Denman

As described in Chapter 6, the difficulty of standardizing test scenes for video has meant that thus far there are no well-established metrics for assessing the performance of a video camera in terms of perceptual quality. However, the video compression codec, described in Chapter 4, is a crucial component of any video camera, and considerable research has been done into metrics for quality assessment for video compression. These metrics are full-reference, digital video quality metrics, that is, the digital images presented as input frames to the codec are assumed to be available for direct, numerical comparison to the output frames.

Numerous subjective studies have been undertaken to determine the perceptual validity of these metrics, some of which incorporate HVS modeling. These metrics measure the quality of a particular rendition of a source video sequence, and therefore the desideratum of perceptual validity is that for all renditions having the same metric value, the average observer would subjectively assess their quality, relative to the source, as equivalent. This is very difficult to attain, due to the complexities of video perception, outlined in Chapter 2. Nevertheless, metrics that do not have this quality in theory do find use in practice. It is important to note that these metrics quantify the quality of a particular video relative to the source, pre-compression. While codecs can be compared using metric scores obtained on their renditions of identical input videos, this is only an indirect comparison of the codecs, and does not afford a true comparative characterization. In particular, the fact that one codec is outperforming another on one particular source video, whether in subjective assessment or by some metric, is not necessarily predictive of superior performance of that codec over the other, over all video sequences.

The most significant of these video metrics are discussed below. In this discussion, intensity-valued (i.e., gray) images are assumed for simplicity, $\hat{I}_n(i,j)$ denotes the intensity value at pixel i, j of input frame n, and the corresponding value in the output, that is, after compression by the codec and decompression by some suitable decoder, is denoted $I_n(i,j)$.

7.8.1 PSNR

The most commonly encountered metric for evaluating video encoding quality is the *Peak Signal to Noise Ratio* (PSNR). This metric is based on quantification of the error, in the sense of numerical discrepancy, introduced to a signal by some processing, for example compression and decompression. Consider an intensity-valued image $I(i,j)$ of dimension $M \times N$, which is a rendition of some original image $\hat{I}(i,j)$. Subtraction of these images yields an *error image*, $E(i,j)$:

$$E(i,j) = I(i,j) - \hat{I}(i,j) \tag{7.42}$$

For the purposes of PSNR computation, this error image is summarized by the mean squared error (MSE), computed using:

$$MSE(I) = \frac{1}{MN} \sum_{i=0}^{M-1} \sum_{j=0}^{N-1} E(i,j)^2 \tag{7.43}$$

The purpose of the summation is to quantify the overall magnitude of the error. The elements of the error image are squared in the summation because the values at different sites may be positive or negative, and if these values were summed without squaring, the positive and negative values would partially cancel each other out.[4] The sum of squared signal values is an estimate of the *power* of a signal. If the mean of the error image is 0, the power is equal to the variance of the error image. An important special case is when the errors are due to additive, zero-mean, white Gaussian noise, in which case the MSE is an estimate of the Gaussian noise distribution. In perceptual terms, the squaring is equivalent to applying a higher weight to higher error values—see also the discussion about the Minkowski metric in Section 5.6.5. For example, a single site having an error value of 10 contributes 10 times more to the overall MSE than 10 sites each with error values of 1.

The PSNR is just a re-expression of the MSE, relative to the peak attainable signal power. For 8-bit digital images, the peak signal value is 255, and the PSNR is calculated as follows:

$$PSNR(I) = 10\log_{10}\frac{255^2}{MSE(I)} \tag{7.44}$$

MSE constitutes an error metric, in that larger values indicate a higher magnitude of error. PSNR converts the MSE to a fidelity metric, wherein larger values indicate a better reproduction of the input image. PSNR expresses the ratio of the peak signal power to

---

4 Summing the absolute value of the signal would resolve this problem, but results in a mathematically awkward entity, as the absolute value is not continuously differentiable.

the power of the error (termed noise) in decibels (dB), as the ratio itself has a very large dynamic range, from a theoretical minimum of 1, up to arbitrarily large values for small MSEs, for example, 260 000 for an MSE of 0.25. The PSNR is described as infinite if the output signal is entirely error free.

PSNR is very widely used in video coding. Despite its well known limitations and theoretical unsuitability with respect to video coding applications, it is a useful metric within a well-constrained context, and is popular as it can be computed very cheaply and because most practitioners have considerable experience with it.

The limitations of PSNR are due to its simplicity. As is clear from the description of the implementation above, all that is measured is the power of the error in a signal transmission. This is appropriate for the original use case of PSNR, which is to characterize the quality of a transmission channel by means of measuring the error in a transmitted signal, rather than to directly assess the quality of the transmitted signal itself. If the error process is one of additive noise with some assumed distribution, as in transmission of electrical signals over wire, then PSNR gives a stable assessment of the quality of the channel in terms of the magnitude of the error process. All signals passing through the channel will be subject to the same noise and will result in the same PSNR.

In the case of video processing and video compression, the error process is not additive noise and the signal processing is highly nonlinear, and so the degradation is signal dependent. Thus, PSNR cannot be used to characterize the quality of a compression "channel" in general, it can only describe the quality degradation suffered by a particular example.

Even when understood as a measure of received signal quality, rather than channel quality, there is little reason to expect that PSNR values should correspond to perceptual quality for video compression. There is clearly no explicit model of human vision or quality at work within the metric. The only implicit quality model in play is that many small-magnitude errors are preferable to fewer large-magnitude errors, brought into consideration by the use of the squared error measure.

As PSNR compares the actual, measured power to the theoretical peak signal power, it does not describe the effect of a given level of noise or other error on the particular signal measured, in terms of the actual power or dynamic range of this signal. The information content, in the sense of the significant structure, of a low power signal may be obliterated by noise at a given level, while the same noise would not affect the intelligibility of a higher power signal. In the case of video assessment, for example, 8-bit images degraded by a noise process introducing intensity deviations with variance $\sigma^2 = 64$ will be assessed at a fixed PSNR of 30 dB, but the quality of the resulting image will depend on the dynamic range of the original image content, as low-contrast text or texture may be obliterated by this noise, while higher contrast content will remain easily discernible.

Furthermore, as PSNR summarizes the error image $E(i, j)$ by summing the squares of its individual values, it can take no account of the visibility of the errors due to structure or correlations in the error process. In the case of non-white (correlated) noise, characterization of noise purely through its power is inadequate for assessing the quality impact of that noise on an image or video, because the spatial frequency of the noise plays a significant role, as described in Section 7.4.3 on visual noise above. More generally, any structure in the error image has a strong effect on the visibility of the errors and thus the validity of PSNR as a quality metric. For example, consider an image of dimensions 1920×1080 pixels with a mean intensity of 128. If this image is corrupted with additive white Gaussian noise with a standard deviation $\sigma = 2.5$, the expected absolute error, given by $\sigma\sqrt{\frac{2}{\pi}}$, is 2.0, yielding an MSE of 4.0, and so the PSNR is 42 dB. Such a low level of noise would be hard to perceive, and this is reflected in the high PSNR value. If the same image were corrupted by a drop-out error, whereby a 180×180 pixel block of data was set to zero, the PSNR would be the same (as $180 \times 180 \times 128 = 1920 \times 1080 \times 2$), but the visibility of the error would be very much greater.

As well as failing to account for the visibility of the errors in themselves, the PSNR metric does not account for the visibility of the errors relative to the particular scene content. For example, noise at a certain power may be highly visible in flat regions of a video frame, but imperceptible in textured regions, depending on the spatial frequency power spectrum of the noise relative to that of the texture. Thus again, images having the same PSNR may have very different perceptual quality, depending in this case on the scene content.

Both of these factors in the signal degradation—the spatial frequency spectrum of the noise in itself, and in relation to that of the scene—require a model of the HVS to account for their impact on the image quality and thus obtain a quality assessment more closely correlated to subjective impressions. This is the approach taken by more sophisticated metrics, considered below.

It is also clear that the PSNR metric contains no temporal component since each frame of a processed video is assessed individually, and typically the average of the PSNRs for each frame is presented as the metric for the video sequence as a whole. As discussed in Chapter 5, temporal factors and biases play an important part in quality perception for video. More sophisticated metrics employ a model of the HVS incorporating a spatio-temporal contrast sensitivity function to take these temporal factors into account. Where PSNR is being used to assess a video, it is prudent to consider the worst of the per-frame PSNRs, or another order statistic such as the first decile of the PSNR values, so that severe but temporally brief drops in PSNR are not masked by the averaging process.

These theoretical deficiencies of PSNR pertain to a general consideration of possible errors, and illustrations of the deficiencies of PSNR rely on contrived examples. In practice, PSNR remains useful as a metric for diagnostic purposes in video coding. Where a stream of "natural" (i.e., not contrived) videos is being processed by a particular codec operating with set parameters, a drop in PSNR can be taken to correspond to reduced quality, and in this context a practical threshold of acceptability or concern can be established by experience. In the simplest scenario, a drop in PSNR to a value below this empirical threshold can be taken as a cue to re-encode the video in question using a higher bitrate or more CPU-intensive parameters. More generally, the choice and interpretation of any metric should be informed by the processing in question and the nature of the artifacts expected.

The misapplication of PSNR occurs outside of this diagnostic scenario, in the use of PSNR to compare different codecs by compressing a fixed set of videos, or to make inference about the relative quality of renditions of different source videos using a fixed codec or set of codecs (Huynh-Thu and Ghanbari, 2008). As well as the deficiencies outlined above, it is important to note that modern codecs often do not seek to optimize PSNR, and may deliberately encode a source using another error metric to obtain lower artifact visibility, rather than the highest possible PSNR, at a given bitrate. In this scenario,

PSNR remains useful for the detection of unacceptable errors, but clearly is not a valid measure for the performance of the codec.

Figure 7.11 shows an example image with the effect of several degradations of different types and degrees of severity. The three types of degradation are the addition of white Gaussian noise (leftmost column), quantization of intensity values (center column), and truncation of the blockwise DCT (rightmost column). The DCT truncation process is intended to result in blocking artifacts similar to those introduced by DCT quantization in aggressive video compression, and so targets the highest spatial frequency components. For example, in Figure 7.11d, the truncation factor of 84% is applied by computing the corresponding passband, which, for the 32×32 DCT blocks used here, turns out to be 12.8×12.8 samples, as $\sqrt{1 - 0.84} = 0.4$ and $0.4 \times 32 = 12.8$. To apply this passband, the 12×12 square of lowest spatial frequency components was preserved, the frequency components adjacent to this square were attenuated by 20%, and all higher-frequency components were set to zero.

The level of each degradation was chosen so that each row of images has approximately equal PSNR. It is very clear that equal PSNR does not correspond to equal image quality, particularly in the last row. In the leftmost column of images, the noise is more visible in the bright flat areas to the top right than in the shadows or in the textured region depicting cloth at the bottom of the image. Thus, a metric intended to approximate the visual quality of an image should account for noise in such different areas differently, unlike PSNR. In Figure 7.11g, the degradation could be described as comprising two qualitative components: a severe blur, and the introduction of discontinuity artifacts organized on a grid. These discontinuity artifacts are particularly visible because of their coherent structure, and therefore, a metric intended to approximate the visual quality of an image should take particular account of the spatial organization of the error signal, again, unlike PSNR.

7.8.2 Structural Similarity (SSIM)

The Structural SIMilarity (SSIM) index is a relatively recent full-reference digital image quality metric, published in 2004 (Wang *et al.*, 2004a). It has had a very great impact in academia and is well-established in industrial contexts.

The metric assumes that at a high level, the HVS is concerned with the extraction of structure from visual stimuli. Rather than computing an error image by subtracting the processed image from the original, and then considering the visibility of this error image in terms of its inherent structure and its structure relative to that of the scene, SSIM seeks to compare the structure of the original and processed images directly, and to give a high value, indicating good quality, when the structure is preserved. This notion of structure is defined by the authors of SSIM as corresponding to the image data after normalizing for luminance and contrast, that is, mean intensity and variance of intensity. Preservation of structure then corresponds to the possibility of straightforwardly reconstructing the original image from the processed image.

As structure is intended to be independent of luminance and contrast, and luminance and contrast vary over the image, the SSIM index is computed for each pixel site by comparison of a window, or patch, centered on that pixel site in each of the two images. The resulting set of values is the *SSIM index map*. Assume that the two corresponding patches in the source and processed image are denoted $\mathbf{x}$ and $\mathbf{y}$, respectively. SSIM is

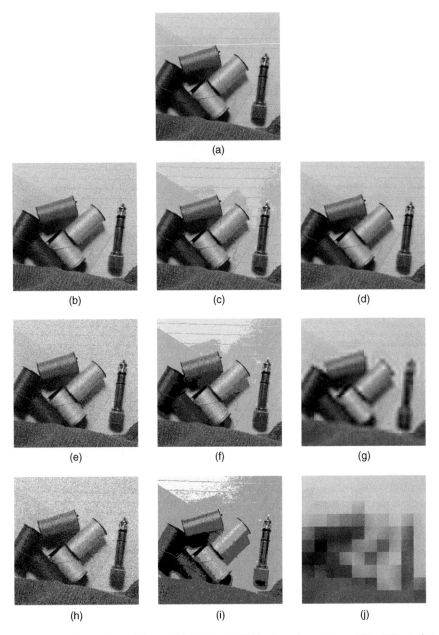

Figure 7.11 Frame degradations, with PSNR and SSIM values, for an image 320 × 320 pixels in size. The degradation processes, from left to right, are additive white Gaussian noise (AWGN), quantization of intensity values, and truncation of the 2D DCT for 32 × 32 pixel blocks. (a) Uncorrupted image. (b) AWGN, $\sigma = 8$. PSNR = 30 dB; SSIM = 0.696. (c) 11 intensity levels. PSNR = 30 dB; SSIM = 0.846. (d) 84% DCT truncation. PSNR = 30 dB; SSIM = 0.831. (e) AWGN, $\sigma = 16$. PSNR = 24 dB; SSIM = 0.453. (f) 7 intensity levels. PSNR = 25 dB; SSIM = 0.737. (g) 98.5% DCT truncation. PSNR = 25 dB; SSIM = 0.640. (h) AWGN, $\sigma = 26$. PSNR = 20 dB; SSIM = 0.290. (i) 4 intensity levels. PSNR = 19 dB; SSIM = 0.561. (j) 99.8% DCT truncation. PSNR = 19 dB; SSIM = 0.530.

based on comparison of the luminance, contrast, and structure of these patches. The luminance comparison is formulated as

$$l(\mathbf{x}, \mathbf{y}) = \frac{2\mu_x\mu_y + C_1}{\mu_x^2 + \mu_y^2 + C_1} \tag{7.45}$$

where μ_x and μ_y are the mean intensity values of the two patches, and C_1 is a constant included for numerical stability.

The contrast comparison takes the form

$$c(\mathbf{x}, \mathbf{y}) = \frac{2\sigma_x\sigma_y + C_2}{\sigma_x^2 + \sigma_y^2 + C_2} \tag{7.46}$$

where σ_x and σ_y are the standard deviations of the intensity values in each patch. C_2 is, again, a constant included for numerical stability.

The structure comparison for the patches is

$$s(\mathbf{x}, \mathbf{y}) = \frac{\sigma_{xy} + C_3}{\sigma_x + \sigma_y + C_3} \tag{7.47}$$

Here σ_{xy} is the covariance of the patches, and C_3 is included for numerical stability, as before.

SSIM is then a weighted geometric combination of these comparative measures:

$$\text{SSIM}(\mathbf{x}, \mathbf{y}) = l(\mathbf{x}, \mathbf{y})^\alpha c(\mathbf{x}, \mathbf{y})^\beta s(\mathbf{x}, \mathbf{y})^\gamma \tag{7.48}$$

A number of parameters are involved in these formulations. While various choices of parameter values yield interesting variations in the metric, all popularly used implementations, to the authors' knowledge, use the fixed values proposed by the authors in the original paper. The combination weights are set to be equal, that is, $\alpha = \beta = \gamma = 1$, which yields an arithmetical simplification of the combined formula. The constants relating to numerical stability are required to be very small relative to the peak signal value, but the specific values are not critical. In the original paper, the authors propose determining these values using $C_1 = (K_1 L)^2$, $C_2 = (K_2 L)^2$, and $C_3 = C_2/2$, where L is the peak signal value (255 for 8-bit images), $K_1 = 0.01$ and $K_2 = 0.03$, and C_3 is determined as a function of C_2 to enable a further simplification of the final, combined formula.

Another crucial parameter is the configuration of the window. While a square, equally weighted window may be used, this can yield a blocky, discontinuous SSIM index map. To avoid this, the authors recommend the use of a Gaussian window function, specifically an 11×11 tap, circularly symmetric Gaussian with a standard deviation of 1.5 samples.

The SSIM index may be computed at every pixel site, or at some subset of sites, for example at a coarsely sampled grid or with a variable sample density corresponding to some putative measure of interest or saliency. In any event, the estimates must be pooled to form an overall score for the quality of the image as a whole. This is achieved by computing the mean SSIM, or M-SSIM, over all the patches in the image pair. In one of the earliest descriptions of an application of SSIM to video as video, rather than to video as a collection of still images, the authors propose applying a lower weight to frames undergoing large motion, as well as applying a lower weight to darker regions of each frame, on the basis that defects are less visible in both these contexts (Wang et al., 2004b). More recent work has shown that the correspondence of SSIM scores to subjective assessments can be improved by determining weights using a fixation model, to predict the areas of an image to which the eye is drawn, and by applying order statistics to give higher significance to poor-quality areas in a frame (Moorthy and Bovik, 2009).

Conceptually, the comparisons of luminance, contrast, and structure making up SSIM are cascaded, such that each component is independent. Specifically, the contrast between two patches is compared after subtraction of the means, so that both patches have a mean of zero in the contrast comparison, and the structure is compared after dividing by the contrast, so that both patches have unit variance. Each of the comparison functions is designed to satisfy the three desiderata of symmetry, boundedness (having a defined maximum value), and of attaining this maximum if and only if the patches being compared are identical. The geometric combination ensures that the overall SSIM index also meets these criteria.

The comparison functions also incorporate some perceptual considerations. The luminance comparison function is sensitive to relative luminance change, rather than absolute change, in accordance with Weber's law (see Chapter 5). The contrast comparison function becomes less sensitive as the base contrast (i.e., that of the reference patch) increases, which is consistent with contrast masking behavior of the HVS. The structural comparison function is the correlation between the windows, which corresponds in some intuitive sense to the preservation of information between the two windows.

This metric seeks to address the principal deficiencies of PSNR, but without incorporating an explicit model of the HVS. Metrics incorporating such a model, such as the VDP metric described below, typically operate in a "bottom-up" fashion, that is, starting with a contrast sensitivity function and incorporating approximations for psychophysical phenomena such as contrast masking. The authors of SSIM suggest that these models involve simplified approximations to the operation of the HVS, specifically in that such models rely on linear or nearly-linear operations derived from psychophysical experiments using simple stimuli, such as sinusoidal gratings, and calibrated according to experiments establishing the threshold of perception. As the HVS involves highly non-linear processing, and real-world visual stimuli are complex and depend on suprathreshold perception, the authors argue that in some sense a "higher-level" of HVS function should be targeted, and that extraction of structure, or information, from an image, is such a high level function and thus a suitable basis for image quality assessment.

As SSIM does not incorporate any reference to the CSF, it cannot factor viewing distance into account. This suggests that comparisons of image quality on the basis of SSIM must be made with the assumption that the source images are of comparable resolution and that the viewing distance is constant. A multi-scale application of SSIM has been proposed to account for viewing distance (Wang *et al.*, 2003). In this approach, the contrast and structural comparison functions are applied at multiple scales, where each scale is obtained by the repeated application of lowpass filtering and downsampling operations. The luminance comparison is applied only at the coarsest scale. The ratings thus obtained are then combined, with a different weighting applied to each scale. The weights are calibrated for a particular viewing distance, and differences in viewing distance may be accounted for by varying the set of weights used accordingly.

Any application of a spatial degradation metric, such as SSIM, to video, relies on the pooling of ratings derived for each frame, and therefore the temporal characteristics of any degradation are lost. For example, consider a case of intensity flicker, in which the intensity of each frame of the processed video is offset by +10 levels in even-numbered frames and -10 levels in odd-numbered frames, contrasted with a fixed intensity shift of +10 levels affecting every frame of the processed video. The flicker will have a much greater impact on perceived quality than the fixed offset, but any quality assessment based on pooling of spatial metrics will rate these degradations as equal.

There is a variant of SSIM that incorporates temporal analysis to address this deficiency, namely "spatio-temporal video SSIM" (Moorthy and Bovik, 2010). This involves an extension to spatial SSIM to consider three-dimensional volumes of pixels, rather than two-dimensional windows. Motion estimation is used to obtain a weight applied in the pooling of the metric values for each level.

7.8.3   VQM

The National Telecommunications and Information Administration (NTIA) General Model for estimating video quality, along with its associated calibration techniques, is generally known as the Video Quality Model, or VQM (Pinson and Wolf, 2004). It is a video quality metric with some similarities in approach to SSIM, in that it is based on feature extraction and comparison without the use of an explicit HVS model. However VQM does involve more explicit emulation of features of the HVS than SSIM, as described below. Due to the strong performance of this metric in comparative tests (organized by the Video Quality Experts Group, described below), it was adopted in ANSI standard ANSI-T1.801.03 in 2003, as well as in the ITU-T Recommendation J.144 (International Telecommunication Union, 2004). Although standardized as a full-reference metric, it is actually a reduced-reference metric.

VQM proceeds in four phases. First, a calibration step is applied to normalize for modifications to the processed video that do not affect the subjective quality. Then *quality features* are extracted from corresponding spatio-temporal blocks of data in the reference and processed videos. Next, the quality features for corresponding blocks are compared, yielding *quality parameters* describing the difference in quality between the processed and reference blocks. Lastly, the quality parameters are pooled, or "collapsed" in VQM terminology, to yield an overall rating.

The calibration step aligns the processed video with the reference, correcting for any spatial, temporal, or intensity shift which may have been introduced in processing. The calibration seeks to estimate and correct only shifts that are fixed over the entire sequence. Variable shifts are not corrected as these certainly do degrade the perceived quality of the processed video.

This step also computes differences in signal gain or amplitude (image contrast). Any offset or change in contrast in the luminance channel is compensated to reconcile the processed sequence with the original, but any offset or change in gain in the chroma channels is considered a degradation and factored in to the quality assessment.

The calibration step also incorporates detection of the "processed valid region," or PVR, which is the region excluding any non-picture content at the edges of the frame corresponding to overscan or letterboxing. These edge regions are not included in the quality assessment.

The feature extraction step involves the division of the processed and reference video into spatio-temporal blocks. As the calibration step has aligned the processed and source videos, each pair of corresponding source and processed video blocks is expected to contain the same scene content. Different, putatively optimal block sizes are used for each quality feature. The quality features are computed by taking either the mean or the standard deviation of the data within each block, after the application of some perceptual filter.

The quality parameters are extracted by the comparison of features, using functions intended to replicate masking effects of the HVS. In particular, features are most often compared using either ratio comparison

$$p = \frac{f_p - f_o}{f_o} \tag{7.49}$$

or a logarithmic comparison

$$p = \log_{10}\left(\frac{f_p}{f_o}\right) \tag{7.50}$$

where p is the quality parameter, f_o is a feature value extracted from a block in the original video, and f_p is the corresponding feature value extracted from the corresponding block in the processed video. In both of these schemes, the quality parameter computed for a given level of feature discrepancy will be lower for higher feature values in the original video. The features typically represent spatial or temporal activity within a block, and so these comparison functions mimic, to some degree, the spatial and temporal masking behavior of the HVS.

While a feature comparison may indicate an increase or decrease of the feature level in the processed video relative to the original, within VQM an extracted quality parameter is always restricted to the description of either an increase in feature level or a decrease in feature level. This is motivated by the requirement of applying a different weighting to increases in the feature level than to decreases, as opposed to applying a single weighting to the change in feature level irrespective of its sign, in the computation of the final quality score. Where the feature corresponds to edge activity, for example, the addition of edges such as those corresponding to blocking artifacts, has a greater impact on the perceived quality than the loss of edges, for example in the case of blur, and so increases in the feature level should be given a greater weight than decreases.

VQM uses six features, from which are extracted seven quality parameters. Four of the features used are variants on the standard deviation of the edge-enhanced luminance data within each spatiotemporal block. The edge enhancement is performed by the application of a filter designed to have a peak response at a spatial frequency of 4.5 cycles per degree, when standard-definition video is viewed at a distance of 6 picture heights. Thus, this metric does include some consideration of the HVS CSF and the expected or presumed viewing conditions. The four variants involve the detection of (1) any loss of edge content in the horizontal and vertical directions combined, (2) any gain of such edge content, (3) any increase in diagonal edge content relative to horizontal and vertical edge content, and (4) any increase in horizontal and vertical edge content relative to the diagonal edge content. These features are used in the calculation of quality parameters identifying blur, blocking artifacts, and any quality improvement introduced by sharpening.

The color content of the video is characterized by one feature, the mean of each color channel in each spatiotemporal block. Two quality parameters are extracted using this feature, one targeting color shifts sustained over a long duration, and the other identifying sudden, large-magnitude color errors affecting a block. These parameters analyze the two chroma channels in combination, rather than separately. The perceptual filter consists of a higher weighting being applied to the red-cyan chromaticity (the Cr channel), in comparison to the blue-yellow chromaticity (the Cb channel).

The final quality feature is a product of the contrast (standard deviation of intensity values) and motion level (sum of the frame-to-frame absolute intensity differences) within a spatiotemporal block. Comparison of this feature in the source and processed blocks yields a quality parameter identifying artifacts affecting edges in motion.

The pooling of quality parameters is in most cases based on order statistics, for example finding the average value of the worst 5% of the values within each block and then taking the tenth percentile value of these values.

The final quality rating is then a weighted linear combination of the quality parameters after pooling. VQM values are clipped or crushed so as to range from zero, indicating perfect fidelity, to a nominal maximum of one, indicating severe degradation. The value may exceed one in cases of extremely high degradation.

7.8.4 VDP

The visible differences predictor, or VDP metric, belongs to a class of metrics categorically different from PSNR, SSIM, and VQM. PSNR assesses fidelity on the basis of the digital data as a vector of values, with no incorporation of their constituting a two-dimensional image. SSIM and VQM do assess the data as two dimensional images, and include elements inspired by or related to the HVS. VDP, by contrast, is among those metrics that include an explicit model of elements of the HVS, and apply this model to the image data in order to determine the visibility of artifacts to a human observer. VDP is among the earliest published of such metrics, published by Scott Daly in 1992 (Daly, 1992). A comprehensive description with additional explanatory text was published as a book chapter the following year (Daly, 1993).

VDP was designed to overcome the limitations of earlier quality metrics, which relied on characterizing the system MTF and noise power spectrum. These approaches imposed the assumption of linearity on the model of the system behavior, and as such they were more suited to quality assessment of analog systems than digital systems. Furthermore, these spectral approaches could not capture phase-dependent effects, and therefore could not model the masking effects of the HVS. Because VDP assesses the digital image data itself, rather than the transfer function of the system, it can incorporate non-linear processing and phase effects.

At the core of the VDP metric is the HVS model. The reference image and an image to be assessed are passed through this model, and the output is a two-dimensional map containing the (putative) probability that a difference will be visible, at each pixel site. There are a number of calibration parameters, pertaining to the display conditions, which must be established before the HVS model can be applied. These include the display size and pixel spacing as well as the viewing distance. The electro-optical transfer function of the display is also required. The inputs to the VDP are not two-dimensional arrays of intensity values, but rather two-dimensional arrays describing the luminance value, in cd/m^2, of each pixel.

The HVS model is focused on lower-level vision processes, specifically those that limit visual sensitivity. Sensitivity is defined in this context as the inverse of the minimum visible (threshold) contrast, and is similar to system gain, albeit with nonlinear aspects. Thus, this metric is oriented toward the detection and demotion of artifacts to which the HVS is not sensitive and which therefore will not influence the perceived quality.

Figure 7.12 presents an overview of the VDP algorithm. In this figure, $I(i,j)$ is the reference, or original, image, and $I'(i,j)$ is the received image to be assessed. Each of these images is passed through the early vision stages (shown outlined in red) and the cortex filtering stages (shown outlined in blue—the diagram omits the cortex filtering of the received image). The early vision stages include an amplitude non-linearity filter (labeled "ANL"), which generates the retinal response images $R(i,j)$ and $R'(i,j)$, and a CSF model. The cortex filtering generates 31 *cortex band images*, selecting different combinations of spatial frequency and orientation. These are labeled $B^{k,l}(i,j)$ for those derived from the reference image and $B'^{k,l}(i,j)$ for those derived from the received image.

Each pair of cortex band images, $B^{k,l}(i,j)$ and $B'^{k,l}(i,j)$, is compared in terms of contrast difference (outlined in magenta) and in terms of *masking* activity (outlined in cyan). For the contrast comparison, each member of the pair is converted to a contrast image, $C^{k,l}(i,j)$ via a global contrast conversion filter, labeled "CNT." These are then differenced using subtraction (the filter labeled "DIFF") to generate the contrast difference image, $\Delta C^{k,l}(i,j)$. The masking comparison uses the masking function, "MSK," to generate a threshold elevation image $T_e^{k,l}(i,j)$ for each image in the pair. These are combined by taking the minimum threshold elevation value at each pixel site (the filter labeled "MIN") to generate a final masking image $\hat{T}_e^{k,l}(i,j)$.

The final stage, outlined in green in the diagram, is the computation of the probability that the difference between the reference and received images will be detected by an observer, for each cortex band, at each pixel site. This probability is computed using the psychometric function, represented by Ψ, which takes the contrast difference and threshold elevation images as inputs, and is labeled $P^{k,l}(i,j)$. These probability maps are then combined by the probability summation operation Π to obtain a map of the overall probability of detection at each site, represented by $VDP(i,j)$.

7.8.4.1 Further Considerations

The VDP metric results in a map describing the probability, at each pixel site, that any discrepancy between reference and processed versions of an image will be detected. Its strength is in modeling the interaction of near-threshold errors with suprathreshold image content. However, all suprathreshold errors result in a VDP value (probability of detection) of 1.0, and thus the VDP does not describe the relative severity of such errors. It is in this sense that VDP is, as described, a visual fidelity metric rather than a visual quality metric.

The VDP is not readily reducible to a single value describing the overall fidelity of the image. One approach, described by Daly, is taking the maximum value of the VDP, the *peak probability of detection*; if this value is less than 0.5, the reference and received image may be assumed to be visually equivalent.

Modifications to the VDP to enable assessment of high dynamic range (HDR) images have been developed (Mantiuk *et al.*, 2004, 2005; Narwaria *et al.*, 2015). These

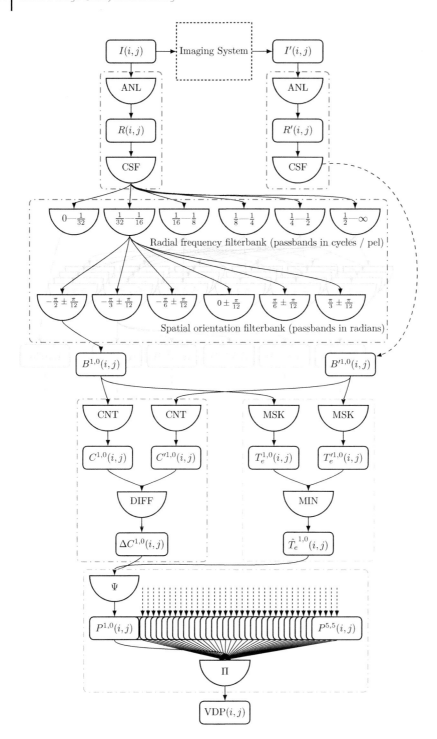

Figure 7.12 An overview of the VDP algorithm. Rectangles represent images, and semicircles represent filtering operations. See the main text for details.

modifications chiefly target the retinal response and CSF aspects, accounting for local variations in contrast sensitivity and the combination of scotopic and photopic vision at play in the perception of these images.

7.8.5 Discussion

The set of metrics described in this section broadly covers the available categories of full-reference digital image metrics. Considering that a digital image is a set of numerical values, destined for presentation as a two-dimensional array on a display device, to be viewed by a human observer, the metrics presented are in an order such that each aims to characterize progressively more of the elements in this chain, and in consequence range in complexity from extremely simple (PSNR) to highly complex (VDP).

Numerous other quality metrics have been developed, but most of these are refinements of the ideas contained in the survey here. The Sarnoff JND Vision model is notable, which is a CSF-based metric incorporating explicit analogs of HVS processes, similar to VDP (Lubin, 1998). The Sarnoff model uses spatial domain, wavelet based processing rather than the spatial frequency-domain processing used by VDP, which reduces the computational complexity at the cost of increased working memory requirements. As an approach to vision modeling with application to image quality, the Sarnoff model has advantages and disadvantages relative to the VDP (Li *et al.*, 1998).

Another important metric is the MOtion-based Video Integrity Evaluation (MOVIE) Index, presented by Seshadrinathan and Bovik (Seshadrinathan and Bovik, 2010). This is essentially a feature-based metric similar to SSIM and VQM, but recalling the VDP in its use of a frequency-domain decomposition of the input. The MOVIE metric aims to more explicitly account for temporal artifacts, and the effects of motion on quality perception, than preceding metrics, and therefore uses a bank of spatiotemporal Gabor filters, over multiple scales, to obtain this decomposition. Spatial and temporal artifacts are assessed separately at each pixel site, each pooled over each frame and then over the video as a whole, and then combined to obtain a single overall score.

The S-CIELAB metric, described above, is also interesting as an example of a quality metric that, while originally intended specifically for color quality assessment, can be applied to general image and video quality, though there has not been a great deal of research of this application (Tong *et al.*, 1999; Fonseca and Ramirez, 2008).

The model developed by Winkler (2000) is also worth mentioning as an example of an HVS-based model that also incorporates color.

Objective comparisons of video quality metrics are periodically performed by the Video Quality Experts Group (VQEG) (Brunnström *et al.*, 2009; Huynh-Thu *et al.*, 2015). Proposers of a metric submit it for evaluation, and VQEG prepares a set of degraded video sequences, conducts subjective experiments to establish a difference mean opinion (DMOS) score for each degraded sequence, and then correlates these DMOS scores with those obtained from the experiments. VQEG also publishes the set of evaluation videos and the subjective scores, when this is licensed by the providers of the video sequences, and so other researchers can evaluate their metrics against the VQEG data. The various reports describing full-reference metrics for standard definition television (Video Quality Experts Group (VQEG), 2000, 2003) and HDTV (Video Quality Experts Group (VQEG), 2010) give excellent accounts of the assessment and performance of various metrics. It is notable that in the early stages of VQEG metric development, PSNR was found to be equivalent to the proposed metrics in several tests; this is testament to the difficulty of developing visual metrics that are robust and general with respect to the range of distortions assessed (Rohaly *et al.*, 2000).

As well as assessing full-reference metrics, the VQEG effort accepts reduced-reference (RR) and no-reference (NR) metrics for assessment. No-reference metrics are potentially of particular interest to the camera designer as they do not require a digital input image and therefore can be used to assess the camera performance as a whole, rather than just the codec in isolation. However, the difficulty of developing a robust, practical, no-reference metric is perhaps evident in the fact that in each of the VQEG assessments conducted to date, any NR metrics submitted were withdrawn by their proponents before the assessment was completed.

While the HVS is a logical starting point for the development of a video quality metric, most recent research into metrics has been primarily concerned with the severity of transmission errors rather than with the characteristics of human vision. This dichotomy and trend was recognized by Winkler and Mohandas (2008), who named the two approaches the *vision modeling approach* and the *engineering approach*. Part of the reason for this is the high computational cost of vision modeling approaches, relative to the engineering metrics. As more results and datasets from subjective evaluation studies become available, engineering metrics continue to improve in terms of correlation with subjective assessments, while remaining computationally relatively cheap (Zhang and Bull, 2016). Indeed, there are now sufficient assessment data to deploy machine learning to quality assessment, potentially bypassing the need for vision modeling altogether.

Another reason for the prevalence of engineering metrics is that vision modeling metrics must take into account the properties of the expected viewing scenario, such as the electro-optical transfer function, display conditions, and the viewing distance. Often these cannot be predicted for a consumer camera, because of both the diversity of use cases prevailing and the speed with which new displays and consumer contexts come into consideration.

On the other hand, the vision modeling approach does provide the most comprehensive framework for assessment of video in circumstances where the viewing conditions, camera, and display are all part of the system being developed, for example medical imaging. Furthermore, such metrics can be updated with more recent or more comprehensive models containing aspects of the vision system as technology evolves, as in the case of the HDR VDP mentioned above. An engineering or machine learning metric, by contrast, may be tuned or calibrated to the resolutions and displays common at the time of deployment, but it will be difficult to predict how new technologies will influence such metrics as their properties (e.g., the dynamic range of the display) are not explicitly modeled.

In choosing a metric for video camera applications, it is important to note that correlation with subjective scores, for example as reported by VQEG, is not necessarily in itself the sole criterion for selection. For example, a small, fixed spatial offset applied to a video does not greatly affect the subjective ratings of the video quality, and the VQM metric devotes considerable computation to shift-invariance so as to replicate this aspect of subjective assessment. However, in a camera codec system, such a shift does indicate a problem, and the shift-invariance computations of VQM should be disabled for this scenario.

Summary of this Chapter

- For color vision, three types of light sensitive elements, cones, are found in the retina: long, medium, and short.
- It is possible to perform reasonably accurate linear transformations between the CIE standard observer and the spectral sensitivities of the cones. One well-known such transformation is the Hunt–Pointer–Estevez approximation.
- The trichromatic theory of human color vision is complemented by an opponent representation, with one luminance channel and two chrominance channels for red-green and blue-yellow color vision.
- The sensitivity to detail in luminance and chrominance is manifested in the contrast sensitivity functions, which are separate for the luminance and chrominance channels.
- Due to the fact that the contrast sensitivity functions are expressed as functions of angular frequency, that is, cycles per degree, the perception of detail will vary as a function of viewing distance and size of images. Therefore, a set of viewing conditions must be specified for perceptually correlated image quality metrics.
- Spatial image quality metrics include sharpness, visual noise, and texture blur.
- Color appearance models may be used to obtain perceptually correlated values for hue, brightness, colorfulness, lightness, chroma, and saturation.
- The color appearance model that is presently recommended by CIE is called CIECAM02.
- Color appearance models may be extended by using image appearance modeling, which also incorporates spatial aspects of color imaging. Examples are S-CIELAB and iCAM.
- Perceptually correlated image quality metrics obtained using the methods in this chapter are expressed in units of Just Noticeable Differences (JNDs).
- In order to obtain a comprehensive metric of image quality, individual metrics have to be combined. The proposed way of combining metrics is through a Minkowski sum.
- For video quality assessment, standardized, well-established metrics are scarce. However, for video codec evaluation, a wide variety of full reference image quality metrics exist. The most widespread is the PSNR metric.
- The SSIM metric was designed to account for some of the deficiencies of the PSNR metric by taking into account the structural similarity between reference and test image. This metric has become quite widespread in industry and academia.
- Other notable full-reference metrics for video quality assessment are VQM and VDP.
- The Video Quality Experts Group (VQEG) performs comparisons and develops objective video quality metrics. The results from their studies frequently become used in standards.

References

Alakarhu, J. (2007) Image sensors and image quality in mobile phones, in *International Image Sensor Workshop*, Ogunquit, Maine, USA.

Barten, P.G.J. (1989) The square root integral (SQRI): A new metric to describe the effect of various display parameters on perceived image quality. *Proc. SPIE*, **1077**, 73.

Barten, P.G.J. (1999) *Contrast Sensitivity of the Human Eye and Its Effects on Image Quality*, SPIE Press, Bellingham, WA, USA.

Bartleson, C.J. (1985) Predicting graininess from granularity. *J. Photogr. Sci.*, **33**, 117–126.

Baxter, D., Cao, F., Eliasson, H., and Phillips, J.B. (2012) Development of the I3A CPIQ spatial metrics. *Proc. SPIE*, **8293**, 829302.

Baxter, D.J. and Murray, A. (2012) Calibration and adaptation of ISO visual noise for I3A's camera phone image quality initiative. *Proc. SPIE*, **8293**, 829303.

Bonnier, N. and Lindner, A.J. (2010) Measurement and compensation of printer modulation transfer function. *J. Electron. Imaging*, **19**, 011010.

Brunnström, K., Hands, D., Speranza, F., and Webster, A. (2009) VQEG validation and ITU standardization of objective perceptual video quality metrics [standards in a nutshell]. *IEEE Signal Process. Mag.*, **26**, 96–101.

Campbell, F.W. and Robson, J.G. (1968) Application of Fourier analysis to the visibility of gratings. *J. Physiol.-London*, **197**, 551–566.

Daly, S.J. (1992) Visible differences predictor: an algorithm for the assessment of image fidelity. *Proc. SPIE*, **1666**, 2–15.

Daly, S.J. (1993) The visible differences predictor: an algorithm for the assessment of image fidelity, in *Digital Images and Human Vision* (ed. A.B. Watson), MIT Press, Cambridge, MA, USA, chap. 14, pp. 179–206.

Ebner, F. and Fairchild, M.D. (1998) Development and testing of a color space (IPT) with improved hue uniformity. *IS&T/SID Sixth Color Imaging Conference: Color Science, Systems and Applications*, pp. 8–13.

Fairchild, M.D. (2013) *Color Appearance Models*, John Wiley & Sons Ltd, Chichester, UK, 3rd edn.

Fairchild, M.D. and Johnson, G.M. (2002) Meet iCAM: A next-generation color appearance model, in *Proceedings of the IS&T 10th Color and Imaging Conference*, Scottsdale, AZ, USA, pp. 33–38.

Fairchild, M.D. and Johnson, G.M. (2004) iCAM framework for image appearance, differences, and quality. *J. Electron. Imaging*, **13**, 126–138.

Farrell, J.E., Okincha, M., Parmar, M., and Wandell, B.A. (2010) Using visible SNR (vSNR) to compare the image quality of pixel binning and digital resizing. *Proc. SPIE*, **7537**, 75370C.

Fonseca, R.N. and Ramirez, M.A. (2008) Using SCIELAB for image and video quality evaluation, in *2008 IEEE International Symposium on Consumer Electronics*, pp. 1–4.

Fujii, T., Suzuki, S., and Saito, S. (2013) Noise evaluation standard of image sensor using visual spatio-temporal frequency characteristics. *Proc. SPIE*, **8667**, 86671I.

Granger, E.M. and Cupery, K.N. (1973) An optical merit function (SQF), which correlates with subjective image judgments. *Photogr. Sci. Eng.*, **16**, 221–230.

He, Z., Jin, E.W., and Ni, Y. (2016) Development of a perceptually calibrated objective metric for exposure. *Electronic Imaging: Image Quality and System Performance*, **2016**, 1–4.

Hubel, D. (1995) Eye, brain, and vision. URL http://hubel.med.harvard.edu/book/bcontex.htm, (accessed 29 May 2017).

Hunt, R.W.G. and Pointer, M.R. (2011) *Measuring Colour*, John Wiley & Sons Ltd, Chichester, UK, 4th edn.

Huynh-Thu, Q. and Ghanbari, M. (2008) Scope of validity of PSNR in image/video quality assessment. *Electron. Lett.*, **44**, 800–801.

Huynh-Thu, Q., Webster, A.A., Brunnström, K., and Pinson, M.H. (2015) VQEG: Shaping standards on video quality, in *1st International Conference on Advanced Imaging*, Tokyo, Japan.

IEEE (2017) IEEE 1858-2016, IEEE Standard for Camera Phone Image Quality. IEEE.

Imatest (2015) Imatest color saturation. URL http://www.imatest.com/docs/colorcheck_ref/, (accessed 29 May 2017).

International Telecommunication Union (2004) Recommendation J.144: Objective perceptual video quality measurement techniques for digital cable television in the presence of full reference, ITU-T Rec. J.144.

ISO (2013) ISO 15739:2013 Photography – Electronic Still Picture Imaging – Noise Measurements. ISO.

Jin, E.W., Keelan, B.W., Chen, J., Phillips, J.B., and Chen, Y. (2009) Soft-copy quality ruler method: implementation and validation. *Proc. SPIE*, **7242**, 724206.

Johnson, G.M. (2003) *Measuring Images: Differences, Quality, and Appearance*, Ph.D. thesis, Rochester Institute of Technology, USA.

Johnson, G.M. and Fairchild, M.D. (2003) A top down description of S-CIELAB and CIEDE2000. *Color Res. Appl.*, **28**, 425–435.

Johnson, G.M. and Fairchild, M.D. (2004) The effect of opponent noise on image quality. *Proc. SPIE*, **5668**, 82–89.

Keelan, B.W. (2002) *Handbook of Image Quality – Characterization and Prediction*, Marcel Dekker, New York, USA.

Keelan, B.W., Jenkin, R.B., and Jin, E.W. (2012) Quality versus color saturation and noise. *Proc. SPIE*, **8299**, 82990F.

Keelan, B.W., Jin, E.W., and Prokushkin, S. (2011) Development of a perceptually calibrated objective metric of noise. *Proc. SPIE*, **7867**, 786708.

Kelly, S.C. and Keelan, B.W. (2005) ISO 12232 revision: Determination of chrominance noise weights for noise-based ISO calculation. *Proc. SPIE*, **5668**, 139.

Koopipat, C., Tsumura, N., Fujino, M., and Miyake, Y. (2002) Effect of ink spread and optical dot gain on the MTF of ink jet image. *J. Imaging Sci. Technol.*, **46**, 321–325.

Koskinen, S., Tuulos, E., and Alakarhu, J. (2011) Color channel weights in a noise evaluation, in *International Image Sensor Workshop*, Hokkaido, Japan.

Kriss, M.A. (1998) Tradeoff between aliasing artifacts and sharpness in assessing image quality. *Proc. IS&T PICS Conference*, pp. 247–256.

Lam, K.M. (1985) *Metamerism and Colour Constancy*, Ph.D. thesis, University of Bradford, England.

Li, B., Meyer, G.W., and Klassen, R.V. (1998) Comparison of two image quality models. *Proc. SPIE*, **3299**, 98–109.

Lubin, J. (1998) A human vision system model for objective image fidelity and target detectability measurements, in *9th European Signal Processing Conference (EUSIPCO 1998)*, pp. 1–4.

Mantiuk, R., Daly, S., Myszkowski, K., and Seidel, H.P. (2005) Predicting visible differences in high dynamic range images – model and its calibration. *Proc. SPIE*, **5666**, 204–214.

Mantiuk, R., Myszkowski, K., and Seidel, H.P. (2004) Visible difference predicator for high dynamic range images, in *Proceedings of IEEE International Conference on Systems, Man and Cybernetics*, pp. 2763–2769.

Moorthy, A.K. and Bovik, A.C. (2009) Perceptually significant spatial pooling techniques for image quality assessment. *Proc. SPIE*, **7240**, 724012.

Moorthy, A.K. and Bovik, A.C. (2010) Efficient motion weighted spatio-temporal video SSIM index. *Proc. SPIE*, **7527**, 75271.

Narwaria, M., Mantiuk, M., Silva, M., and Callet, P.L. (2015) A calibrated method for objective quality prediction of high dynamic range and standard images. *J. Electron. Imaging*, **24**, 010501.

Phillips, J.B. and Christoffel, D. (2010) Validating a texture metric for camera phone images using a texture-based softcopy attribute ruler. *Proc. SPIE*, **7529**, 752904.

Phillips, J.B., Coppola, S.M., Jin, E.W., Chen, Y., Clark, J.H., and Mauer, T.A. (2009) Correlating objective and subjective evaluation of texture appearance with applications to camera phone imaging. *Proc. SPIE*, **7242**, 724207.

Pinson, M.H. and Wolf, S. (2004) A new standardized method for objectively measuring video quality. *IEEE Trans. Broadcast.*, **50**, 312–322.

Reinhard, E., Khan, E.A., Akyuz, A.O., and Johnson, G.M. (2008) *Color Imaging – Fundamentals and Applications*, A. K. Peters, Wellesley, MA, USA.

Rohaly, A.M., Corriveau, P.J., Libert, J.M., Webster, A.A., Baroncini, V., Beerends, J., Blin, J.L., Contin, L., Hamada, T., Harrison, D., Hekstra, A.P., Lubin, J., Nishida, Y., Nishihara, R., Pearson, J.C., Pessoa, A.F., Pickford, N., Schertz, A., Visca, M., Watson, A.B., and Winkler, S. (2000) Video Quality Experts Group: current results and future directions. *Proc. SPIE*, **4067**, 742–753.

Seshadrinathan, K. and Bovik, A.C. (2010) Motion tuned spatio-temporal quality assessment of natural videos. *IEEE Trans. Image Process.*, **19**, 335–350.

Tong, X., Heeger, D.J., and den Branden Lambrecht, C.J.V. (1999) Video quality evaluation using ST-CIELAB. *Proc. SPIE*, **3644**, 185–196.

Video Quality Experts Group (VQEG) (2000) Final report of the FR-TV Phase I validation test, *Tech. Rep.*, VQEG. URL http://www.its.bldrdoc.gov/vqeg/projects/frtv-phase-i/frtv-phase-i.aspx, (accessed 29 May 2017).

Video Quality Experts Group (VQEG) (2003) Final report of the FR-TV Phase II validation test, *Tech. Rep.*, VQEG. URL http://www.its.bldrdoc.gov/vqeg/projects/frtv-phase-ii/frtv-phase-ii.aspx, (accessed 29 May 2017).

Video Quality Experts Group (VQEG) (2010) HDTV phase I final report, *Tech. Rep.*, VQEG. URL http://www.its.bldrdoc.gov/vqeg/projects/hdtv/hdtv.aspx, (accessed 29 May 2017).

Wandell, B.A. (1995) *Foundations of Vision*, Sinauer Associates, Sunderland, MA, USA.

Wang, Z. and Bovik, A.C. (2006) *Modern Image Quality Assessment*, Morgan & Claypool Publishers, USA.

Wang, Z., Bovik, A.C., Sheikh, H.R., and Simoncelli, E.P. (2004a) Image quality assessment: From error visibility to structural similarity. *IEEE Trans. Image Process.*, **13**, 600–612.

Wang, Z., Lu, L., and Bovik, A.C. (2004b) Video quality assessment based on structural distortion measurement. *Signal Process. Image Commun.*, **19**, 121–132.

Wang, Z., Simoncelli, E.P., and Bovik, A.C. (2003) Multiscale structural similarity for image quality assessment, in *Signals, Systems and Computers, 2004. Conference Record of the Thirty-Seventh Asilomar Conference*, vol. **2**, vol. 2, pp. 1398–1402.

Winkler, S. (2000) *Vision models and quality metrics for image processing*, Ph.D. thesis, Ecole Polytechnique Fédérale de Lausanne, Switzerland.

Winkler, S. and Mohandas, P. (2008) The evolution of video quality measurement: From PSNR to hybrid metrics. *IEEE Trans. Broadcast.*, **54**, 660–668.

Wyszecki, G. and Stiles, W.S. (1982) *Color Science: Concepts and Methods, Quantitative Data and Formulae*, John Wiley & Sons Ltd, Chichester, UK.

Zhang, F. and Bull, D.R. (2016) A perception-based hybrid model for video quality assessment. *IEEE Trans. Circuits Syst. Video Technol.*, **26**, 1017–1028.

Zhang, X., Farrell, J.E., and Wandell, B.A. (1997) Applications of a spatial extension to CIELAB. *Proc. SPIE*, **3025**, 154.

Zhang, X. and Wandell, B.A. (1997) A spatial extension of CIELAB for digital color image reproduction. *J. Soc. Inf. Display*, **5**, 61–63.

Zhang, X. and Wandell, B.A. (1998) Color image fidelity metrics evaluated using image distortion maps. *Signal Processing*, **70**, 201–214.

8

Measurement Protocols—Building Up a Lab

The content of the book so far has been describing and defining aspects related to components of camera image quality benchmarking. In the next two chapters, we will set forth the important elements of how one might utilize this knowledge to actually obtain the objective and subjective data and subsequently combine it to generate benchmarking comparisons. Chapter 8 details lab needs and protocols; this is followed by descriptions of benchmarking systems and presentation of extensive examples of actual benchmarking images and data in Chapter 9.

When it comes to performing accurate and repeatable measurements, it is absolutely critical to establish and define the so-called *protocols*. The protocols provide a full description of the testing conditions that are required when performing image quality measurements. Important elements include the room environment, the lighting, the subject matter (e.g., charts or scenes), the camera (for objective metrics), and the observer and viewing conditions (for subjective metrics).

This chapter will successively go over the protocols to be applied for objective and then subjective measurements. We will show how there are general protocols as well as ones that are specific to individual image quality attributes or parameters being measured. Discussion will include how protocols can vary as test equipment technology, such as lighting and display, evolves.

8.1 Still Objective Measurements

A thorough description of objective metrics for image quality evaluation was provided in Chapters 6 and 7. See the list below summarizing the image quality attributes for which objective metrics were defined. For each attribute below, specific metrics may be devised, which are calculated from still images and videos. In this section, the protocols to obtain these measurements will be defined.

- Exposure and tone
- Dynamic range
- Color
- Shading
- Geometric distortion
- Stray light
- Sharpness and resolution

Camera Image Quality Benchmarking, First Edition. Jonathan B. Phillips and Henrik Eliasson.
© 2018 John Wiley & Sons Ltd. Published 2018 by John Wiley & Sons Ltd.
Companion website: www.wiley.com/go/benchak

- Texture blur
- Noise
- Color fringing
- Image defects

8.1.1 Lab Needs

Setting up an equipped lab for objective metrics is a fundamental step in the benchmarking process. If the lab is not set up properly, reliable benchmarking will be a challenge because there will be difficulties in repeating the same protocols as time passes. For example, a set of cameras could be tested during a given week. Then, in a few months, there could be a desire to test a new set of cameras and compare the new results to the set of previous results. Unless care is taken to enable an identical setup (and record the details), there could easily be systematic errors between the two sets, making it challenging to accurately quantify the differences between cameras captured at different times under different conditions. Aspects to consider include the spectral reflectance of the room walls, ceiling, and floor, the camera hardware mounting equipment, the light fixtures and stands, the light metering equipment, and the charts and real world objects.

8.1.1.1 Lab Space

Because the task of image quality benchmarking is ultimately to measure the camera's response to light, then the light sources in the lab should be controllable. Ideally, there should be no windows in the lab. If present, there should be light-blocking curtains and edge traps to block all exterior lighting. Typically, the lab's interior surfaces outside of the field of view (FOV) of the camera are low reflectance, that is, matte black or dark gray walls, ceiling, floors and doors, while the test chart is effectively spectrally neutral within the visible wavelength region, typically 18% gray within the FOV where there is not active chart content, that is, the elements to measure for image quality. The low reflectance of the room interior is to minimize camera flare, and the spectrally neutral area within the camera FOV is to enable the auto exposure to achieve similar exposure levels between cameras without manual tuning. Depending on the complexity of the chart region, the practitioner is often advised to use the addition of a gray card to modify the autoexposure level to the aim value if necessary. In fact, some charts have components built into them such that the reflectance or transmittance can be modified to tune the auto exposure level precisely.

Another aspect of the lab studio setup is to allow for ample floor space with shooting distances necessary for benchmarking different camera models. For example, cameras with zoom lenses need greater distances from the chart in order to test different zooming positions of the optics. (While collimators are sometimes used in the optical path to compensate for distances that are too short in the lab, these lenses can introduce systematic errors due to their additional optical properties.) In addition, with varying FOV from camera to camera, identical framing of the chart requires changes in camera position. Benchmarking cameras with a wide FOV can be particularly challenging, especially relative to setting up the size of room. As will be discussed later, varying the chart size can aid in this process. However, capturing at closer distances necessitates more care in ensuring that the chart's print quality is not the limiting factor for testing the camera.

Finally, a lab should have temperature and humidity control in order to maintain testing consistency over time and protect valuable equipment and charts. Certain light fixtures and equipment can generate enough heat that cooling options may be necessary during use. Lab practitioners should also be aware that changes in temperature and humidity can impact chart curvature or adhesion, which can impact both the chart quality and measurement for some metrics, for example, geometric distortion. Typically, lab condition aims are ambient temperature conditions of 23 °C and a relative humidity (RH) level of 50%. Acceptable tolerances about those aims can vary, depending on precision needs, though ±2 °C and ±20% RH are an example (ISO, 2006). Regardless of the specificity, recording the temperature and humidity with a hygro-thermometer data logger will provide useful information if lab conditions are suspected for causing erroneous measurements.

8.1.1.2 Lighting

As noted many times in the book, light is a critical element of a scene's definition; without the presence of light, objects in the scene are not visible to the unaided eye. For consumer photography with an emphasis on realism, the aim is to reproduce the appearance of a scene. Thus, the spectral properties, intensity, uniformity, specularity, and directionality are important characteristics of the light; changes in these types of characteristics will change the appearance of the scene. The types of light sources continue to expand both in the real world environment and in selections for the lab setting. While tungsten has been the dominant light source in the home and cool white fluorescent (CWF) in the office in many countries for many decades, energy-saving products such as compact fluorescent lights and LEDs (light emitting diodes) have become commonplace. This is changing what is prevalent in lighting, which in turn changes the way in which cameras should be tested.

As discussed in Chapter 6, light sources differ from illuminants, which are theoretical light sources. Often, test standards refer to illuminants rather than light sources. This leaves some ambiguity regarding what physical sources are appropriate because not all illuminants are physically realizable. For example, ISO (2002) includes the relative spectral power distribution for both daylight (D55) and tungsten illuminants. These are often cited in other standards when specifying the type of lighting to use for illuminating charts (ISO, 2014, 2009, 2006; I3A, 2009b; IEEE, 2017). However, light sources that may have correlated color temperature (CCT) values matching to the ISO 7589 illuminants (i.e., 5500 and 3200 K) may, in fact, have very different spectral properties than the illuminants. And, if the spectral properties of the light sources do not match the illuminants, the acceptable deviations are not typically defined in reference materials. As a reminder, the calculation of the CCT is derived from the chromaticities of the light source and their relationship to isotemperature lines crossing the Planckian locus perpendicularly (Wyszecki and Stiles, 1982; Hunt and Pointer, 2011). This can allow for widely varying light sources with identical CCT values; fundamentally, CCT values are underspecified with respect to spectral properties. Thus, one should be very aware of spectral properties when selecting light sources for the lab setup and not only choose based on light source names or CCT.

Figure 8.1 contains a plot showing comparisons of the spectral power distributions of some commercially available light sources advertised as 2700 K. As seen in the figure, the spectral differences can be significant. Caution should be taken when interpreting the image quality metric results if an objective metric specifies only a CCT for the illuminant in the protocol, allowing for various light sources. This same caution should

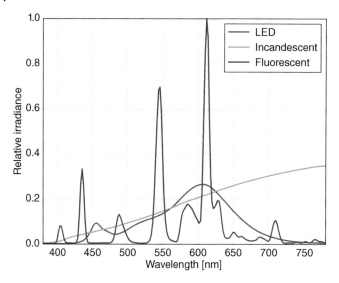

Figure 8.1 Example of three light sources with advertised 2700 K CCT. Note the differences in spectral power distributions for incandescent, compact fluorescent, and LED light sources.

be applied when switching between light sources that may have identical CCT values but different source types. As introduced in Chapter 4, the spectral sensitivities of the color filters in the image sensor are a dominant factor in the color performance of a camera. Thus, changes in spectral content of scene illumination, particularly fluorescent light source variations, can cause measurable changes in camera performance. Note that the precautions taken for illuminating reflective charts should also be taken for illuminating transmissive charts, including uniform flat fields for objective metrics such as luminance and color shading. There are often cases in which usage of "the same" light source without spectral verification has caused conflicting metric results between labs. Again, caution cannot be overstated regarding awareness of spectral properties of the light sources in use.

Another common light source specification having underspecificity regarding spectral information is the color rendering index (CRI). CRI, like CCT, is also a single value light metric, which is thus compressing complex behavior into a simplified numeral. Based on chromaticities of only 8 Munsell colors (or with an additional six), the CRI defines how far away the colors shift under the quantified light source versus a standard illuminant. Like CCT, CRI is also tied to the Planckian locus in that the index is calculated based on the distance in chromaticity space between the light source and the closest Planckian radiator to the CCT of the light source, for those that have CCT values less than 5000 K. For light sources with CCT values greater than 5000 K, the calculated distance in chromaticity space is between the light source and the closest CIE D-illuminant. While CRI values greater than 90 indicate very good color rendering, the index for a particular light source does not mean that a light source with matching CRI will render color in a similar manner (Hunt and Pointer, 2011). Again, the unique spectral properties of a light source will be compressed in obtaining the CRI, and so this metric should be treated as supplemental rather than fundamental to choosing light sources for the lab.

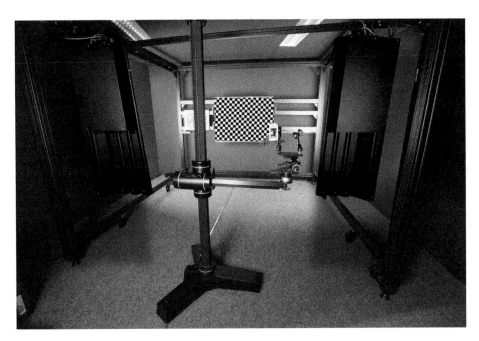

Figure 8.2 Example of lab with lighting at 45° to the normal of the chart surface. This particular setup features multiple light sources and computer-controlled light level adjustment via a DMX interface and also allows for moving the tandem lighting closer or farther away, depending on light level needs. *Source*: Reproduced with permission of DxO Labs.

In addition to the spectral properties discussed above, parameters such as the intensity, uniformity, specularity, and directionality are important properties that are strongly influenced by how the lighting is assembled in the lab environment. The intensity can be impacted by aspects such as the range in brightness of the source, the distance to the chart, and the number of sources. Typically, for reflective chart surfaces, the lighting is set up with two lamps at 45° to the normal of the chart surface. See Figure 8.2 for an example lab setup. This is to minimize any specular reflections off of the chart surface from entering the camera. In addition, baffles are placed between the lamps and the camera to minimize direct lamp light from entering the camera. Additional pairs of light fixtures can provide brighter illuminance levels and better uniformity in most cases. Another option would be to install pairs of different light sources to enable switching between various illuminants. The additional light fixtures would be implemented one on either side, and would also be at 45° to the normal of the chart surface. For example, with four lights, there could be a pair of one light source slightly above or inside a pair of another light source. Each pair would be used separately and adjusted to achieve uniform illumination of the chart. Metric specifications should indicate the necessary uniformity of the illumination, typically more stringent for methods that are measuring uniformity of camera capture, for example, luminance and color shading. Generally, illuminance at any position within the chart area should be within ±10% of the illuminance at the center of the chart. Regardless of the tolerances, the chart luminance and uniformity should be confirmed before capturing images, allowing for lamp warm-up time, for example, 30 minutes or until the measured levels reach a steady state, before measuring the uniformity and proceeding with the testing. Practice has shown that the uniformity should be measured for the entire FOV, including the chart surround. If, for instance, there are bright spots from the light sources outside of the chart area but inside of the camera FOV, these bright spots could impact algorithms such as the autoexposure, causing an unwanted shift.

A common type of LED light source used in the lab is one with blue LEDs, where a layer of phosphorous will shift some of the light from the LED up to longer wavelengths. This produces the characteristic spectrum seen in Figure 6.3 (also, see Chapter 6). RGB LEDs are also starting to become common, where mixing the light from a red, green, and blue LED, respectively, will produce light with the desired chromaticity. For a specialized LED light option for the camera lab, it is also possible to buy LED solutions where a large number of LEDs of varying peak wavelengths are combined to produce a spectrum that can be made to emulate a wide range of illuminants. However, even with a large amount of different LEDs, the resulting spectrum will not be continuous, but will have peaks at discrete wavelengths, and this might affect how different cameras respond to this light. For all LED options, even though it is comparatively simple to control the light intensity of an LED, it is usually done through pulse width modulation (PWM). The LED is here fed with a time varying, square wave voltage. By changing the *duty cycle*, that is, the ratio of time the LED is switched on and off, the apparent light level can be made to change. The voltage frequency is usually relatively high, of the order of tens of kHz, but may still cause flicker in cameras in some situations if the frequency is too low compared to the camera integration time. This factor must therefore be carefully considered before acquiring LED lighting.

Flicker may also be an issue for other light sources, especially fluorescents. All light sources driven by AC power will flicker to some extent at twice the AC line frequency (50 or 60 Hz). However, fluorescent light flickering will generally show much worse behavior compared to incandescents, since the thermal inertia of the filament in the incandescent lamp in effect dampens the amplitude of the flicker. High frequency ballasts should be employed in the light fixture to ensure that no flickering is apparent. There are products on the market that offer "flicker-free" fluorescent light fixtures with appropriate ballasts, for example, 25 to 48 kHz. However, for critical applications where flicker is a factor that needs to be avoided, DC controlled incandescent sources (or even LEDs) may be the best choice.

In order to increase efficiency and reproducibility, automation of the lighting protocols can be implemented. For example, a DMX (Digital Multiplex) controller can be connected either via a port or wirelessly to those light panels that are enabled, and allows for controlling aspects such as illuminance and CCT. Automation can be further assisted by incorporating in-situ light metering with feedback to the controller. Response to the feedback can be especially beneficial for light sources that shift CCT as illumination levels vary, particularly in low-light conditions.

8.1.1.3 Light Booths

While the descriptions above are aimed at describing a room-sized lab for capturing reflective charts ranging in size from small to large, there are compact options, that is, light booths, which are often used for supplemental needs or in place of the room-sized labs when real estate space is limited. Typically, these compact options are designed as boxes with light sources in the perimeter, space on the back wall for placing small to

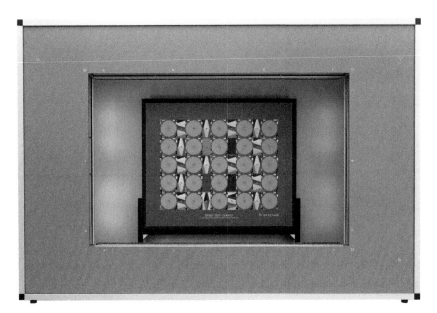

Figure 8.3 Example of light booth designed specifically for illuminating a chart. Multiple light sources are included and all are set up to provide uniform illumination. *Source*: Reproduced with permission of Image Engineering.

medium chart sizes, room on the base for placing real world objects, and a front-facing opening to access the box. For example, products on the market include models with multiple sources in the ceiling behind a diffuser panel. Light sources such as fluorescent, halogen, and LED options are typically included and can be interchanged with other sources as allowed. Many models have dimmable controls. However, be aware that often the CCT of the light source will change with adjustments of the illuminance, particularly at low light (for example, 25 lux or less), especially with incandescent and halogen light sources. There are also models that have ports for connecting peripherals that can be used to automate changes in the light sources of the light box.

Some light booths are specifically designed for illuminating reflective charts uniformly, while others are designed to inspect objects under different light sources, that is, not for camera benchmarking. This means that the latter light booths have not been optimized for necessary light uniformity or light levels, which can be challenging for meeting test specifications when placing charts on the rear wall area. Creative measures such as using two booths, or further diffusing or blocking light strategically from the default configuration may provide improved illumination. However, light booths specifically designed for chart capture as shown in Figure 8.3 provide an objective lab with a more convenient approach. Some of these types of light booths made for camera testing include neutral density (ND) filters to reduce the illumination levels to values less than 5 lux.

8.1.1.4 Transmissive Light Sources

Transmissive charts are also part of camera image quality benchmarking, and light boxes are often the source of the light. See Figure 8.4 for an example of this type of light source. A critical component of using transmissive charts is making sure the

Figure 8.4 Example of a portable light box designed for illuminating a chart. The type of lighting can have various sources such as LED or fluorescent lights. *Source*: Reproduced with permission of Imatest.

backlighting is uniform and controlled. Traditionally, light boxes have comprised fluorescent or incandescent backlighting and a diffuser sheet to provide increased uniformity, though newer options incorporate LED illumination, which has an advantage of adjustable CCT and substantially higher peak luminance. The maximum luminance of the light box will to some extent determine the dynamic range of the scene when external factors such as flare are taken into account. Therefore, a high peak luminance is desirable for measuring image quality attributes such as exposure, tone, and dynamic range of the camera. Some newer light boxes also have capacity for wireless or USB support, enabling automatic control of the light. Typically, light boxes are not large, so the chart size is typically small. Another fundamental element is that a dark surround for the light box is necessary, such as dark walls in the room, or a light tunnel.

Another light source to use with transmissive charts is an integrating sphere or hemisphere as shown in Figure 8.5. Fundamentally, the integrating sphere provides a uniform illumination source because the light is reflected throughout the inner surface of the sphere or hemisphere prior to exiting the aperture. Light sources are typically tungsten halogen or LED. As with the light boxes, the different light sources provide different options for CCT and brightness. However, a potential disadvantage of this type of light source is that the luminance becomes lower due to the integrating sphere's configuration. Both manual and automatable options are available.

8.1.1.5 Additional Lighting Options

For obtaining certain luminance levels, particularly lower ones, supplementals such as neutral density (ND) filters, grids, and shutters may be necessary for controlling the light levels. However, care must be taken to choose equipment that does not change the spectral properties in an undesirable way. These types of items are particularly useful for incandescent sources, which shift in CCT with varying brightness, but can also be useful with other light sources. Typically, one can reduce the number of individual light bulbs or increase the distance between the light sources and the charts. But, these options are not always available to the test engineer.

Figure 8.5 Example of an integrating sphere as the source for illuminating a chart. The type of lighting can have various sources such as LED or tungsten lights. *Source*: Reproduced with permission of Image Engineering.

As described in Section 6.6, an integrating sphere can be used to measure veiling glare in addition to illuminating transmissive targets. Typically, the light source may be an incandescent or LED source.

8.1.1.6 Light Measurement Devices

Typical equipment for setting up and monitoring lab lighting includes items such as illuminance (lux) and luminance meters, spectroradiometers, spectrophotometers, and reference white working standards. Illuminance and luminance meters are typically less costly because they provide integrated results using color filters rather than the spectral data provided with the more complex spectroradiometers and spectrophotometers. Spectroradiometers are used to measure spectral properties of light sources, while spectrophotometers are used to measure the spectral properties of materials. More discussion about these types of light measurement devices follows.

The *illuminance meter* is a crucial piece of equipment, used to measure the amount of light illuminating the scene, in units of lux. As it is very common to specify the scene illumination in lux, this instrument is perhaps the most commonly used in camera testing. However, even though using this instrument is quite straightforward, it is often misused. The reason behind this misuse is most likely because of a misunderstanding in what the illuminance value (in lux) reported by the instrument is actually describing. Many times this value is taken as a description of the scene, while in reality it can only be used to quantify the strength of one or several light source(s) placed some distance away from, and directed toward, the illuminance meter. What the camera, and a person, sees will depend not only on the illumination level, but also on the reflectivity of the scene. For example, if a gray card with 20% reflectance is illuminated with 10 lux, the amount of light detected by a camera will be the same as for a gray card with 40% reflectance illuminated with 5 lux, assuming that the reflective properties, such as specularity, are the same for both gray cards. In fact, a more correct quantity for describing what the camera sees is the *luminance*, measured in cd/m^2. This can be measured using a *luminance meter*, which is a type of spot meter with a viewfinder that must be directed toward the object to be measured. With this instrument, the recorded light level will correspond directly to what the camera detects if it is directed toward the same object as the camera.

The illuminance and luminance meters will only provide information about the light level emitted by a light source or reflected by an object. In many situations, the color of the light is also important. A *chroma meter* for characterizing the light source, or a *colorimeter*, for characterizing the reflected light, may be used. These instruments are equipped with three detectors, each covered by a colored filter in order to obtain spectral sensitivities closely resembling the standard CIE observer.

The accuracy of the colorimeter or chroma meter may not be high enough for certain applications. In this case, a spectrometer must be used. This type of instrument is available for measuring reflected light (radiance) in $W/sr/m^2/nm$, or $W/sr/m^3$, in which case it is referred to as a *spectroradiometer*, or, together with a calibrated light source, the spectral reflectance of a surface (spectrophotometer). With a diffuser attached to the photosensitive surface of the spectrometer (a *cosine corrector*), it can also measure the irradiance of a light source in $W/m^2/nm$ or W/m^3.

8.1.2 Charts

Descriptions of objective metrics in Chapters 6 and 7 should have made it apparent that many types of charts are necessary to measure camera image quality when focused on benchmarking. A suite of charts will be necessary, including charts for both global and local attributes. Table 8.1 provides an example of what charts could be used for a list of image quality metrics. Based on the number of example charts required for a list of benchmarking measurements in Table 8.1, the practitioner should be able to realize that it is advantageous to utilize charts with combinations of individual elements such that the number of captures can be minimized while analysis is maximized. Some charts specified by standards organizations are already combination charts, though many are not. For example, the 2014 edition of the ISO 12233 chart contains low contrast edges to measure SFR and 16 or 20 gray scale patches to measure the OECF. Metrics such as ISO or CPIQ visual noise can also be measured from the gray scale patches. Thus, with one camera capture, multiple measures can be made for multiple attributes. This approach of minimizing captures can increase productivity and can also reduce test variability due to multiple captures.

8.1.2.1 Printing Technologies for Reflective Charts

Print quality is an important aspect of charts for image quality metrics because it can be the limiting factor when measuring spatial metrics that depend on the interplay of the minimum spatial frequency content of the printed edge features and the capture distance. The size of the chart, the minimum chart spatial frequency, and the pixel

Table 8.1 For the given image quality attributes, some measurements from example charts are presented. Charts reproduced with permission of DxO Labs and Imatest. ColorChecker Classic chart reproduced with permission of X-Rite, Incorporated.

Image Quality Attribute	Measurement	Chart
Exposure & Tone	SNR, Dynamic range, Tonal range, OECF, ISO sensitivity, Exposure accuracy	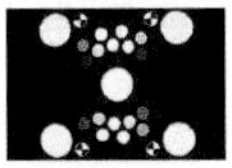
Dynamic Range	Dynamic range, OECF	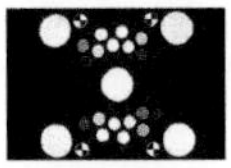
Color	Color Accuracy, White balance	
Shading	Luminance & Color shading	
Geometric Distortion	Optical distortion, TV distortion	
Stray Light	Veiling glare/Flare	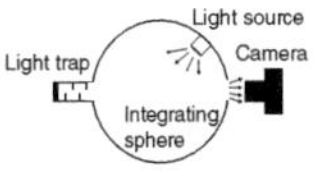
Sharpness & Resolution	MTF (SFR) and derivative metrics: Edge acutance, Limiting resolution, Sharpness uniformity, Longitudinal chromatic aberration	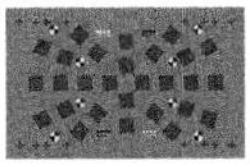
Texture Blur	Texture MTF & Texture acutance	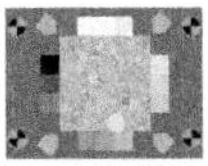
Noise	ISO/CPIQ Visual Noise	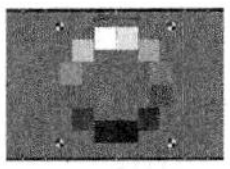
Color Fringing	Color fringing	
Image Defects	Hot pixels	

resolution of the image sensor are interrelated with respect to achieving an optimum lab setup. As an example, the CPIQ document on the SFR metric includes a table that lists the minimum printed target frequency in cycles/mm with 80% SFR for a series of given physical target sizes and image sensor resolutions (I3A, 2009a). As the size of

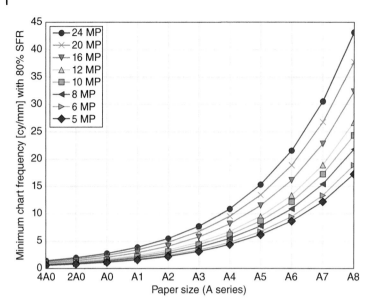

Figure 8.6 For chart printing quality, the plots include the minimum printed target frequency in cycles/mm with 80% SFR for a given chart size (A series formats) and image sensor resolution *Source*: Data for graph from I3A (2009a).

the SFR chart decreases and the image sensor resolution increases, the minimum chart frequency increases exponentially as shown in Figure 8.6.[1]

Thus, it is important to pay attention to the chart print quality, being aware that various technologies have different limitations. Not only is the native printer resolution important, but also the impact of the substrate, for example, the type of media used for inkjet printing. Glossier photo papers tend to have higher edge quality and achieve darker blacks, but can be problematic due to specular highlights when illuminated. For edge quality with high resolution, digital film recording and contact silver halide printing has traditionally been the technology of choice. Inkjet printing can also achieve high resolution, though typically lower than silver halide options. These generalizations apply to both reflective and transmissive prints.

An example of the impact of print quality and capture distance follows (Koren, 2016). The results from measuring SFR acutance and CPIQ visual noise metrics are shown. To determine the degree to which measurements were impacted by print quality, two sizes of a combination chart with OECF and SFR components printed with inkjet were captured at different distances with a 12 MP camera at 255 lux as shown in Table 8.2. The purpose was to determine if, at closer distances, the SFR acutance would decrease and the visual noise would increase due to the camera image sensor having more spatial resolution than the chart. Based on the data in Figure 8.6, the minimum printed target frequency with 80% SFR should be 1.68 and 3.36 cycles/mm for the 4x and 2x charts, respectively. The SFR acutance and CPIQ visual noise values were calculated for a use case of a 4k UHDTV (30 inches, 146 ppi) viewed at 50 cm and appear in Table 8.3.

---

1 For reference, A8 is a typical business card size (52 × 74 mm), A4 is a common sheet size (210 × 297 mm), and A2 (420 × 594 mm) and A0 (841 × 1189 mm) are common for larger lab test chart sizes.

Table 8.2 Capture distances (cm) for two sizes of a combination chart with OECF and SFR components. The distances in **bold** were captured with the specified chart framing as per ISO (2014). The 4x chart was only captured at a far distance. The 4x size and 2x size are closest to A series formats A0 and A2, respectively (Koren, 2016). *Source*: Data from Koren, 2016.

Chart Size	Far (cm)	Near (cm)
4x (800 mm × 1220 mm)	**96.3**	–
2x (400 mm × 610 mm)	124.4	**56.5**

Table 8.3 Results for SFR acutance and visual noise at different capture conditions. The results for the capture of the 2x-sized chart (400 mm × 610 mm) at the closest distance of 56.5 cm, which was captured with the specified chart framing as per ISO (2014), are compromised compared to the results for captures at farther distances. This provides an example of how measurements can be impacted by, presumably, the print quality of the chart. The results were calculated for a use case of a 4k UHDTV (30 inches, 146 ppi) viewed at 50 cm using Imatest Master 4.5.7 (Koren, 2016). *Source*: Data from Koren, 2016.

Condition	SFR Acutance	CPIQ Visual Noise
4x, Far	93.82	0.67
2x, Far	94.70	0.74
2x, Near	85.26	0.77
2x, Near vs. 2x, Far	−11.07%	3.90%
2x, Near vs. 4x, Far	−10.04%	12.99%

As seen in the results, the measurements were impacted by the print quality at the closest capture distance of 56.5 cm compared to the results from captures at farther distances. Specifically, the SFR acutance decreased by over 10% and the visual noise increased by over 12%, implying that the chart print quality can impact measurements if minimum frequencies are not achieved. Note that the 2x chart captured at 56.5 cm was captured with the specified chart framing as per ISO (2014), indicating that practitioners should be cognizant of the print chart quality, particularly with increasing image sensor resolution and closer distances. The same cognizance should be applied as chart size decreases due to the increased minimum frequency requirement of the print quality with decreasing chart size.

Some vendors provide additional means to compensate for printed chart limiting resolution for sharpness measurements as described in the following example (used by DxO Labs). The process entails performing a micro-analysis of the edge profile of a resolution chart (e.g., checkerboard or low-contrast SFR) and then deriving a compensation function that is applied to the MTF measurements. Note that this technique enables the measurement of, for example, edge acutance at much closer distances to the chart than without the described calibration. This compensation can be particularly useful when making measurements at very long focal lengths.

8.1.2.2 Technologies for Transmissive Charts

Specifications for using transmissive charts are generally similar to using reflective charts, though there can be differences in aspects such as size and luminance. Regardless, the requirements on minimum spatial frequencies for transmissive charts are as stringent as for reflective charts, and, because transmissive charts are generally smaller in size and require closer capture distances, requirements may be stricter than technologies such as inkjet are able to provide. Thus, transmissive targets are generally printed with light valve technology (LVT) film recording or comparable film recording with the highest of resolutions. Understandably, electronic displays may not provide high enough minimum resolution to be used as sources of test charts for spatial metrics. But, depending on testing needs, this option may prove to be applicable, particularly as display resolutions continue to increase over time.

Existing film recording techniques have a limitation with respect to the uniformity of the density patches (as required for OECF or noise charts, for example). Indeed, film as a material contains micro-structure inherent in the light sensitive layer of the film. Even with very fine-grain films (down to micron-level structure), it may still be possible that, for high resolution image sensors and very sharp lenses, the camera will have imaged some of the chart film structure in capturing a transmissive chart. Therefore, a standard deviation analysis on a uniform patch in a still image may be biased by the "noise" of the chart itself. Such a bias is very challenging to compensate for by computation. Thus, vendors of film-based charts recommend defocusing the camera to avoid this phenomenon. However, the issue then becomes the repeatability of the defocus and, unfortunately, it is difficult to ensure all tested cameras are defocused similarly when benchmarking a set of cameras.

As an alternative to using film for transmission charts, some vendors use optical neutral density (ND) filters, which have the benefits of being perfectly uniform, reasonably spectrally uniform, and very stable in density over time. For example, an OECF chart can be designed with a set of densities of a given distribution of steps complying with the required standard. Typically, the patches are arranged evenly around a centered circle. The measurement protocol typically specifies that the camera optical axis has been aligned with the center of the chart circle, so as to limit camera light fall off due to shading. See Figure 8.7 for an example of a chart made with ND filters to measure dynamic range of a camera, among other image quality attributes.

8.1.2.3 Inhouse Printing

Generally, it is not difficult to have custom charts manufactured by any of the major test chart sources specializing in image quality charts in the marketplace. However, for increased flexibility it may be necessary to produce inhouse test charts. This can be done using a high quality inkjet printer, preferably large format with pigment inks in order to ensure long print life. However, the printer drivers that are supplied with the printer were never intended to be used to make prints with such specific image quality requirements as for camera test charts. Therefore, specialized software that allows detailed control of each of the individual inks is needed. Such software is called a *raster image processor* (RIP) and is made for commercial printing in prepress and other non-technical applications. As such, it might be somewhat cumbersome to linearize the printer output, and so on, for demanding test chart quality. For this reason, it may be even easier to make a completely custom printer driver using open source software, for example, Gutenprint

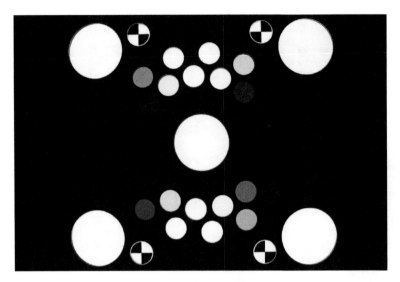

Figure 8.7 Example of a transmissive chart made with neutral density (ND) filters of varying densities. The ND filters are more uniform than other technologies, for example, printing with film recording. *Source*: Reproduced with permission of DxO Labs.

(The Gutenprint Project, 2016). This allows complete control of the individual ink amounts, and therefore makes it possible to obtain spectrally true output, which can be used for many advanced purposes (McElvain *et al.*, 2009). These approaches to printing can be applied to printing on both reflective and transmissive materials.

8.1.2.4 Chart Alignment and Framing

Due to aspects such as the nature of automatic camera settings and chart constituents, proper alignment and framing in the capturing step is very important. Typically, individual metrics will specify aims and tolerances for these aspect of camera capture. Alignment of the chart to be orthogonal to the camera's optical axis is necessary for robust measurements across the FOV. A typical alignment approach is to use a mirror in the plane of the chart and ensure that the camera's lens reflection in the mirror is centered in the viewfinder FOV. See Figure 8.8 for more details. This assumes that the chart is planar, a fundamental necessity for objective measurements such as sharpness uniformity. In fact, some metrics specify the degree to which nonplanarity is tolerable and the subsequent impact on the measurements (ISO, 2015b). On the topic of properly framing a chart with the camera, some charts, for example, the ISO 12233 resolution chart, include indicators in the chart content to assist the photographer. However, other times the procedure will provide guidance for including the proper FOV. For example, the current approach for using the dead leaves chart for the CPIQ texture blur acutance is to frame the target such that the dead leaves pattern itself is centered and filling 1/4 to 1/3 of the vertical FOV (IEEE, 2017). This type of framing protocol for a given metric should be properly followed as results may change with change in FOV. Although the dead leaves pattern is designed to be scale invariant, variation in vertical FOV has been shown to impact the results (Nielsen, 2017). Table 8.4 contains data from captures with a smartphone in which the vertical FOV of the dead leaves pattern varies. The results

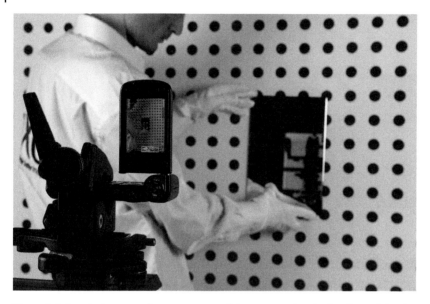

Figure 8.8 A typical camera alignment approach is to use a mirror in the plane of the chart and ensure that the camera's lens reflection in the mirror is centered in the camera viewfinder. *Source*: Reproduced with permission of DxO Labs.

Table 8.4 Comparison of CPIQ texture acutance for varying vertical FOV of the dead leaves pattern (Nielsen, 2017). The results are for the use case of viewing the image at 100% magnification on a 100 ppi monitor from a distance of 60 cm.

	CPIQ Texture Acutance		
Illuminance Level	1/4 Vertical FOV	1/3 Vertical FOV	1/2 Vertical FOV
25 lux	73.4	69.3	60.8
1000 lux	90.0	79.3	74.9

show that CPIQ texture acutance can vary, even within the specified vertical FOV. For example, the data shows that the CPIQ texture acutance with a use case of viewing the image at 100% magnification on a 100 ppi monitor from a distance of 60 cm decreases in value from 90.0% to 79.3% for captures at 1000 lux for the specified range of 1/4 to 1/3 of the vertical FOV, respectively. At 1/2 of the vertical FOV, the CPIQ texture acutance decreases further to 74.9%.

Some vendors have designed charts with repeated patterns, for example, a chart of a regularly spaced grid of dots or checkerboard pattern, to make framing the chart much less constrained than as with, for instance, the ISO 12233 resolution chart. Indeed, the protocol here simply specifies the range of number of patterns that should fit within the sensor height (e.g., 7 to 14 dots per vertical height). This type of approach gives much greater flexibility in obtaining test captures. Coupled with the chart-based micro-analysis edge profiling method described earlier, the protocol is also resilient to a broader range of camera-to-chart distances.

8.1.3 Camera Settings

As presented in Chapter 4, the camera's functional blocks, that is, the lens, image sensor, and image signal processor (ISP), all impact image quality attributes. For example, the lens aperture affects the amount of light entering the camera and subsequently impacts exposure level, among other attributes. The camera flash also impacts image quality. As such, the camera settings will impact the outcome of the image quality benchmarking. Thus, the settings should be carefully chosen and noted to fulfill the purpose of the benchmarking scope and to ensure measurement repeatability, respectively. For most cameras, available settings of influence include the aperture, exposure time, focal length, ISO sensitivity, white balance, and flash. These parameters should be set and known for manually based camera benchmarking. However, often the goal of a benchmarking task is to compare the automatic settings of cameras. In that case, these types of settings should be contained in the image Exif (Exchangeable Image File Format) tags. Note, however, as camera image processing continues to evolve, for example, become more localized, the divergence between image quality with automatic settings in the lab and in real world captures continues to increase. For example, an automatic setting with HDR image processing turned on as the default often treats images of charts in the lab with typically limited dynamic range in a manner noticeably different from the processing of real world scenes with significantly higher dynamic range. For this reason, benchmarking is often performed with a variety of camera settings, depending on what options might be available.

8.1.4 Supplemental Equipment

A tripod is a fundamental tool for the lab, providing stability for image captures. Tripods are available in various materials, for example, aluminum and carbon fiber, as well as various sizes and configurations, for example, table-top, telescoping for portability, and with a variety of heads. Some features that provide pragmatic elements are adjustable center stems, ball heads, heads with coarse and fine adjusters, and heads with built-in bubble levels. Each of these assist in the ease and accuracy of camera alignment relative to the chart. Having various options available is beneficial for benchmarking a variety of devices with a variety of measurements.

In camera benchmarking, many sizes of cameras are typically encountered, from modules on development boards to camera phones to DSLRs and more. As the typical metrics require captures with camera stability such as with a tripod, various mounting hardware will be necessary to accommodate the setup of devices without onboard tripod sockets. Selection of tripod adapters with holders for devices such as camera phones and tablets continues to grow, including spring loaded options as well as those with tightening screws. It is important to verify that the camera fit in the mount is stable with respect to the holder and the tripod, that external device controls and screen are not depressed by the holder mount itself, and that necessary access points on the device to set up the camera and review images are readily accessible. Holders with built-in bubble levels are useful, though these levels can be purchased separately and attached as needed to ensure accurate leveling. Tripods with ball heads can be useful for ease of adjustment, for example, switching a device between portrait and landscape mode.

Labs needing short-term or modular solutions will find traditional tripods to be sufficient. But, for rigorous and repetitive testing needs, a more permanent lab configuration with railings for moving the camera mounting system along a path normal to the chart center will provide for easier and faster test setup. As for the test device mounting, there are tripod adapters and sliding plates with coarse and fine adjusters, which can ease the process of accurate camera-to-chart alignment. In addition to setting up the mounting, a critical element is the alignment of the camera with the chart without blocking the existing lighting setup; once the uniform lighting has been set up, it is important to avoid interference of the light path due to the camera device and mounting equipment. Automation of the measurement protocols is possible with additional equipment and software. These include accommodating the communication with the test devices and equipment, which may have wireless communications options. Tasks such as capturing a given number of images with a prescribed settling period, downloading image files, renaming images, and providing information in the metadata will allow for a streamlined test protocol flow. Controlling the lighting systems, position of the camera mounting system, and chart selection can be part of the automation design with use of communication mechanisms such as USB to serial base unit.

8.1.4.1 Real World Objects

While not always affiliated with setting up a lab, there are reasons why an objective measurements lab should also have real world objects to capture. Before delving into some of these reasons, let us first observe components in charts that are present but not used specifically in the objective measurements of image quality attributes. For example, there are typically components in charts that serve as fiducials for automated chart detection in analysis software and that can also assist the autofocus mechanism of the camera. These elements can be objects with high contrast such as crosshairs contrasting to the background, or barcode-like structures. For another example, consider objects such as hyperbolic wedges and Siemens stars, which appear in some charts for the purpose of visual assessment of resolution (though these can also be used for objective measurements if set up to do so). There are some charts that contain 2D images of real world objects. These, too, are typically used for visual assessment beyond the elements that are in the chart for objective measurements. However, none of these are real world objects.

Returning to the reasons for including real world objects as part of the toolkit in an objective measurements lab, let us make some observations about which aspects of real world objects are fundamentally represented in the image quality measurements of some example charts. Regarding color charts, most typically the objective measurements are calculated from flat field (uniform) patches of a collection of single colors. Recall from Chapter 1 that low-level characteristics of objects in a scene include color, shape, texture, depth, luminance range, and motion. If color always occurred in the real world as uniform areas, then capturing a color chart in the lab should provide image quality results with high correlation to color image quality in an image. However, often color appears as a gradient, with texture elements, or with specular highlights rather than simply a solid area. This means that measurements from a color chart in the lab will diverge from how colors often exist in the real world.

Another chart example is a chart of the dead leaves pattern or the Siemens star to quantify texture reproduction. The former is a randomized pattern that can represent natural textures while the latter is representative of structured texture. Recall that there are both neutral and randomized color versions of the dead leaves pattern currently in use and that the Siemens star is typically neutral, though colored versions have been suggested (Hubel and Bautsch, 2006). With respect to real world objects and how image processing pipelines are applied to images, research has shown that the objective metrics correlate to various degrees with observations of real world captures (Phillips *et al.*, 2009; Phillips and Christoffel, 2010; Artmann and Wueller, 2010). However, because the dead leaves pattern and Siemens star configuration are limited representations of textures in real world objects and because image processing is often localized in nature, there is value in incorporating actual objects into the chart or collection of images for objective metrics such as texture acutance.

As mentioned above, there are some charts that do incorporate 2D images of real world objects. The elements in the chart do allow for visual assessments, but the real world content is still limited to that presented in the chart and does not contain low-level aspects such as extreme depth and luminance range as with 3D objects and scenes. For example, a 2D photo containing a metallic object may be part of a test chart. But, the luminance range of that object is limited to the dynamic range of the technology and media used to produce that chart rather than the dynamic range of an illuminated scene with a metallic object, which could have a much higher dynamic range due to shadow regions and specular highlights on the object. Thus, photographs

Figure 8.9 Example of real world objects to capture in lab conditions. Note differences in characteristics such as color, texture, gloss, and reflectance. *Source*: Reproduced with permission of Image Engineering. ColorChecker Classic chart reproduced with permission of X-Rite, Incorporated.

included in a chart may have shortcomings when being representative of real world objects in a scene. If there are physical objects present in the lab, then they can be set up like a still life scene, for example, and be captured under the same illumination as the charts. This will provide a direct comparison of how the benchmarking results from objective measurements from the lab conditions relate to observations of objects captured under similarly controlled conditions.

Figure 8.9 contains an example set of real world objects that represents a collection having useful content with respect to image quality benchmarking. The objects vary in characteristics such as color, texture, gloss, and material and can be arranged as the practitioner desires. Once captured, the image quality of the objects can be observed for attributes that were measured objectively. Objects with high spatial frequency content such as the text in the newspaper print, the fur on the stuffed animal, and the details in the twisted rope can be inspected to see if the local attributes correspond to the levels predicted from the objective metrics. Similarly, the full image can be observed to determine if the global attributes correspond to those predicted from the objective metrics.

Figure 8.10 is an example of real world objects and chart components that are pre-organized in a fixed manner such that testing over time is repeatable and consistent. The chart components in this collection allow for a comparison between what can be observed in an image of this still life collection and that in an image of solely a chart with similar content to these components. Note that this scene also contains objects that can be observed for assessing image quality attributes as with those in the previous figure. Note, though, that these objects are shallower in depth, enabling illumination with the lighting setup in the objective measurements lab used for charts; shallower depth prevents strong shadows from the lighting at 45° to the still life backing.

Figure 8.10 Example of real world objects combined with chart components to capture in lab conditions. This collection is permanently fixed and limited in depth which functions well for photographing in an objective measurements lab using existing chart lighting. *Source*: Reproduced with permission of DxO Labs. ColorChecker Passport chart reproduced by permission of X-Rite, Incorporated.

8.2   Video Objective Measurements

Contributed by Hugh Denman

Frequently, basic video objective measurements are an adaptation of still image quality measurements applied to individual frames from video streams. For example, a video clip can be taken of an ISO 12233 chart, a frame is grabbed from the video stream, and then the frame is analyzed, for example, for sharpness. But, as pointed out in Chapter 6, to fully characterize the video acquisition pipeline, objective metrics must include temporally varying features in the test scene. The chapter also highlighted that these types of metrics have limited affiliated standards. Recall that the following temporal video objective metrics were defined and described in that chapter:

- Frame rate and frame rate consistency
- Frame exposure time and consistency
- Auto white balance consistency
- Autofocusing time and stability
- Video stabilization performance
- Audio-visual synchronization
- Temporal noise
- Fixed pattern noise

8.2.1   Visual Timer

In order to obtain measurements of the first two items, related to frame rate and frame exposure time, a high-speed timer or clock is a necessary lab tool. This provides ground-truth information for the temporal aspects of video capture, such as the elapsed time between frames. The temporal frequency presented by the clock should greatly exceed the temporal frequency of the characteristic being assessed, for example, the frame rate of the camera. This is analogous to the requirement that still image test charts have a spatial resolution beyond that resolvable by the lens and image sensor.

The ISO standard 15781 (ISO, 2015a) sets out requirements for an LED-panel clock suitable for evaluating temporal aspects of camera performance such as shutter lag, autofocus time, and so on. These devices offer accurate time measurement to increments as low as 20 microseconds, and can be used as a reference for video timing.

These timers are typically constructed as a number of LEDs, organized in a grid or a set of horizontal lines. See Figure 8.11 for an example of a timing box. The LEDs are configured to light in succession, so that the time elapsed can be read from the grid. For example, consider a device with 5 horizontal lines, each containing 10 LEDs. Such a device may be configured so that the LEDs in the top line light at a rate of 1 every 10 seconds, and in each lower line the time period between LEDs is successively one-tenth of the period of the line above. With 5 lines, such a configuration could record elapsed time with millisecond precision, for a duration of 100 seconds. When the timer is being used to record very short intervals, for instance exposure time, with a camera using an electronic rolling shutter, readout should be confined to one line to eliminate the influence of line-to-line delay.

The timer can be read automatically if it includes fiducial markers, and vendors of proprietary systems for camera assessment typically sell both the timer hardware and a software module to extract the timings from the video. An account detailing the subtleties

Figure 8.11 The pictured example timing box can be used to measure metrics such as frame rate and frame exposure time. The multiple tracks of LED arrays allow for moving, visible traces of light to be captured by the camera. The video stream is subsequently analyzed. Note that the fiducials are present to assist autofocus and framing. *Source*: Reproduced with permission of DxO Labs.

involved in extracting the timings, written in relation to one such proprietary system, is presented in Masson *et al.* (2014).

In some cases the video recorded by the camera will be output in a file format that includes a timestamp for each frame. This information should not be used for characterization of the frame rate, as its accuracy may not be assured.

For use of the timing stream to measure metrics such as shutter lag and electronic rolling shutters, the timing box should be configured to support such metrics and linked to a shutter trigger. Figure 8.12 (left) contains an example of a mechanical finger that can be set up to depress a mechanical camera shutter button in a manner such that the initiation is directly linked to timing box operation. An alternative type of shutter trigger can be used for cameras with displays that operate based on capacitive sensing. In this case, the touchscreen serves as a conduit for activating the camera capture, for example via a shutter button displayed on the touchscreen in the camera application. Figure 8.12 (right) contains an example of a capacitive trigger that operates by activating the camera via a location of choice on the touch screen region.

8.2.2 Motion

To obtain a broad, representative characterization of the video system's behavior, such as objective metrics of auto white balance consistency and autofocusing time and stability, the test setup should permit camera motion, motion of elements in the scene (object motion), and changes in lighting while the video is being recorded. These dynamic elements should have repeatable behavior if performance of different cameras is to be compared, should be quantifiable if performance of a camera is to be quantified, and should be reproducible if assessments are to be carried out at multiple labs.

These desiderata are only partially realizable. Camera motion can be introduced to a video assessment setup by placing the camera on a motorized track running perpendicular to the test scene where the object distance is to be varied. Camera

Figure 8.12 Shutter trigger examples. Left: Mechanical finger example. *Source*: Reproduced with permission of Image Engineering. Right: Capacitive trigger example. *Source*: Reproduced with permission of DxO Labs.

motion can also be introduced by placing the camera on a motorized, articulated platform with six degrees of freedom where the camera's orientation and local position is to be varied, for example, to emulate shake. See Figure 8.13 for an example of this type of platform. Both of these motorized components are available from specialist suppliers in high-precision, computer-controlled implementations, and therefore satisfy the requirement for repeatable, quantifiable behavior. This permits the use of quantitative metrics for autofocus adaptation speed and stabilization performance, respectively.

As mentioned earlier in the chapter, lighting that can be varied in an automated way is becoming more common. This enables computer-controlled adjustment of brightness, color temperature, and in some cases light spectrum, in a predetermined sequence. The sequencing is assisted by in-situ light meters. As such, this allows for the use of quantitative objective metrics such as auto white balance and auto exposure adaptation speed. The test scene should contain an objective reference within the scene to indicate the timing of illumination modifications, for example, an LED which illuminates when a lighting condition change begins.

For audio-visual synchronization measurements, protocols necessarily need to detect both the motion and the sound of the video streams. There are specialist products that are designed to measure the latencies between audio and video signals, allowing calculations of the degree of synchronization by comparing, for example, visible LED blinking patterns and audible beeping patterns. Recall from Chapter 3 that it is recommended that sound should not be more than 40 ms advanced of the corresponding video frames

Figure 8.13 Example of a motorized, articulated platform with six degrees of freedom which can vary a camera's orientation and local position. *Source*: Reproduced with permission of DxO Labs.

or more than 60 ms delayed (European Broadcasting Union, 2007). As discussed in Chapter 6, the synchronization should be measured at both the start and end of video capture to account for any drift that may occur.

A necessary aspect of a video test scene is the incorporation of moving elements. Having a means of introducing camera motion does not obviate the need for moving elements in the scene, principally because of the fact that the video compression system will deal differently with global and local motion, as described in Chapter 4. Localized motion can be introduced by incorporating a motorized toy or ornament, for example. This local motion is sufficiently repeatable for comparison of different cameras, but is difficult to reproduce across labs, and it is not quantifiable—in other words, the ground truth describing this motion is not available. The performance of the video compression system must therefore be assessed subjectively, by inspection of the resulting video. A more quantitative approach may be obtained by mounting a chart with spatial elements, for example, a series of edges or a dead leaves pattern, on a spindle, so that the chart rotates while the video is being recorded. The rotating chart will exercise the video compression system in a way more representative of local motion in the real world, and the performance can be assessed by computing the chart metric in each frame. Computer control of the spindle permits repeatable assessments.

While different test scenes, or ad-hoc video recording exercises, may be used to assess different aspects of camera performance, in some cases it is necessary to ensure that multiple aspects may be assessed simultaneously. For example, spatial and temporal noise is difficult to compress and, where present in a scene, can consume a significant portion of the available bitrate. In a generally static scene, noise can be satisfactorily compressed, but when motion is present, the codec will generally prioritize allocation of bits to the area containing local motion in the scene, and the visual appearance of the noise will degrade, becoming blurry or exhibiting blocking artifacts. Therefore, the test scene should include a flat patch for noise assessment, as well as a local motion element so that noise performance can be assessed in the presence of motion.

8.3   Still Subjective Measurements

A thorough description of psychometrics for image quality evaluation was provided in Chapter 5. See the list below of the subjective methodologies that were described for still images. These subjective approaches allow the practitioner of image quality benchmarking to either validate the objective metrics or provide a starting point for determining how attributes perform relative to other cameras. In the following section, the protocols to obtain subjective data from these subjective methods will be defined.

- Rank order
- Categorical scaling
- Acceptability scaling
- Anchored scaling
- Forced-choice comparison
- Magnitude estimation

8.3.1   Lab Needs

Unlike the objective lab setup with black interior to the room, subjective evaluation of still images is best administered in a room with spectrally neutral gray interior. This allows for a more comfortable viewing environment when observers are making image quality judgments and provides a neutral surround that is ideal for photographic assessment, particularly for attributes such as color and tone, and for evaluation repeatability. For judging images on a monitor, that is, softcopy viewing, the task is made less fatiguing due to the extended gray surround compared to a dark surround with bright stimulus. In addition to gray walls, a surround lighting setup with lighting matching the white point of the monitor (e.g., D65) behind the monitor can provide ambient light that can be adjusted to be similar to the luminance of an average still image rendered on the monitor. Peak luminance of monitors continues to increase. Therefore, the ambient light level will vary depending on the particular monitor in use. Ideally, direct illumination falling on the monitor should be avoided, as light on the front surface of the monitor can generate flare, reducing image contrast. Some monitors have optional shading hoods to block stray light, similar in fashion to camera lens hoods.

A specific example of softcopy viewing requirements for anchored scaling is described with supplemental materials to ISO 20462-3 included with the digital reference stimuli (DRS) (ISO, 2012), and this relates well to general lab needs. The requirements document describes setting up the viewing station, including a high-quality monitor to display the subjective test images rendered via a computer with a high-quality graphics card, and the viewing environment, which includes a room painted neutral gray, that is, an extended gray surround for the monitor. For this particular lab setup example, pairs of light fixtures with fluorescent bulbs are suggested as a light source for generating,

Figure 8.14 Example of a softcopy viewing lab setup. Note the lighting used to illuminate the spectrally neutral wall behind the monitor without generating front surface reflections on the monitor. The monitor and illuminated wall have similar brightness and the desk surface is a neutral gray. *Source*: Used with permission by Maria Persson (Persson, 2014).

for example, D65 ambient conditions. As example lab setup can be seen in Figure 8.14. The recommended luminance of the background wall is one that is similar in brightness to the monitor, in this particular description being one with peak brightness of 400 cd/m^2. The document also describes a headrest used for controlling the viewing distance between the viewer and the display, which is attached to a table that is described as being gray to match the rest of the room. Because there are calibrated quality JND values based on sharpness level associated with the reference images being used for ISO 20462-3, the viewing distances are to be ensured, for example, 86 cm. Note, however, that the calibrated monitor is being used as a psychophysical tool in this instance; viewing distances for desktop monitors for typical usage are 50 to 60 cm.

8.3.2 Stimuli

As charts and cameras are fundamental to objective measurements, images and observers are fundamental to subjective measurements. This section will present some insights regarding stimuli generation and presentation followed by a section on observer needs. Understanding the utilization of both is an important aspect to obtaining solid subjective measurements.

8.3.2.1 Stimuli Generation

The task of generating and choosing images for subjective evaluation should not be assumed to be trivial. In fact, stimuli generation may be the most time-consuming portion of subjective measurements. As mentioned earlier in the chapter, in the

Objective Measurements section, physical objects are useful to include in images captured in a lab set up for objective metrics. But, the real world includes a much broader range of objects and lighting than can be reproduced in a typical objective measurements lab. This necessitates capturing images beyond what was discussed earlier. A typical approach for benchmarking a set of cameras is to capture a set of images with similar framing of the scene. See Figure 8.15 for an excerpt of an example set from four different cameras, which includes outdoor, indoor, and macro scenes. Note the differences in aspect ratio of the cameras and that the scenes are captured to ensure that the vertical FOV and center point of the image are equivalent for each camera when capturing landscape format images.

An approach for creating a collection of landscape format images with a set of cameras having various aspect ratios is to choose a notable, repeatable center point of the scene and to note the upper and lower bounds of the scene in the camera vertical FOV of the camera with the widest FOV. Then, to capture the scene with additional cameras, the center point and upper and lower bounds can be matched, adjusting distance to the objects in the scene as needed. This allows for direct comparison of the full image from each camera. For portrait format images, the practitioner should ensure that the horizontal FOV and center point are equivalent. Note, however, that if one wished to compare equivalent scene content for, say, 100% magnification of the images, then the practitioner would need to ensure that the scene content remained equivalent across cameras of varying image sensor resolution by adjusting camera-to-subject distance accordingly.

For many scenes, a best practice is to take the images at the same time in order to ensure optimal matches between cameras. However, for some scenes, for example, an outdoor scene with clouds moving across the sun's position, or other moving objects such as swaying trees in the wind or people walking by, capturing similar images with each camera can be challenging if not impossible. Additionally, as one might realize, if the set of cameras to be benchmarked needs to be expanded at a later date, the challenge is to be able to capture the scene as it was, including object placement and illumination. This is typically not possible unless means have been taken to assure such stability in the scene environment. One example of how to accomplish a scene with longterm stability is to create a life-sized scene similar to a stage setup in a theater, for example, furniture and props with fixed positions and controllable room lighting. Such a setup allows for expansion of a camera benchmarking set to accommodate more cameras in the subjective measurements as needed. However, this limits the scenes for the subjective evaluations to those that can be assembled as space allows. A seemingly more flexible, though complex, approach has been taken by McElvain *et al.* (2009). The system design consists of a large light box (48″ × 36″, 1.2 m × 0.91 m) composed of multiple xenon arc lamps with diffusers and a set of inkjet transparencies that have been printed using a spectral printer model such that photographic scenes can be printed with spectral matches to the real world spectra. The combination of the uniform light source and the acrylic-mounted transparencies allows for a maximum contrast ratio above 100 000:1 with a peak brightness over 11 000 cd/m^2. This system can thus mimic real world scenes with dynamic ranges matching those in natural scenes. One can envision camera stimuli being generated by photographing these transparency scenes with a collection of cameras to be benchmarked. Both the life-size room setup and the transparencies offer solutions to having repeatability and longterm availability when needing to generate stimuli for subjective measurements.

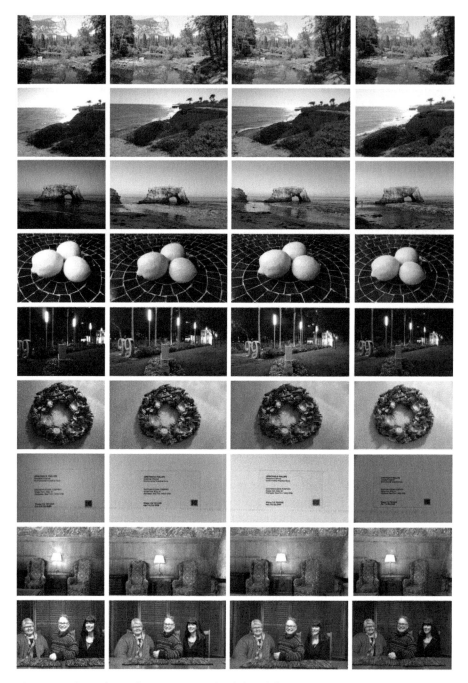

Figure 8.15 Example set of scenes, captured with four different cameras, to use for subjective measurements. Note the various scene content such as outdoor, indoor, and macro. Differences in global attributes can be noted, while apparent differences in local attributes would require larger magnification.

Just as multiple charts and multiple components in the charts are necessary to make objective measurements for benchmarking an array of image quality attributes, so too are multiple scenes necessary to provide representative subjective measurements. An important component of generating stimuli relates to populating photospace, the probability distribution of subject illumination level and subject-to-camera distance in photos taken by consumers (Rice and Faulkner, 1983). The collection of stimuli for subjective measurements should be broad enough to fill the photospace of interest as well as specific enough to contain scenes that will surface the image quality attributes to be evaluated. For example, photographs of people are an important part of consumer photography. Thus, stimuli with faces are typically an important part of subjective measurements. More details about stimuli selection will be discussed in Chapter 9.

8.3.2.2 Stimuli Presentation

Recall the discussion on metamerism in Chapter 7, which relates to how observers detect and interpret image colors. Because displays used to present the subjective stimuli are made up of a limited set of color primaries, the majority of colors produced on the display are metamers of the colors in the real world. This means that the typical encoding of images using sRGB values will produce colorimetric rather than spectral match of colors. In order to optimize these colorimetric matches for subjective viewing, color calibration tools are beneficial. Without color calibration, there is the possibility that the display's color primaries have shifted over time, producing inaccurate colors, or that the default color tables are not optimized for color accuracy, which is beneficial for camera image quality benchmarking. The calibration process consists of using the tool's software to generate a series of known colors on the display, which are measured using the colorimeter supplied with the tool. For some monitors on the market, the colorimeter is built into the display itself. The software then generates a monitor color profile with the display measurements and known sRGB values, which is applied to images to result in minimization of color errors, resulting in more accurate color presentation on the display. In this manner, the representation of camera colors will be more accurately presented in the subjective evaluation.

As mentioned earlier, it is important to not have direct light illuminating the front surface of the monitor used for subjective testing in order to reduce flare for the viewing conditions. More capable color calibration tools incorporate measurement of the ambient lighting conditions, generating room-specific monitor profiles. Note that the color calibration process is typically run with measurements at one location of the monitor surface, for example, the center, and will not correct for any display nonuniformities. This means that the native uniformity of the monitor should be verified. To help reduce the impact of existing nonuniformities in a display, image presentation should be randomized such that images are presented and judged on both the left and right sides of the display within the design of a given experimental set.

Regarding the six subjective methodologies listed at the start of this section, an important aspect of obtaining the subjective assessments by observers is consideration of the layout organization of the stimuli on the display with respect to the benchmarking questions being asked, for example, ranking local versus global attributes. For example, simultaneous rank ordering of 36 images for sharpness level could be challenging and fatiguing for an observer due to the lack of apparent resolution for what would necessarily be small-sized images presented on a display, for example, a 6 × 6 matrix of thumbnail-sized stimuli to rank. But, it may in fact be reasonable to ask an observer to rank those same images for global attributes such as white balance, color accuracy, or exposure preference.

The other subjective methodologies, that is, categorical scaling, acceptability scaling, anchored scaling, forced-choice comparison, and magnitude estimation, typically have fewer simultaneous assessments. For example, forced-choice comparison usually involves two to four simultaneous alternatives, for which the layout would allow much more real estate per image on the display. However, the same consideration of what is feasible for local and global attributes applies to these methodologies; for local attributes, the benchmarking practitioner must still be vigilant in understanding to what degree the details of interest are apparent to observers. As an example, consider how one might compare the texture blur level of a set of cameras being benchmarked. If considering how observers would compare the image quality impact of texture blur on their camera phone display with "zoom-to-fit" images, then the subjective experiment could involve multiple camera phone devices that would be compared in a side-by-side fashion or simulated on a single, color calibrated display. With typical smartphone display resolutions surpassing 300 ppi and brightness of several hundred cd/m^2, this type of presentation could provide meaningful results. As long as the observers do not magnify the images on the display, for example, a "pinch-and-zoom" expansion, then this type of presentation would result in image quality assessment that would most likely be judged as higher than that of a presentation of magnified images, for example, 100% (1:1), on the same set of displays. This is because the degree of texture blur, a local attribute, would be more apparent in the magnified images, particularly for images with higher resolution and greater degree of possible magnification. Thus, the practitioner should be aware that differences in stimuli presentation may significantly impact image quality results in subjective evaluations, particularly for local attributes.

8.3.3 Observer Needs

Subjective testing as defined in this book would not be possible without the inclusion of observers. Thus, a critical aspect of subjective measurements is the pool of observers involved in the judgment of images. Important aspects to consider include what type of observers to recruit, how to screen observers, and experimental design and duration.

8.3.3.1 Observer Selection and Screening

As has been mentioned in the book, the process of camera image quality benchmarking requires the practitioner to determine the purpose of the comparisons. The main scope of this book addresses benchmarking overall image quality for the consumer market. But, certainly, benchmarking other aspects of photography exist, for example, comparison of individual portions of the image pipeline such as autofocus, noise reduction, and color preference. More specifically, a benchmarking aim might be to compare the image quality assessed by experts versus naïve observers, which requires surveying the observers' experience level prior to administering a study. An important fundamental is that observers familiar with the specific preparation of test images will have a biased understanding of the subjective study and, thus, are typically not included in the pool of collected observer data. As noted in Chapter 5, subjective evaluation will differ between observer classes, for example, those intimately associated with the experimental task, experienced observers, and naïve observers (Keelan, 2002; Engeldrum, 2000). Therefore, before running a benchmarking assessment, the class of observer to be quantified should be determined.

Once the class of observer has been defined, a pool of potential observers can be surveyed regarding the match to this desired observer class. The observers must also be screened for color vision and visual acuity to ensure that the subjective evaluation is being performed by a known quality of observers. Typically, observers with normal color vision and high visual acuity are the targeted participants. Example means of administering the color vision screening include the collection of test charts by Ishihara (Ishihara, 1990) and the Farnsworth–Munsell 100 Hue Test (Farnsworth, 1943). The former includes a set of colored plates of colored dots with depicted numbers for the observer to identify in order to provide assessment of whether that observer is color normal or color anomalous. The latter includes four sets of colored chips for the observer to order based on hue relationship between the chips.

Visual acuity can be measured with a process as simple as using the Snellen or LogMar charts (Snellen, 1965; Sloan, 1959). For this approach, the test administrator needs to ensure that the proper distance between the observer and the chart is maintained. More advanced testing can include a vision screening instrument, which can test for a broader set of tests such as visual acuity, binocularity, depth perception, and peripheral vision.

Remember from discussions in Chapter 5 that, as image quality degrades, variability in the subjective data increases. In order to accommodate this, more observers and images should be used when the practitioner knows or expects that the image quality range of a particular camera or set of cameras includes poor image quality as measured objectively. A typical experiment includes a minimum of 20 observers. The number of recommended observers and images can vary depending on the subjective methodology. For example, recall from Chapter 5 that the recommendations for the quality ruler method using ISO 20462-3 are a minimum of 10 observers and 3 different scenes, but preferably 20 observers and 6 scenes (ISO, 2012). Engeldrum (2000) discussed a range of 4 to 50 observers, though typically 10 to 30, for scaling methodologies, depending on scale precision requirements. For paired comparisons, Montag (2006) demonstrated the relationship between number of observations for a given number of observers and observer standard deviation, which can be used to assist with trade-offs for modifying experimental designs.

8.3.3.2 Experimental Design and Duration

The purpose of the experiment should be stated and obvious to the observers. Thus, the practitioner must have clearly prepared this purpose, be it evaluation of individual image quality attributes, overall preferences, or other purpose related to camera image quality benchmarking. As one calibrates an instrument, for example, a lux meter, prior to making measurements to ensure accuracy, so should a process be included to calibrate observers. How the instructions are written and conveyed are key to this observer calibration. To enable consistent relay of the instructions, they should be the same written instructions for each observer. If a proctor is present, a useful approach can be to read the instructions to the observer. Care should be taken to only read the instructions as written and to not modify the instructions in any way: each observer should be told the same set of instructions to ensure inter-observer consistency. For example, typical benchmarking studies are for the purpose of comparing overall image quality of images. This should be stated clearly in the instructions. Also useful to the observer is the knowledge of how many sets of images and how many total comparisons will be administered. An example of observer instructions is given in Chapter 5. A visual progress indicator will be appreciated by the observer.

In setting up the observer data collection, items to obtain include observer identification, age, gender, and experience level. A useful detail for analyzing the data is the duration of each judgment. This allows for more understanding, for example, of whether there was any assessment that varied in length from the mean evaluation time.

A practical approach to validating an experimental design and the effectiveness of the instructions is to run a pilot study with a small number of observers. This process typically reveals issues with parameters such as clarity of the purpose to the observer (both in the instructions and in the presentation of images), appropriate range of the image quality attributes for the subjective methodology, and typical duration of the experiment. The data can also be analyzed to predict if the purpose of the experiment will be fulfilled. For example, the Thurstonian scaling from a forced-choice comparison cannot be calculated if judgments from all observers for a given pair are unanimous. If this is the result from the pilot study, then an alternative method such as rank order should be selected or the images themselves should be changed, for example, differences in quality between images could be decreased by modifying the scene selections to those that were less discerning.

A general understanding is that observer fatigue will begin to impact results if the duration of the experiment exceeds one hour. Multiple sessions separated in time, however, allow for extended observations for a given observer. Another aspect to consider is the number of individual assessments and the complexity of the experimental design. For example, if a benchmarking study includes comparison of 6 cameras that were used to capture 10 scenes, then it may be that an experimental order with randomization of the 6 cameras for each of the 10 scenes is easier to judge compared to presenting randomized scenes for each camera. This is because the observer will be judging the same scene six times in succession rather than judging a completely randomized order, which may require more time to reach an assessment.

Chapter 5 contained information on topics such as including null images in the sets of images to check for observer consistency and variability, judging rates, and impact of observer sensitivity. The reader interested in more information should refer back to that chapter as well as the included relevant references.

8.4 Video Subjective Measurements

As with subjective measurements for still images, an extensive description of psychometrics for video image quality was provided in Chapter 5. The list below contains the three general categories of subjective methodologies that were described to obtain video quality measurements. Many of the protocols presented above in the Still Subjective Measurements section relate directly to the methodologies below. Therefore, the contents of this section will be concise in nature.

- Categorical scaling
- Continuous scaling
- Functional assessment

Returning to topics discussed in Chapter 3, video quality attributes that are more easily quantified with subjective evaluation include jitter, impact on realness due to high frame rate (HFR), and temporal aliasing. In addition, for the no-reference condition of camera benchmarking, subjective evaluation can be applied to assessment of image quality aspects such as blocking artifacts, ghosting, and temporal pumping. Finally, noise attributes such as temporal, fixed pattern, and mosquito noise can be evaluated subjectively.

Topics related to video subjective measurements that were detailed in Chapter 5 include observer selection, the viewing setup, video display and playback, and clip selection. Other topics include presentation protocols (e.g., single-stimulus and double-stimulus variants), the three assessment methods in the above list, and sources of video clip libraries with affiliated subjective ratings to consider when running subjective evaluation comparisons.

Summary of this Chapter

- Measurement protocols are necessary to perform accurate and repeatable image quality measurements.
- Protocols for objective measurements include aspects such as room environment, lighting, chart quality, subject matter, and camera settings.
- Protocols for subjective measurements include aspects such as viewing conditions, observer screening, stimulus generation and presentation, and experiment duration.
- Objective image quality metrics can only be as good as the quality of the charts; aspects such as minimum spatial resolution and color accuracy of the chart are critical to obtaining reliable and accurate measurements.
- Subjective image quality metrics are dependent on the assessment capability of the observers; observer screening and training are important protocols to follow.
- Video image quality measurements must include motion of the subject and/or the camera in order to fully characterize the video acquisition pipeline.
- Video image quality measurements are not only dependent on the visual quality, but also the audio quality, for example, audio-visual synchronization.
- Evolving technologies that impact test equipment for image quality measurements include light sources, for example, decline of incandescent and increase of LED prevalence, and display technology used for both camera displays and presentation of subjective stimuli.

References

Artmann, U. and Wueller, D. (2010) Differences of digital camera resolution metrology to describe noise reduction artifacts. *Proc. SPIE*, **7529**, 75290L.

Engeldrum, P.G. (2000) *Psychometric Scaling: A Toolkit for Imaging Systems Development*, Imcotek Press, Winchester, MA, USA.

European Broadcasting Union (2007) The relative timing of the sound and vision components of a television signal, EBU Recommendation R37-2007.

Farnsworth, D. (1943) The Farnsworth-Munsell 100 hue and dichotomous tests for color vision. *J. Opt. Soc. Am.*, **33**, 568–578.

Hubel, P. and Bautsch, M. (2006) Resolution for color photography. *Proc. SPIE*, **6069**, 60690M.

Hunt, R.W.G. and Pointer, M.R. (2011) *Measuring Colour*, John Wiley & Sons Ltd, Chichester, UK, 4th edn.

I3A (2009a) Camera Phone Image Quality – Phase 2 – Acutance – Spatial Frequency Response. IEEE.

I3A (2009b) Camera Phone Image Quality – Phase 2 – Color Uniformity. IEEE.

IEEE (2017) IEEE 1858-2016, IEEE Standard for Camera Phone Image Quality. IEEE.

Ishihara, S. (1990) *Ishihara's tests for colour-blindness*, Kanehara, Tokyo, Japan.

ISO (2002) ISO 7589:2002 Photography – Illuminants for sensitometry– Specifications for daylight, incandescent tungsten and printer. ISO.

ISO (2006) ISO 12232:2006 Photography – Digital Still Cameras – Determination of Exposure Index, ISO Speed Ratings, Standard Output Sensitivity, and Recommended Exposure Index. ISO.

ISO (2009) ISO 14524:2009 Photography – Electronic Still Picture Cameras – Methods For Measuring Opto-Electronic Conversion Functions (OECFs). ISO.

ISO (2012) ISO 20462-3:2012 Photography – Psychophysical experimental methods for estimating image quality – Part 3: Quality ruler method. ISO.

ISO (2014) ISO 12233:2014 Photography – Electronic Still Picture Imaging – Resolution and Spatial Frequency Responses. ISO.

ISO (2015a) ISO 15781:2015 Photography – Digital Still Cameras – Measuring Shooting Time Lag, Shutter Release Time Lag, Shooting Rate, and Start-up Time. ISO.

ISO (2015b) ISO 19084:2015 Photography – Digital cameras – Chromatic displacement measurements. ISO.

Keelan, B.W. (2002) *Handbook of Image Quality – Characterization and Prediction*, Marcel Dekker, New York, USA.

Koren, H. (2016) Personal communication.

Masson, L., Cao, F., Viard, C., and Guichard, F. (2014) Device and algorithms for camera timing evaluation. *Proc. SPIE*, **9016**, 90160G.

McElvain, J., Miller, J., and Jin, E. (2009) Spectral printer modeling for transparency media: toward high dynamic range scene reproduction. *Proc. SPIE*, **7241**, 72410U.

Montag, E.D. (2006) Empirical formula for creating error bars for the method of paired comparison. *J. Electron. Imaging*, **15**, 010502–1–3.

Nielsen, M. (2017) Personal communication.

Persson, M. (2014) *Subjective Image Quality Evaluation Using the Softcopy Quality Ruler Method*, Master's thesis, Lund University, Sweden.

Phillips, J.B. and Christoffel, D. (2010) Validating a texture metric for camera phone images using a texture-based softcopy attribute ruler. *Proc. SPIE*, **7529**, 752904.

Phillips, J.B., Coppola, S.M., Jin, E.W., Chen, Y., Clark, J.H., and Mauer, T.A. (2009) Correlating objective and subjective evaluation of texture appearance with applications to camera phone imaging. *Proc. SPIE*, **7242**, 724207.

Rice, T.M. and Faulkner, T.W. (1983) The use of photographic space in the development of the disc photographic system. *J. Appl. Photogr. Eng.*, **9**, 52–57.

Sloan, L.L. (1959) New test charts for the measurement of visual acuity at near and far distances. *Am. J. Ophthalmol.*, **48**, 807–813.

Snellen, H. (1965) Letterproeven tot Bepaling der Gezigtsscherpte (PW van der Weijer 1862) cited in Bennett AG Ophthalmic test types. *Br. J. Physiol. Opt.*, **22**, 238–271.

The Gutenprint Project (2016) Gutenprint web page. URL http://gimp-print.sourceforge .net, (accessed 29 May 2017).

Wyszecki, G. and Stiles, W.S. (1982) *Color Science: Concepts and Methods, Quantitative Data and Formulae*, John Wiley & Sons Ltd, Chichester, UK.

9

The Camera Benchmarking Process

The first step to building a camera image quality benchmark is to determine the key image quality attributes to be measured and then to establish a method to weight and combine them to obtain a global scale so that one can benchmark all cameras against each other. Answering benchmark scope questions such as "Do I need to benchmark the individual components such as image sensors?" "Is this comparison for the camera as a whole system?" and "Do I need to be able to access raw images?" helps determine the necessary components or devices to gather to be tested and to what extent the comparisons can be made. By now, the reader of the book should understand that utilizing image sensor megapixel count as a single image quality benchmark is an oversimplification and is far from guaranteed to have any correlation with subjectively observed image quality.

In this chapter, we will show how a comprehensive camera benchmark should combine subjective and objective image quality assessment methodologies, and how some can substitute others when correlation is established. We will describe the ideal benchmark and will show that, given the intrinsic subjectiveness of image quality, various approaches nearing the ideal might reach different conclusions. The chapter will also describe a number of existing camera benchmark systems and will point to the ones that are the most advanced. Example benchmarking data will be shared for a collection of cameras, highlighting how various individual metrics can sway results. Finally, we will detail the possible evolution to move even closer to the ideal benchmark and highlight the technologies that remain to be developed to achieve this goal.

9.1 Objective Metrics for Benchmarking

As discussed in Chapter 1, imaging that aims for realism contains the low-level characteristics of color, shape, texture, depth, luminance range, and motion. Physical objects that are illuminated in the scene environment around us are interpreted by the HVS based on these characteristics. As such, faithful reproduction of the physical nature of these properties in an image results in an accurate, realistic reproduction of the objects in a scene. Thus, the goal in benchmarking camera image quality for the perspective of the book (and most typical for those seeking perspective on camera performance comparisons) is to quantify the degree to which the system is able to portray these characteristics in an image. The challenge arrives when one must determine *how* to do this.

Camera Image Quality Benchmarking, First Edition. Jonathan B. Phillips and Henrik Eliasson.
© 2018 John Wiley & Sons Ltd. Published 2018 by John Wiley & Sons Ltd.
Companion website: www.wiley.com/go/benchak

Image quality attributes were specified in terms of global and local attributes as well as video quality attributes in Chapters 2 and 3. Let us revisit these attribute categories and observe how they relate to the objective metrics presented in Chapters 6 and 7. As such, the following lists provide guidance on how to consider measuring the objective components of camera image quality. Example objective metrics have been categorized into appropriate image quality attributes.

Recall that global attributes are those that are mostly independent of magnification and viewing distance such that attributes of exposure, tonal reproduction, flare, color, geometric artifacts, and nonuniformities should be similarly perceived in images for most typical use cases. Thus, for use cases such as viewing small compressed thumbnails through magnification up to full resolution images on a large UHD TV, the attributes should be similar. That is, the calculations of the metrics for these attributes do not change significantly in appearance with different use cases. For example, an underexposed image looks similarly darker than normal exposure in both a thumbnail and on a UHD TV.

However, for local image quality attributes, the use case can greatly impact the perception of the attributes of sharpness and resolution, noise, texture rendition, color fringing, defects, and artifacts. That is, a change in conditions such as viewing distance, magnification of the image, or physical size of the image will strongly impact the perception of image quality for local attributes. As such, the objective metrics for these attributes are necessarily spatially dependent as described in Chapter 7.

Example Metrics for Global Image Quality Attributes

- Exposure, tonal reproduction and flare
    - Exposure index
    - ISO speed
    - Standard output sensitivity (SOS)
    - Opto-electronic conversion function (OECF)
    - Dynamic range
    - Veiling glare index
- Color
    - White balance
    - Color saturation
    - ΔE_{76}
    - ΔE_{94}
    - CIEDE2000
- Nonuniformities
    - Luminance shading
    - Color shading
- Geometrical artifacts
    - Optical distortion
    - TV distortion
    - Rolling shutter distortion

Example Metrics for Local Image Quality Attributes

- Sharpness and resolution
    - Edge spatial frequency response (SFR)

- Sine modulated Siemens star SFR
- Edge acutance
- Subjective quality factor (SQF)
- Square root integral (SQRI)
● Texture rendition
- Texture modulation transfer function (MTF)
- Texture acutance
● Noise
- Signal to noise ratio (SNR)
- ISO visual noise
- CPIQ visual noise
- vSNR
- SNR 10
● Color fringing/Lateral chromatic displacement
- Lateral chromatic aberration
● Image Defects
- Dead pixels
- Hot pixels

Example Metrics for Global Video Quality Attributes

● Frame rate and frame rate consistency
● Frame exposure time and consistency
● Auto white balance consistency
● Autofocusing time and stability
● Video stabilization performance
● Audio-visual synchronization

Example Metrics for Local Video Quality Attributes

● Temporal noise
● Fixed pattern noise (FPN)

9.2 Subjective Methods for Benchmarking

As discussed in Chapter 5, there are many psychophysical methods for both still and video image quality evaluation. As has also been mentioned throughout the book, benchmarking is only as good as the metrics that are included in the approach. Additionally, the breadth of the objective metrics typically dictates the comprehensive nature of the benchmark. When subjective evaluation, either informal or formal, is not included in the benchmarking process, often this means that image quality shortcomings arise that were not part of the initial objective assessment. In addition, with localized image processing, the difference between what is measured using images captured in lab environments and what occurs in image processing of real world scene content can be significant. For example, measuring noise with flat field patches captured in lab conditions may provide a different conclusion from what is perceived in processed images of real world scenes with spatially varying content. Incorporation of subjective evaluation as a necessary component of benchmarking will increase the probability that the image quality benchmarking results will be an accurate assessment of a particular camera.

Revisiting methods presented in Chapter 5, there follows a list of approaches to consider for quantifying the subjective impact on image quality in a benchmarking process. The reader should reference that chapter for more details, including the pros and cons of each methodology. What should be taken into consideration in the benchmarking process is that the time necessary to run organized, repeatable subjective experiments is typically more than that needed for obtaining objective metrics and necessarily involves multiple observers, with more being better.

Psychophysical Methods for Image Quality

- Rank order
- Categorical scaling
- Acceptability scaling
- Anchored scaling
- Forced-choice comparison
- Magnitude estimation

Psychophysical Methods for Video Quality

- Category scaling
- Continuous scaling
- Functional assessment

9.2.1 Photospace

An important aspect when using subjective metrics in benchmarking is determining the subset of salient scene content to evaluate from what may seem like endless scene possibilities a consumer may capture. Certainly, needs will be different for different types of cameras and applications, but general understanding of photospace as mentioned in Chapter 1 provides guidance. Recall that photospace is based on the probability distribution of subject illumination level and subject-to-camera distance in photos taken by consumers (Rice and Faulkner, 1983). Early work on this topic utilized a three-dimensional map with axes for the scene luminance, the distance, and the frequency of occurrence. This analytical approach provided a quantitative means of summarizing what began with manual inspection for both camera usage and image quality of tens of thousands of prints from customer rolls of film. The distributions in this three-dimensional space vary for different usage. For example, use of a camera for indoor illumination conditions has high occurrence at 4 to 8 feet (1.2 to 2.4 m) and lower brightness compared to high occurrence at greater than 30 feet (9.1 m) and high brightness for outdoor illumination conditions (Segur, 2000).

One can map out the expected camera usage on the plot of subject illumination level and subject-to-camera distance, known as the System Coverage Space (SCS), in order to specify the aim conditions in the scene content for the subjective evaluation (Rice and Faulkner, 1983). This space provides useful boundaries to the regions to consider when quantifying the subjective image quality and acceptability used in benchmarking. Many of the boundaries are the most challenging regions for obtaining excellent image quality from a camera, for example, capturing an object greater than 30 feet (9.1 m) away under low brightness or capturing an object at a very close distance. When designing camera systems, this mapping activity is an important step for selecting hardware components and planning for needs in the capabilities of the ISP.

By collecting subject illumination and subject-to-camera distance from collections of images, one can gain knowledge of the types of images that are important to the consumer. The capabilities of the camera system influence what types of images people take because the capabilities impact the breadth of coverage in the SCS (Segur, 2000). For example, early camera phones were not optimized for macro images due to constraints of the hardware, so consumers were not apt to take closeup shots that were all-too-often very blurry. But, smartphones have readily accessible capabilities of taking macro images and this part of the SCS is now commonplace. Hultgren and Hertel (2008) have shown evidence that the photospace distributions have changed over time for consumers, particularly as photography has shifted from film cameras to digital cameras to smartphones, and will continue to shift with technology advances in areas such as image processing and connectivity. Thus, an important part of determining the breadth of images included in subjective evaluations is to ensure that one has an understanding of the relevant photospace. In fact, this understanding can also influence the prioritization of designing such aspects as capture distance and illuminance for the lab evaluations of objective metrics.

9.2.2 Use Cases

Recall from Chapter 1 that a benchmarking approach can be described by a matrix of scene content and application use cases (see Figure 1.15). For a comprehensive understanding of a camera's image quality, the full matrix would be tested: all relevant combinations for both still and video performance would be assessed. This would include understanding how each type of scene content capture would appear in each use case. If a less thorough assessment were desired, specific combinations could be selected. For example, a benchmark could include image quality quantification and comparison of images or videos such as those taken with macro photography viewed on a camera phone display, an outdoor landscape viewed on a tablet, and an outdoor sporting event viewed on a 4k UHD TV. These types of combinations provide a sampling and partial assessment of what could be a full assessment if all combinations were analyzed.

Regardless of the fullness of the matrix design, calculating image quality for multiple use cases provides better understanding. A keen observer would notice that various scene examples (e.g., macro, dim bar, indoor portrait, sports, and landscape) make up elements of photospace and that use cases (e.g., phone display, tablet display, 4k TV, and enlarged wall art) are both spatially and media-dependent. As presented in Chapter 7, perceptually correlated image quality metrics incorporate models that account for spatial and media components. Metrics for local attributes such as sharpness, texture, and noise include those that can be modified to calculate image quality for varying distances. These types of metrics are those that will potentially modify the image quality results for varying use cases. For example, Figure 2.1 demonstrated that global attributes are apparent in both larger and smaller images, while local attributes are generally more difficult to discern as images become smaller. Thus, as the angular subtense changes for objects in the scene as magnification changes, the appearance of local attributes will change.

Just as wine experts are found to have superior olfactory recognition compared to novices for wine-related odors (Parr *et al.*, 2003), one can expect to find experts in the imaging field to have more sensitivity to image quality evaluation than novices. Indeed, in a study with colored textile samples, results have shown a statistically significant difference in capability of naïve and expert observers to identify subtle color differences in paired comparisons (Cárdenas *et al.*, 2006). Expert observers were found to be more sensitive. To approach an understanding of this from a broader perspective, consider the task of training observers in a psychophysical experiment prior to starting the data collection. The training is used in order to familiarize an observer with the experimental task, reducing data variability. From experience, it is known that this process stabilizes subjective results and avoids a startup phenomenon, that is, a break-in period for an observer's judgments in an experiment (Engeldrum, 2000; Bartleson and Grum, 1984). This indicates that even a short training process can impact the observer's judgment, regardless of whether novice or expert. Similarly, the common practice of randomization of sample presentation order between observers to avoid any judgment order biases indicates that the startup phenomenon is real. This difference between novices (untrained) and experts (trained) is one element of observer impact that should be considered when designing the subjective aspect of a benchmarking process. Most often the experienced observers are more sensitive to detecting an attribute. This leads to differences in results from average observers, particularly for preference studies (Engeldrum, 2000). So, if the benchmarking intent is to obtain understanding of the average observer, then experts should be avoided (unless specifically trained to judge as average observers). However, often the experimental variability is larger than the differences between expert and naïve observers. Keelan (2002a) points out that the 95% confidence intervals for the overall experimental data to judge the image quality impact of misregistration between color channels are significantly larger than the difference between the expert and novice results.

Another example of observer impact relates to regional cultural differences between various collections of observers. For example, research over several years by Fernandez *et al.* studied the impact of cultural differences in preferences for color reproduction of pictorial images (Fernandez and Fairchild, 2001, 2002; Fernandez *et al.*, 2005). Observers rank ordered printed scenes that were modified globally by individual attributes, for example, lightness, contrast, chroma, and hue. Each study revealed that there were statistically significant differences in color preferences between cultures they studied, including American, Japanese, European, and Chinese. However, the analysis also indicated that those differences were not a dominant factor in the experimental variability. Specifically, in Fernandez *et al.* (2005), the results indicated that the color preferences due to image content and interobserver differences were greater contributors to the experimental variability than the cultural differences. In research by Töpfer *et al.* (2006), the aim of the studies was very specific: exploring cultural and regional preferences of color reproduction of skin tones. Whereas the Fernandez studies had scenes with and without faces, all of the images in the Töpfer *et al.* (2006) study contained faces. Observers in the United States, China, and India rank ordered prints that varied globally in both lightness and chroma. Results indicated that each culture had varying aims and tolerances for skin tone reproduction, both for color and lightness. Of note, the preferred aims of Caucasian skin tones by observers in the United States were more saturated and narrower in hue compared to the similar preference aims of Indian and Chinese skin tone by their respective cultures. The analysis showed that the Chinese observers had the smallest tolerance in their preferred skin tone aim.

It is also known that the relationship of the observer to the content of the image will impact judgment. Terms such as first party, second party, and third party help define this type of relationship. First party observers are those who were present at the time the scene was captured, second party observers are those familiar with the subject matter though not present at the time of image capture, and third party observers are those with no direct relation to the scene itself or subject matter in the scene, though able to relate to the scene via knowledge of similar content in other contexts. Typically, first party observers are more sensitive to changes because of their direct knowledge of the scene. For example, perspective distortion impacting the face of a person would be detected at a lower level of distortion because that observer would know the shape of the face of the person compared to an observer who had never met the person in the image. An example related to color saturation can be observed in how typical marketing images are presented with overexaggerated colors to catch attention, such as on packaging or advertising posters. When this type of color saturation level is applied to images used in first-party studies, typically the observers will not rate the image quality as high due to perceiving color saturation that has surpassed the preferred level for content such as familiar objects and complexions.

These three types of example observer impact should provide understanding with respect to the importance of choosing observers specific to the benchmarking task at hand. As mentioned, if wanting to obtain the perspective of average observers, then expert observers should be avoided, particularly if preferences are sought. If one wants to optimize color aims for a product in which memory colors such as skin tones are important subject matter, then incorporation of cultural preferences could be advantageous. Finally, products that are used directly by the consumer may require different aims than broader applications where first-party preferences do not play a role. Beyond the three types of impact described, recall that screening observers for color vision and acuity is a necessary step to ensure that data being collected is coming from observers who are accurate "detectors." In addition, replicates in the presentation protocol will allow an understanding of the reliability of each observer, regardless of sensitivity.

9.3 Methods of Combining Metrics

One of the challenges of benchmarking is determining how to mathematically combine metrics from individual measurements into an overall score summary. Publications in the field of computer performance engineering have provided some fundamentals related to this benchmarking challenge (Fleming and Wallace, 1986; Smith, 1988; Lilja, 2000). Averaging approaches using arithmetic, harmonic, and geometric means have been compared for combining computing metrics that have the same unit of measurement, for example, time in seconds, or are normalized. The geometric mean has been identified as appropriate for combining measurements of normalized numbers or with a wide range of values (Lilja, 2000). This averaging approach has been utilized in combining measurements for camera systems that have both image quality and performance metrics (Peltoketo, 2014). However, typical image quality benchmarking studies have used weighted arithmetic averages of normalized metrics to address the complexities of image quality (Farnand *et al.*, 2016; DxO, 2012c). For example, in addition to consideration of the comprehensive list of global and local image quality metrics, images are typically captured under multiple illuminants with a range of illuminances and spectral properties as well as capture distances. This can result in a large collection of image quality measurement data, which can become unwieldy. Typically, there are experimental conditions that are more or less important for the intended benchmarking assessment. Thus, a higher weighting can be applied to those conditions that are more important and lower weighting for those less important.

9.3.1 Weighted Combinations

As mentioned, a weighted combination of image quality metrics is a common approach used in the type of benchmarking discussed in this book. Firstly, weighting does not necessitate metrics with varying magnitudes and ranges to be converted into a common scale, though normalization is often applied; instead, weights are adjusted as desired for each metric as calculated. Secondly, the benchmarking practitioner can choose which metrics to consider and how to combine individual metrics in such a manner as to optimize for the application of choice. For example, the purpose of benchmarking may be to compare the low-light performance of a collection of cameras. In this case, the practitioner could determine at what illuminance to capture images, for example, 5, 10, and 25 lux, and generate objective metrics for each condition. Then, if the captures at 5 lux are most important, those results could be weighted more than the other low-light results prior to applying simple summation or more complex arithmetic averaging. Another example would be in how to consider the impact of image quality across the image array. Take sharpness as an example. Here, the practitioner may wish to incorporate a benchmarking approach that accounts for center and edge sharpness.

9.3.2 Minkowski Summation

Minkowski summation has been shown to be a valid mathematical approach for combining a set of image quality metrics of individual, independent attributes to obtain a prediction of overall image quality (Keelan, 2002a,b). In this approach of combining metrics, previously described in Sections 5.6.5 and 7.7, we see the importance of using a calibrated JND scale for evaluating multiple, yet individual and orthogonal, metrics and then summing them to obtain a total quality loss prediction as a benchmarking system.

Examples of conversions of individual objective metrics into calibrated overall quality loss JNDs using the ISO 20462-3 softcopy ruler methodology (see Chapter 5) have been published (Phillips *et al.*, 2009; Keelan *et al.*, 2011, 2012; Baxter *et al.*, 2014; He *et al.*, 2016; I3A, 2009b,e,f). While this set of published metrics is not complete to the extent of obtaining a comprehensive benchmark, the process to perform an experiment to obtain perceptual correlation for additional attributes of interest is comparatively straightforward to set up. For example, conversions to overall quality loss JNDs for objective metrics such as white balance and tonal reproduction have not been published as of yet, but subjective experiments could be designed to obtain such values to incorporate into the multivariate formalism.

Recall from Chapter 5 that the following Minkowski metric has been noted as an approach (Keelan, 2002a; Wang and Bovik, 2006):

$$\Delta Q_m = \left[\sum_i (\Delta Q_i)^{n_m} \right]^{1/n_m}$$ (9.1)

where ΔQ_i is the image quality degradation in JNDs for attribute i. A special case of the Minkowski metric is the root mean square (RMS) sum, with $n_m = 2$. Even for such a small exponent, one can see that the smaller number will have less influence on the result. For instance, for two measurement values of 1.5 and 4.0, the average will become 2.75, and the smaller value has therefore had an unreasonably large effect on the result. The RMS value, on the other hand, is 3.0, and the smaller value has therefore contributed less to the end result.

The question is now which exponent to choose. According to Keelan (2002a), the following variable exponent has previously been used with very good results:

$$n_m = 1 + 2\tanh\left(\frac{\Delta Q_{max}}{16.9}\right)$$ (9.2)

Essentially, the hyperbolic tangent function allows the attribute with the highest quality loss to dominate the total, as such quality loss becomes more differentiated from the rest of the quality loss attributes in the set. As noted in Chapter 7, the loss from the dominant attribute will impact the image quality to the extent that any other smaller losses from other attributes will not be important. Using this formalism, it is important that the individual metrics produce values with the same units, for example, JND, otherwise combination metrics like the above will not provide relevant results.

9.4 Benchmarking Systems

As mentioned above, camera benchmarking systems can vary widely in their breadth and depth of image quality coverage. The following subsections provide a survey of some key benchmarking systems and perspectives on their respective utility. Note that some are limited to objective or subjective elements, while others contain both.

9.4.1 GSMArena

GSMArena provides an online tool to visually compare up to three camera outputs in their database for identical still or video captures of controlled lab content. Specifically, GSMArena provides online information and reviews on the topic of products using GSM (Global System for Mobile Communications), particularly image quality of mobile devices (GSMArena, 2016b). The website interface allows visitors to inspect camera images and video frame grabs from a vast selection of consumer devices with brands from A to Z. Specifically, the database has three images captured with each device—the year 2000 edition of the ISO 12233 chart with added gray scale and color charts illuminated under bright (unspecified lux) conditions, and a collection of real world objects on a gray board illuminated with either the bright or dim (28 lux) conditions (GSMArena, 2016a). These captures for each device can be viewed at various scaling levels from downsampled low resolution, that is, 2MP (stills) and 720p (video frames), up to 100%

magnification. The latter magnification allows for convenient close inspection, or "pixel peeping," while the former is more indicative of device display presentation. However, though this camera content is useful and informative subjectively, the objective component of the site is limited to listings of camera specifications. Thus, this benchmarking system is practical for directly assessing subjective attributes of still and video captures, but does not provide objective image quality content.

9.4.2 FNAC

In contrast to GSMArena's strongly subjective benchmarking capacity, FNAC (Fédération Nationale d'Achats des Cadres) provides a strongly objective benchmarking system example for still images. FNAC is a chain store in France that sells cameras and mobile phones among many other products. The store provides image quality assessment for the digital camera products it sells. The site notes that the camera's automatic mode is utilized, though for measurements of sensitivity and image stabilization, manual mode is utilized (FNAC, 2016). FNAC's assessment includes optical quality (maximum resolution, geometric distortion, chromatic aberration, and vignetting), color (under daylight, tungsten, and cool white fluorescence), noise and texture, autofocus (threshold determination for minimum illuminance and contrast levels), flash, image stabilization, and latency. Three categories of benchmarking are provided for a given phone: (1) an overall technical star rating on a 4-point scale; (2) scores on a 5-point scale for the main categories of resolution, optics, color, sensitivity, latency, and autofocus; and (3) scores on a 10-point scale for each individual test category. Note that the optical testing is provided for wide angle as well as telephoto conditions, where applicable.

While FNAC indicates that they utilize metrics from the company DxO for their analysis, the process for obtaining the overall star rating, the category scores, and individual scores is not fully specified. However, FNAC reports include information regarding illumination and chart types, and raw data is shown for some tests such as the texture acutance and SNR at various ISO speed levels. In this way, there is accessibility to some of the objective image quality evaluation. But, the relationship of the scoring to perceptual elements is implied rather than specified. In fact, no images are available for comparison.

9.4.3 VCX

The VCX (Valued Camera eXperience) assessment for smartphones is another example of a benchmarking approach based on objective image quality metrics. The VCX camera score for still images, comprising 70% image quality metrics and 30% performance metrics, provides summary information to customers interested in purchasing smartphones (Image Engineering, 2016). The white paper explaining the scope, metric details and procedures, and score generation was written by Vodaphone in cooperation with Image Engineering, an independent test lab and supplier of image quality testing supplies (Vodafone and Image Engineering, 2016). Image quality metrics include those in categories of resolution, texture loss, sharpening, colors, dynamic range, shading, and distortion. The image quality portion comprises five conditions for testing each camera phone: low (63 lux) light, mid (250 lux), and bright (1000 lux), flash turned on under low light, and zoom under bright light. Performance metrics include two categories: (1) response time metrics such as shooting time lag, autofocus failure rate, and startup time; and (2) metrics such as visual noise and edge width with camera shake applied. VCX's version 1.5 scoring for each phone consists of 100-point scales for the VCX (total) score, for each of the five image quality conditions and for the performance. The report contains a breakdown for the image quality of the five conditions and the total for the performance metrics.

While this benchmarking approach lacks any subjective scoring component, the white paper indicates that the conversion from objective metrics to the scoring is based on use-case studies by Vodafone, which implies observer data usage. Though not specified for all metrics, conversions are indicated to be a continuous function with respect to the metric or to have a local maximum for those metrics that have an optimum value in the inner portion of the scale. No specifics are publicly provided. Despite this lack of transparency, care is taken to specify the score weights for each image quality condition, which is based in part on a case study of how consumers use camera phones. As such, the bright light condition is weighted the most, followed by mid and low light, and, finally, zoom and flash with the least weight.

As discussed earlier in the chapter, multiple means of conversion of objective metrics to scores are possible. VCX's identification of conversions to scores, such as linear functions and functions with local maxima, offers an objective benchmarking system with more insight into its score obtainment for still photography. In addition, their website provides the visitor with an interactive comparison of 100% magnification of an image quality chart for two selectable cameras and any of the five capture conditions. This allows the visitor with a means of providing his or her own subjective assessment relative to the objective VCX assessment.

9.4.4   Skype Video Capture Specification

Microsoft's documents, Skype for Business Video Capture Specification for Conferencing Devices and Skype for Business Video Capture Specification for Personal Solutions, are, in actuality, requirements documents for webcam certification rather than examples of image quality benchmarking systems (Microsoft, 2016b,c,a). However, the publications contain significant content relevant to benchmarking video image quality. Included are both subjective assessment related to general usage of webcams and objective assessment specific to video attributes, categorizing detail quality, tone response and noise, color, geometry, shading, and timing. The document provides specification limits for each metric in order to classify a webcam as "standard" or "premium," which represent "good" and "exceptional" video quality, respectively. Many of the metrics are based on standards from ITU, ISO, and IEEE CPIQ. Several of the thresholds were chosen based on the IEEE CPIQ metrics, which established useful and relevant JND predictions.

There are several aspects of a screening process in the specifications that provide a pragmatic perspective on video image quality benchmarking. Fundamentally, the screening process is used to determine if the webcam system to be tested is achieving basic operation. Particularly for engineering devices, new products, and technologies, taking time to assess suitability for comprehensive testing with this type of prerequisite screening prevents the need to redo a test suite due to a camera system that is not fully functioning. For example, the specifications define aspects such as the range of image resolutions, minimum frame rates at different illuminance levels, and color spaces for acceptable webcams to test. This type of detail is something to consider when determining which aspects of a camera's video image quality will be useful for a given benchmarking process. Another aspect of screening includes ensuring that the camera's vertical field of view and center position remain consistent when switching between different resolutions and aspect ratios.

A fundamental difference between the two specifications is that they define video quality based on different use cases—one being the use of the camera in a conference room and the other as a personal webcam (integrated or peripheral). One illustration of this is that the specifications describe unique testing distances for the two types of use cases, with the personal webcams tested at closer distances. Another example is that the personal webcam specification has tests for image quality metrics under 20, 80, 160, 200, and 1000 lux with an emphasis on 20 and 80 lux, whereas the conference room specification has tests under 80, 200, and 1000 lux with an emphasis on 80 and 200 lux. Differences in standard and premium aim values specific to the two use cases are also noted for various metrics, for example, oversharpening, spatial and temporal SNR, and white balance error.

While the video capture specifications do not provide a continuous integrated assessment output scale, the bimodal specification aims can be used to determine the overall quality category of "good" or "exceptional" for a camera under testing. All metrics must achieve the aims of either standard of premium in order to achieve these categories, respectively. A strength is that the documentation is thoroughly detailed, including specifying necessary charts, lab equipment, lab setup and procedures. That these specifications are publicly available is an advantage, too. Note that most of the metrics utilize one or two frame grabs from a video stream rather than analysis of video clips, which is typical for basic video objective metrics in general.

9.4.5 VIQET

VIQET, the open-source VQEG (Video Quality Experts Group) Image Quality Evaluation Tool, is a no-reference image quality assessment model to evaluate still consumer images, having a primary output of predicted MOS (mean opinion score) values (VIQET, 2016). As described in earlier chapters, these values provide subjective evaluation on an intervallic quality scale from excellent to bad. For the VIQET evaluation, a camera's overall predicted MOS comprises four categories of images: outdoor day landscape, indoor wall hanging, indoor arrangement, and outdoor night landmark. Five images for each category are captured with a camera and subsequently evaluated by the open source code individually, by category, and overall. VIQET can be used as a camera benchmarking tool if images of the same scene with similar fields of view are taken with each camera.

The model used in the tool was built from two datasets of images: (a) the Consumer-Content Resolution and Image Quality (CCRIQ) camera image database of 392 images of 16 different scenes captured with 23 cameras and mobile phones (Saad *et al.*, 2015b) and (b) a non-public internal set of photos from Intel (Saad, 2016). Two analyses of the database images have been incorporated into the objective component of the model in VIQET: (1) natural scene statistics using DCT parameters that account for frequency and orientation aspects of the images; and (2) image quality attributes, that is, sharpness and detail, noise, dynamic range, quality of scene illumination, color saturation, and color warmth (Saad *et al.*, 2015a,b). The source of the subjective MOS data used to train the VIQET model via a linear kernel support vector regressor consists of MOS values obtained from thousands of observers rating the CCRIQ images (Saad *et al.*, 2016). Specifically, the database was rated for MOS under controlled viewing conditions in a lab with each image rated by at least 45 observers and also rated via crowdsourcing with variable viewing conditions by at least 300 observers per image. The resultant model generates links between calculable attributes of images and ground truth MOS values such that objective evaluation of an image can produce a predicted MOS.

The scoring output for a given phone comprises an overall predicted MOS with standard error. In addition, scores and standard errors are provided for outdoor day, indoor (combined wall hanging and arrangements), and outdoor night categories as well as for the individual images. Additional camera analysis provided by VIQET includes values with standard error for the following image quality attributes: multi-scale edge acutance, noise signature index, color saturation, illumination, and dynamic range. For individual images, the values include additional attributes such as flat region index, multi-scale texture acutance, and color warmth. While these image quality attributes are related to global and local attributes described in this book, note that the VIQET attributes are not calculated with the objective metrics spelled out earlier. Thus, there would be challenges to directly relate the objective components of VIQET to actual camera hardware and tuning. Also, the relationship between individual objective components and predicted MOS are not independently linked, so a goal of improving the predicted MOS becomes more challenging and perhaps elusive. Additionally, the images are resized so the smaller dimension is 1080 pixels and the aspect ratio is maintained (1080 resize) prior to analysis, so there are limitations as to which use cases can be applied to this resizing. However, use of VIQET as a benchmarking tool does provide an approach consisting of both objective and subjective components, which is an important characteristic of a good benchmarking approach.

9.4.6 DxOMark

DxOMark benchmarking has two scoring systems: the first system used for DxOMark Sensor and DxOMark Lens scores based on open scales and objective metrics from raw captures of DSLRs and DSCs, and the follow-on system used for DxOMark Mobile scores based on both objective and subjective metrics from the rendered images and processed videos of mobile cameras. Both systems have been developed by DxO, a company experienced with camera hardware and ISP pipelines. As such, DxO metrics are strongly applicable to camera and image quality tuning engineers who want to carefully examine hardware and software influences on image quality.

DxO designed its initial camera component benchmarking process, DxOMark, as an evaluation of raw files to quantify the performance of lenses and camera sensors. As such, DxO notes that the evaluation does not quantify image signal processing impact such as denoising or JPEG compression (DxO, 2017b). The scores are objective only and are obtained from captured charts. DxO utilizes specific protocols to capture a set of images of targets taken under various conditions. Analysis is run twice in total on different days by different lab technicians in order to ensure reliable results. The DxO-Mark Lens results include an overall score and subscores in the categories of sharpness, distortion, vignetting, transmission, and chromatic aberration. The DxOMark Sensor results comprise an overall score and subscores evaluating maximum color sensitivity, maximum dynamic range, and low-light ISO speed. The overall Lens and Sensor scores are combinations of the respective subscores. DxO notes several key aspects of the DxO-Mark Lens scores as follows (DxO, 2016a, 2017a):

- Test protocols are run for each available focal length and aperture combination. The highest scores for each focal length are weighted to obtain the overall score.
- The testing is limited to 150 lux and 1/60 s exposure time.
- The overall Lens score is linear, increasing as the size of the assumed print size increases. For example, doubling each print dimension should quadruple the overall score.

Launched in 2012, DxOMark Mobile is a camera benchmarking system for mobile phones and tablets that is an assessment of both the still and video components of the rear-facing camera (DxO, 2012b). As mentioned previously, the overall score is for rendered images and processed videos, not raw as in DxOMark Sensor and Lens scores described above. About 1000 photos and 20 videos are captured for each device under test to generate the perceptual analyses, which are combined with the objective metrics to obtain the overall score, a weighted average of the subscores from the separate processes for photo and video (the photo score is weighted more). DxOMark Mobile Photo consists of the following image quality categories, each on a scale of 1 to 100:

- Exposure and contrast
- Color
- Autofocus
- Texture
- Noise
- Artifacts
- Flash

DxOMark Mobile Video has similar categories as DxOMark Mobile Photo except flash is replaced by stabilization. The scores for these respective seven categories are non-linearly combined to obtain the respective photo and video subscores. The results are supplemented with descriptions of pros and cons for both subscores as well as image examples.

The current published DxOMark Mobile scores for mobile phones range from 50 for the 2010 Apple iPhone 4 to the upper 80s for several 2016 phones, for example, 89 for the 2016 Google Pixel, and will continue to increase over time. At the 2012 launch of DxOMark Mobile, DxO provided some examples of the new scoring system applied to DSLR and DSC devices to show how the image quality of these camera types compared to mobile phones, even though DxOMark Mobile is used exclusively for mobile phones. The following comparisons of the photo portion of the scores were made: the Nikon D7000 DSLR has a 95 on the DxOMark Mobile Photo scale (which is higher than the best camera phone on the market as of early 2017), the Canon Powershot S100 DSC has an 88, and the Canon Powershot G9 DSC has a 77 (DxO, 2012a). Note, however, that the video image quality assessment of the Canon Powershot S100 DSC is only 66 on the DxOMark Mobile Video scale. This DSC video performance is significantly lower than the DxOMark Mobile Video scores of current high-end mobile phones, for example, the Google Pixel, which has an 88 in the video portion of its DxOMark Mobile scoring.

DxOMark Mobile is an example of a benchmarking system that is accessible in part by usage of the DxO Analyzer software and hardware. DxO uses its own full and purchasable system to obtain the objective metric component of their scoring. This includes five modules: optics, photo, video and stabilization, timing, and array and dual 3D. What is not available to a practitioner with their tools is the proprietary formulation of the DxOMark Mobile score, including the incorporation of the subjective evaluations performed by DxO staff.

9.4.7 IEEE P1858

The IEEE P1858 initiative is developing the Standard for Camera Phone Image Quality (CPIQ) based on Phase 1 and Phase 2 documents written under the jurisdiction of I3A (International Imaging Industry Association) (I3A, 2007, 2009g,a,b,d,e,f,c; IEEE, 2017). The fundamental building blocks of the construct include a set of objective metrics (OM) with psychophysically obtained formulations for calculating associated predicted quality loss (QL) values for each individual metric and, ultimately, a total quality loss using Minkowski summation of the individual QL values. These QL values are in units of secondary Standard Quality Scale (SQS_2) JNDs (see Chapter 5 for a discourse). Thus, benchmarking using the IEEE CPIQ standard contains both objective and subjective components, providing a numeric value for the objective metrics and a "to-what-extent" understanding of the objective value's impact in consumer imaging applications via the predicted JND values. Ultimately, the Minkowski summation can predict an overall quality level of the camera on the intervallic quality scale of excellent, very good, good, fair, poor, and not worth keeping (also described in Chapter 5).

The specific metrics are spatial frequency response, lateral chromatic displacement, chroma level, color uniformity, local geometric distortion, visual noise, and texture blur. For the local metrics, that is, the spatially dependent metrics, the standard provides guidance for selection from a set of given use cases such as viewing on a computer monitor at 100% magnification, on a camera phone display, or via a 4R 4 × 6 inch (10 × 15 cm) print. For each of these use cases, critical components are the resolution of the image input, the cutoff spatial frequency of the output system, and the observer's viewing distance to the image. With these parameters one can obtain the cycles per degree of the observed image on the human retina, which enables modeling of perception by incorporating achromatic and chromatic CSFs as described in Chapter 7.

Chart images can be captured under different light levels, from dim to bright, and under various correlated color temperatures, from low to high, in order to establish performance levels that can represent scene content categories. For example, dim light and low color temperature can represent images taken in a dimly lit bar or restaurant, while images taken with bright light and high color temperatures can represent images taken outdoors during the day. While the specific lab conditions are not specified for the complete set of objective metrics, there are recommendations for individual metrics.

As a benchmarking system, the IEEE CPIQ standard has advantages of including both objective and predicted subjective metrics as well as open accessibility. Because the predicted quality losses have already been established with psychometrics, subjective experimentation is not necessary to obtain benchmarking results. A current shortcoming is that the seven metrics are an incomplete set. For example, metrics such as color reproduction, white balance, exposure index, and image artifacts are absent. Thus, if images from a given camera were to have reduced quality due to one or more of such absent metrics, the total quality loss value would be underpredicted.

9.5 Example Benchmark Results

In order to allow for more understanding of camera benchmarking approaches, this section provides examples using approaches described in the previous sections. Beginning with four different cameras, a flip phone circa 2006 with 1.3 MP camera, generations 5 and 6 of a 16 MP smartphone, and an 18 MP basic DSLR, a VIQET analysis will be presented and discussed. Following this, an analysis of these four phones using the IEEE CPIQ metrics will be described, including the resulting comparisons with various use cases. The discussion will include examples of the objective metric charts and real world images used in the analysis and demonstrate important elements to consider, especially in light of local attributes. Both VIQET results and CPIQ metrics will be compared to existing DxOMark scores. Additional comparisons will include CPIQ metric insights from a current 21 MP high-end DSLR representing high image quality.

9.5.1 VIQET

Table 9.1 contains VIQET scores for the four cameras calculated with VIQET version 2.3.117.87 for desktop computers (VIQET, 2016). These scores were obtained by capturing five scenes in the four categories of outdoor day landscape, indoor arrangement, indoor wall hanging, and outdoor night landmark as described in the previous section of the chapter. Outdoor day and indoor images were captured handheld. Outdoor night images were captured on a tripod. Discussion of the scores follows.

Clearly, the flip phone is in a different class with its distinctly lower MOS predictions being in the poor quality category with 2s, compared to the fair and good categories with 3s and 4s for the other cameras. However, remember that all images are 1080 resized prior to analysis. Thus, for the flip phone with height of 960 pixels, some upsampling occurs and, for the other cameras, the images are downsampled by 3-fold or more. This downsampling would tend to favor the image quality of the three higher resolution cameras. Note that the variability for each predicted VIQET MOS value is ±0.1 standard error. Therefore, the Generation 5 camera phone and basic DSLR are predicted to be equivalent for overall MOS, outdoor day, and indoor, while the basic DSLR has

Table 9.1 Comparison of VIQET scores for flip phone, camera phones, and basic DSLR. Note that the variability for each value is ±0.1. Outdoor day and indoor images were captured handheld. Outdoor night images were captured on a tripod.

Camera	Overall	Outdoor Day	Indoor	Outdoor Night
Flip phone, 1.3MP	**2.2**	2.4	2.1	2.2
Generation 5 Smartphone, 16MP	**3.9**	4.4	3.5	3.7
Generation 6 Smartphone, 16MP	**4.1**	4.3	3.9	4.3
Basic DSLR, 18MP	**3.9**	4.3	3.4	4.0

advantage for outdoor night. However, the Generation 6 camera phone is advantaged overall and even more significantly advantaged for indoor and outdoor night. It is noteworthy that the outdoor day MOS predictions are equivalent for the camera phones and basic DSLR, but not in the dimmer indoor and outdoor night. In conditions with sufficiently large illuminance levels such as in daylight, the camera image sensors are not photon-starved, so the image quality is at its peak. However, in the dimmer conditions of indoor and outdoor night, the amount of light reaching the sensor is reduced. These challenging conditions then pass from the sensor to the camera ISP chip, which needs to apply processing such as noise cleaning and frame combining to compensate for the artifacts introduced by the photon-starved image sensor. For the Generation 5 and 6 comparison, we note that one year's development cycle can have a significant impact on improving image quality in a product, especially in heavier image processing of the lower-light capture conditions; Generation 6's VIQET scores for indoor and outdoor night increased significantly over those of Generation 5. The higher MOS values for the outdoor night sets compared to the indoor are most likely due in part to the use of a tripod for the former.

Figures 9.1 through 9.3 contain a sample of the images captured to calculate the VIQET scores. Examples of outdoor day landscape, indoor arrangements, and outdoor night landmark are presented. Because the flip phone, smartphones, and basic DSLR cameras have different aspect ratios, they are shown with equal heights but different widths in landscape and equal widths but different heights in portrait mode. Note that with the size of the images shown in the book, the differences in image quality appear to be minimal. Recall that the image quality attributes evaluated for VIQET include local attributes of sharpness/detail and noise as well as global attributes of dynamic range, goodness of scene illumination, color saturation, and color warmth. There are some global image quality differences such as exposure levels, white balance, and dynamic range, which are apparent in some scenes. For example, the sunset warmth in the top outdoor day landscape scene varies in both warmth of the sunset and the exposure level between phones. Additionally, the neutral appearance of the background in the top indoor arrangements scene and the pavement in the top outdoor night landmark varies between phones.

Cropping the images to take a closer look at the scene content is necessary to see differences in local attributes. Figure 9.4 shows crops from an outdoor day landscape scene and outdoor night landmark scene. Note that the images were all 1080 resized as described above for the VIQET analysis, prior to cropping similar regions with the same crop dimensions (here 472 × 354 pixels). As expected based on the VIQET scores, there are local image quality differences in both sharpness/detail and noise in these example cropped images.

9.5.2 IEEE CPIQ

IEEE CPIQ analysis of the four phones of interest were carried out by capturing a set of camera test charts under three conditions in a light box: U30 fluorescent light (3000 K) at 10 lux, TL84 fluorescent light (4000 K) at 100 lux, and D65 filtered halogen light (6500 K) at 500 lux. Recall that the spectral properties of fluorescent sources with several strong, narrow spectral spikes (such as the mercury lines) differ from the smoother spectral response of daylight, here simulated by filtered halogen light. Thus, the three sources

Figure 9.1 Example scene selections from four cameras with varying imaging quality for the outdoor day landscape category of VIQET (VQEG (Video Quality Experts Group) Image Quality Evaluation Tool) analysis. Note that the differences in aspect ratio are native to each camera. Counterclockwise from upper right for each scene: (a) Flip phone; (b) Generation 5 smartphone; (c) Generation 6 smartphone; (d) Basic DSLR. Images were captured handheld.

and three illuminance levels provide a sampling of the real world, which consists of many types of light sources and light levels. The seven CPIQ objective metrics and quality loss values were calculated for each capture condition using Imatest Master version 4.4.12 (Imatest, 2016). A total quality loss prediction was obtained from the individual QL values by using Minkowski summation as described in Eq. (9.1).

Figure 9.2 Example scene selections from four cameras with varying imaging quality for the indoor arrangements category of VIQET (VQEG (Video Quality Experts Group) Image Quality Evaluation Tool) analysis. Note that the differences in aspect ratio are native to each camera. Counterclockwise from upper right for top scene and left to right for bottom scene: (a) Flip phone; (b) Generation 5 smartphone; (c) Generation 6 smartphone; and (d) Basic DSLR. Images were captured handheld.

9.5.2.1 CPIQ Objective Metrics

Tables 9.2 to 9.4 contain the objective metrics for the four cameras. For the spatial metrics, that is, sharpness, texture blur, and visual noise, the chosen use case was for 100% magnification (1:1) of the image on a 100 ppi (pixels per inch) monitor from an 86 cm viewing distance. Note this relates to "pixel peeping" of the images, which is often used for critical inspection of spatial image quality. Discussion of the results follows.

There are several objective metric results that are notable for a given camera with respect to the behavior of the other cameras in the set. Firstly, in the global attributes, the

Figure 9.3 Example scene selections from four cameras with varying imaging quality for the outdoor night landmark category of VIQET (VQEG (Video Quality Experts Group) Image Quality Evaluation Tool) analysis. Note that the differences in aspect ratio are native to each camera. Counterclockwise from upper right for each scene: (a) Flip phone; (b) Generation 5 smartphone; (c) Generation 6 smartphone; and (d) Basic DSLR. Images were captured on a tripod.

flip phone has a noticeably high color uniformity error of 12.1 when captured under U30 fluorescent lighting at 10 lux. As shown in Figure 9.5, this can be seen in the thumbnail of the flat-field capture in which the color shifts from a yellowish cast at the top of the image to a bluish cast at the bottom. Compared to the other phones, the color uniformity is noticeably worse. This trend is true for the other two light conditions. Secondly, for the chroma level, the basic DSLR has similar values for the three light conditions, whereas

Figure 9.4 Cropped scene selections from four cameras for the outdoor day landscape (top) and outdoor night landmark (bottom) categories. Images were 1080 resized as with the VIQET process, prior to cropping. Counterclockwise from upper right for each scene: (a) Flip phone; (b) Generation 5 smartphone; (c) Generation 6 smartphone; (d) Basic DSLR.

Table 9.2 Comparison of OM results captured under U30 light at 10 lux. Metrics include chroma level (CL), color uniformity (CU), local geometric distortion (LGD), spatial frequency response (SFR), texture blur (TB), visual noise (VN), and lateral chromatic displacement (LCD).

Camera	CL	CU	LGD	SFR	TB	VN	LCD
Flip phone, 1.3MP	74.4	12.1	3.13	77.0	62.0	1.2	0.35
Generation 5 Smartphone, 16MP	83.0	5.03	1.82	60.3	34.4	1.5	0.050
Generation 6 Smartphone, 16MP	86.2	5.04	2.15	91.8	75.4	1.1	0.22
Basic DSLR, 18MP	90.2	2.38	1.30	74.4	46.6	1.3	0.97

Table 9.3 Comparison of OM results captured under TL84 light at 100 lux. Metrics include chroma level (CL), color uniformity (CU), local geometric distortion (LGD), spatial frequency response (SFR), texture blur (TB), visual noise (VN), and lateral chromatic displacement (LCD).

Camera	CL	CU	LGD	SFR	TB	VN	LCD
Flip phone, 1.3MP	106	6.67	2.78	91.6	67.7	1.2	0.40
Generation 5 Smartphone, 16MP	104	2.50	1.45	80.6	57.7	0.72	0.043
Generation 6 Smartphone, 16MP	114	4.02	2.97	103	93.8	0.90	0.19
Basic DSLR, 18MP	92.1	1.56	0.930	69.3	49.9	1.0	1.6

Table 9.4 Comparison of OM results captured under D65 light at 500 lux. Metrics include chroma level (CL), color uniformity (CU), local geometric distortion (LGD), spatial frequency response (SFR), texture blur (TB), visual noise (VN), and lateral chromatic displacement (LCD).

Camera	CL	CU	LGD	SFR	TB	VN	LCD
Flip phone, 1.3MP	105	4.93	2.90	92.0	71.4	1.0	0.48
Generation 5 Smartphone, 16MP	104	2.40	1.74	129	86.4	0.86	0.11
Generation 6 Smartphone, 16MP	108	3.53	2.36	95.5	108	0.74	0.21
Basic DSLR, 18MP	96.4	1.02	3.48	73.0	52.9	0.77	1.5

the other phones are noticeably desaturated at 10 lux compared to 100 and 500 lux, due in part to underexposure at low illumination. Figure 9.6 shows the example of the Generation 6 smartphone in which the colors are more chromatic in the 100 and 500 lux captures.

In the local attributes, the objective metric results between phones are more differentiating. Illuminance levels can have significant impact. For example, the SFR values vary from 60.3% edge acutance for the Generation 5 smartphone at low illuminance to 129% for the same phone at 500 lux, which is from visibly soft to extremely oversharpened. Figure 9.7 shows a crop of these images to show this significant difference in edge quality. While other phones do not vary as much from low to high illuminance, it is noteworthy that the basic DSLR has noticeably lower edge acutance values, especially compared to other phones at brighter conditions. This will be explored in the next section.

Figure 9.5 Flat field capture with flip phone for color uniformity metric captured under U30 10 lux illumination. Notice the color shifts from a yellowish cast at the top of the image to a bluish cast at the bottom. The objective metric value of maximum chrominance variation is 12.1 for this image.

Figure 9.6 Color chart captured with Generation 6 smartphone under U30 10 lux, TL84 100 lux, and D65 500 lux illumination. The objective metric values of mean chroma level are 86.2, 114, and 108, respectively. Notice the lower chroma level in the 10 lux capture, due in part to the underexposure. ColorChecker Digital SG chart reproduced with permission of X-Rite, Incorporated.

Figure 9.7 Crop of SFR chart captured with Generation 5 smartphone under U30 10 lux (left) and D65 500 lux (right) illumination. The objective metric values of edge acutance are 60.3% and 129%, respectively. The significant differences in edge acuity are due to variables such as scene illuminance level and tuning of image processing.

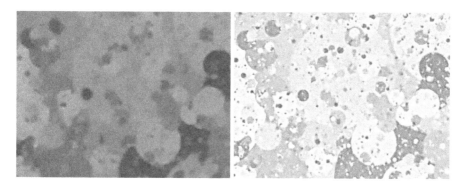

Figure 9.8 Crop of dead leaves chart captured with Generation 5 smartphone under U30 10 lux (left) and Generation 6 smartphone under D65 500 lux (right) illumination. The objective metric values of texture acutance are 34.4% and 108%, respectively. The significant differences in texture acuity are due to variables such as scene illuminance and tuning of image processing.

Not surprisingly, this lower edge acutance also impacts the texture blur objective metric; the basic DSLR is consistently low for texture acutance (46.6%, 49.9%, and 52.9% for 10, 100, and 500 lux, respectively). However, the Generation 5 smartphone achieves an even lower texture acutance value, 34.4%, at 10 lux. The Generation 6 smartphone achieves the highest texture acutance value, 108% at 500 lux, for the phones and conditions tested. Figure 9.8 shows a crop of the dead leaves pattern to demonstrate these two extreme texture acutance values.

With one exception, the visual noise values improve as illuminance levels increase for each phone, which is logical given the increased number of photons hitting the sensor as the light level increases. But results also show how the noise cleaning strength can vary at different illuminance levels. That the noise level is higher for the Generation 5 smartphone at 500 lux (0.86) compared to 100 lux (0.72) is counterintuitive. However, close

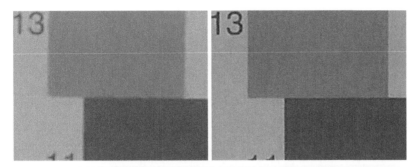

Figure 9.9 Crop of OECF chart captured with Generation 5 smartphone under TL84 100 lux (left) and D65 500 lux (right) illumination. The objective metric values of visual noise are 0.72 and 0.86, respectively. This is unexpected as higher illuminance levels typically have lower visual noise. Presumably, the image processing is tuned differently for each illuminance and the noise reduction appears stronger for the 100 lux capture. However, when converted to JNDs, this objective difference is 0.9 JNDs, that is, perceptually small.

visual inspection shown in Figure 9.9 reveals that the noise cleaning is in fact stronger at 100 lux, thus resulting in the lower noise at this lower light level.

Finally, the lateral chromatic displacement (LCD) results reveal that the basic DSLR camera has the highest levels of aberration compared to low levels for other cameras. As expected, the aberrations increase in size radially. An example of the strongest LCD (1.6 arcminutes) from a corner dot captured with TL84 at 100 lux can bee seen in Figure 9.10.

Before moving on to the CPIQ subjective quality loss predictions from the objective metrics, there is value in looking more closely at the CPIQ texture blur objective metric results compared to a newer, alternative approach mentioned in Chapter 6. We will also look at the CPIQ visual noise results compared to ISO visual noise results.

See Figure 9.11 for a comparison of unmatched crops of the dead leaves chart for the four cameras taken under U30 10 lux conditions. Texture acutance values are 62.0, 34.4, 75.4, and 46.6%, respectively, for the flip phone, Generation 5 smartphone, Generation 6 smartphone, and basic DSLR. Visual inspection would indicate that these values are representative. However, without the reference target present, the fidelity of the texture

Figure 9.10 Crop of dot chart captured with basic DSLR under TL84 100 lux illumination. This particular dot was cropped from the lower right corner and represents the maximum LCD of 1.6 arcminutes.

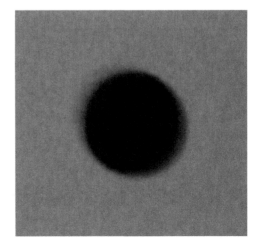

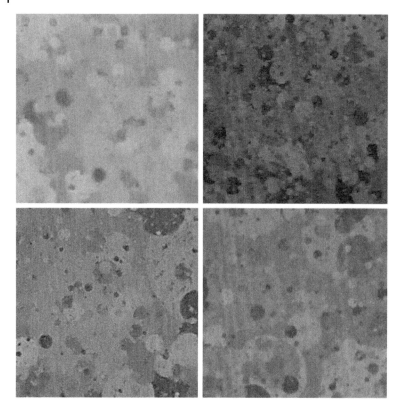

Figure 9.11 Unmatched crop of dead leaves chart captured with the four cameras of interest. Counterclockwise from upper right: (a) Flip phone; (b) Generation 5 smartphone; (c) Generation 6 smartphone; (d) Basic DSLR. Texture acutance values are 62.0, 34.4, 75.4, and 46.6%, respectively.

reproduction is not readily known. For example, "false" texture can be generated due to artifacts from oversharpening. An alternative method, that is, cross correlation with a colored version of the dead leaves (Artmann and Wueller, 2012; Kirk *et al.*, 2014), provides the mathematics that account for the full-reference aspect, incorporating a full transfer function with its phase information instead of solely the power spectrum. Generating a new set of data from the alternative texture acutance metric, we can observe several differences from the CPIQ metric. See Table 9.5 for a comparison of the averaged

Table 9.5 Comparison of CPIQ and cross correlation texture blur (TB) acutance results. For each camera, the value is an average of the results from U30 10 lux, TL84 100 lux, and D65 500 lux captures.

Camera	CPIQ TB	Cross Correlation TB
Flip phone, 1.3MP	67.0	47.1
Generation 5 Smartphone, 16MP	59.5	41.3
Generation 6 Smartphone, 16MP	92.4	70.2
Basic DSLR, 18MP	49.8	48.8

results from U30 10 lux, TL84 100 lux, and D65 500 lux for each texture blur metric. Firstly, the texture degrades more strongly with the colored pattern. As discussed in Artmann and Wueller (2012), this type of colored pattern was specifically chosen due to its advantage over a grayscale pattern regarding correspondence to real world image degradation. Thus, the cross correlation values should be more correlated to image quality in real world images, though the colored content does diverge from most natural textures, which have a narrow range of randomized hues rather than a full range of hues. Secondly, the ordering of the texture acutance changes—the flip phone is no longer strongly differentiated from the other phones and is similar to the Generation 5 smartphone and DSLR. Finally, the range from best to worst level decreases from 42.6 to 28.9, indicating that the basic DSLR and Generation 6 smartphone are not as different as presumed with the CPIQ metric following McElvain *et al.* (2010), though image processing differences between the grayscale versus colored pattern would certainly have impact on this change in range.

For a comparison of the CPIQ and ISO visual noise metrics, one must keep in mind that the source of the subjective data used for the objective-to-subjective conversion is fundamentally different. For the ISO visual noise, the subjective data set was obtained from a scaling experiment of gray patches with applied levels of Gaussian noise, whereas the subjective data set for the CPIQ visual noise was obtained from a quality ruler experiment with noise masks applied to photographic scenes (ISO, 2013; Baxter and Murray, 2012; Baxter *et al.*, 2014). Thus, the ISO visual noise corresponds more to a threshold detection of noise, whereas the CPIQ visual noise corresponds to a suprathreshold detection. Table 9.6 contains a comparison between the CPIQ and ISO visual noise metrics. As can be seen, the ISO visual noise values are at least two times greater in magnitude compared to the CPIQ results. This is understandable due to the nature of the context of the noise assessments as described. However, the more important element is determining the differences in relationship to the respective metric, that is, the normalized visual noise. When that is taken into account, we find that the maximum visual noise is different for each metric. The top images in Figure 9.12 show the two chart images with the highest respective noise from each metric. The bottom of the same figure shows how those same phones perform in nighttime real world images. All are shown at 100% magnification as was the use case for both the CPIQ and ISO visual noise metrics. While the flip phone crop of the OECF chart may appear to have the maximum noise level as indicated by the ISO visual noise metric, the Generation 5 smartphone performance in the context of nighttime scenes has noise that seems more prevalent as indicated by the

Table 9.6 Comparison of CPIQ and ISO visual noise (VN) results. Due to the context of the visual aspect of noise, that is, in context of a color photograph or neutral flat field, respectively, the two metrics have different scale strengths.

	CPIQ VN			ISO VN		
Camera	10 lux	100 lux	500 lux	10 lux	100 lux	500 lux
Flip phone	1.2	1.2	1.0	4.6	3.3	2.5
Gen 5 Smartphone	1.5	0.72	0.86	3.7	1.6	2.1
Gen 6 Smartphone	1.1	0.90	0.74	2.7	2.2	1.4
Basic DSLR	0.48	0.43	0.59	1.8	1.2	1.3

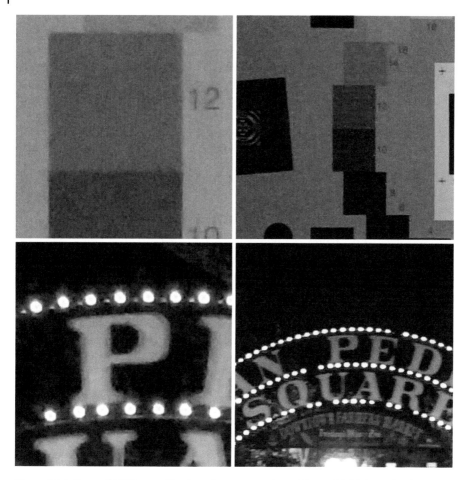

Figure 9.12 Crops of 100% magnification of cameras and conditions with highest visual noise metric results. Content varies due to differences in native resolution of cameras. Top left, worst CPIQ visual noise; top right, worst ISO visual noise; bottom left, worst CPIQ visual noise in context of image; bottom right, worst ISO visual noise in context of image. Left images are Generation 5 smartphone and right images are flip phone. All are taken at low light, for example, 10 lux for the charts.

CPIQ visual noise metric. Therefore, in light of the subjective context of each metric, the choice may be more of a choice of specific use case and interest—that of critical neutral flat field or that of scenic context. A caution should be given regarding the use of negative coefficients in the CPIQ visual noise metric as specified in the standard. The impact of this seems to be present for the flip phone example.

What these comparative examples of texture acutance and visual noise metrics should convey is that image quality metrics do have strengths and weaknesses, which then necessitate continuation in the development of the metrics themselves. For the texture acutance metric, development continues in understanding the content of the dead leaves pattern and the process to obtain the texture MTF. For visual noise, development continues in the understanding of the context in which to characterize the visual perception of the noise, be it critical viewing in flat fields or still images, or moving images. For both metrics, the type and strength of image processing has a significant influence on the resultant visual impact of the structure of texture and noise in the context of still and video imaging. For example, the spectral content of noise changes as the image moves from the raw state through the various image pipeline processes, including the compression and encoding/decoding steps. The task of generating metric calculations that accurately quantify the visual impression of such substructure remains challenging.

9.5.2.2 CPIQ Quality Loss Predictions from Objective Metrics

The individual and total quality loss (QL) results for the use case of 100% magnification on a monitor viewed at 86 cm are shown in Tables 9.7 to 9.9. Note this use case relates to "pixel peeping" of the images, a close inspection of the image structure, rather than a resize-image-height-to-screen use case.

It may seem counterintuitive that a basic DSLR camera is being outperformed by a smartphone with rigorous image quality metrics. But, considering the trends in sales and investments in the imaging sector, one should pay attention to the fact that the number of worldwide cameras in smartphones dwarfs the volume of sales of DSLRs (and DSCs, for that matter). Thus, it should not be surprising that the industry as a whole is investing strongly in hardware and software aimed at mobile imaging markets, with smartphone form factor being a significant component of this sector. Again, the fact that the image quality improvement is significant for the transition from Generation 5 to Generation 6 in one year should provide a glimpse of how quickly camera innovation and development is currently happening in smartphones. However, one should not ignore elements in the

Table 9.7 Comparison of individual and total quality loss (QL) results captured under U30 light at 10 lux. Subjective predictions include chroma level (CL), color uniformity (CU), local geometric distortion (LGD), spatial frequency response (SFR), texture blur (TB), visual noise (VN), and lateral chromatic displacement (LCD).

Camera	CL	CU	LGD	SFR	TB	VN	LCD	Total
Flip phone	3.5	4.1	0.00	1.6	7.8	5.5	0.00	12
Generation 5 Smartphone	2.2	1.3	0.50	8.2	25	7.9	0.00	26
Generation 6 Smartphone	1.7	1.3	0.22	0.00	3.1	4.9	0.00	7.2
Basic DSLR	1.2	0.37	0.37	2.4	16	6.4	0.40	17

Table 9.8 Comparison of individual and total QL results captured under TL84 light at 100 lux. Subjective predictions include chroma level (CL), color uniformity (CU), local geometric distortion (LGD), spatial frequency response (SFR), texture blur (TB), visual noise (VN), and lateral chromatic displacement (LCD).

Camera	CL	CU	LGD	SFR	TB	VN	LCD	Total
Flip phone	0.14	2.0	0.00	0.00	5.6	5.4	0.00	8.9
Generation 5 Smartphone	0.00	0.41	0.00	0.078	9.8	2.3	0.00	10
Generation 6 Smartphone	1.2	0.95	0.70	0.00	0.11	3.4	0.00	4.6
Basic DSLR	0.95	0.12	0.22	4.3	14	4.4	1.5	15

Table 9.9 Comparison of individual and total QL results captured under D65 light at 500 lux. Subjective predictions include chroma level (CL), color uniformity (CU), local geometric distortion (LGD), spatial frequency response (SFR), texture blur (TB), visual noise (VN), and lateral chromatic displacement (LCD).

Camera	CL	CU	LGD	SFR	TB	VN	LCD	Total
Flip phone	0.096	1.3	0.010	0.00	4.3	4.2	0.00	7.1
Generation 5 Smartphone	0.039	0.38	0.61	0.00	0.87	3.2	0.00	3.9
Generation 6 Smartphone	0.33	0.77	0.12	0.00	0.00	2.4	0.00	3.0
Basic DSLR	0.46	0.010	0.014	2.9	12	2.6	1.1	13

DSLR sector that provide superior options and features, such as optical zoom capability, interchangeable lenses for varying needs, and precision manual focus.

More importantly, as mentioned earlier, the DSLR in the test pool is a basic model (and not a new release). In addition, the particular setting chosen in the onboard camera menu was meant to provide the highest integrity for the DSLR captures, that is, "neutral" settings for sharpness, contrast, saturation, and color tone. In hindsight, some screening would have revealed that alternate settings, particularly a higher sharpness setting, would have increased the image quality performance for the rendered JPEG files.

9.5.3 DxOMark Mobile

A comparison in Table 9.10 of the VIQET MOS and the IEEE CPIQ predicted total quality loss values to DxOMark Mobile scores of the Generation 5 and 6 camera phones provides some insights. Ideally, DxOMark Mobile scores would be available for all four cameras included in the benchmarking comparisons. However, this is not the case—DxOMark Mobile scores are only available for the Generation 5 and 6 smartphones because the flip phone is a legacy product and DxOMark Mobile scores are not used for DSLR cameras. If we normalize the VIQET overall scores to a maximum of 5 on the perceptually linear MOS scale, the scores are 78 and 82% for the Generation 5 and Generation 6, respectively. Recall that the score variability is ±2% (±0.1 MOS) for the VIQET values, so these two scores are presumed to be similar. For the IEEE CPIQ, we choose to use a simple average of the total quality loss scores for the three light levels tested. The JNDs of quality loss for the Generation 5 and 6 phones are then 13 and 4.9, respectively, indicating that the Generation 6 camera is better by more than 8 SQS JNDs. This difference is a separation in quality category (see the Category Scaling section in Chapter 5 for more details), so this spread is larger than is implied

Table 9.10 Comparison of DxOMark Mobile scores for the Generation 5 and 6 camera phones. Note that the variability for each value is ±2 for the DxOMark Mobile score. The normalized score for VIQET and the average CPIQ total quality loss values are compared.

Camera	DxOMark Mobile	VIQET (%)	CPIQ QL
Generation 5 Smartphone	79	78	13
Generation 6 Smartphone	87	82	4.9

by the VIQET scores. However, the CPIQ total quality loss values seem to correspond more to the DxOMark Mobile results, which have greater spread. There are more differences in the CPIQ total quality loss values and DxOMark Mobile scores, which may also have an impact on the score comparisons. These differences will be discussed in the final Benchmarking Validation section of this chapter.

9.5.4 Real-World Images

There is value in looking at how the image quality of real world images captured with these four cameras reflect both the CPIQ objective metrics and the predicted quality losses: ultimately, the performance of the image quality in real world settings is usually what matters most to the consumer. Recall that the VIQET evaluation was predicted from real world images under outdoor dim, indoor, and outdoor bright conditions. Also recall that the CPIQ metrics were obtained from lab conditions of 10, 100, and 500 lux in the lab. Thus, we can make some comparisons of the respective sets of VIQET images and CPIQ results because the respective conditions should pair well regarding image quality. As we observed in the VIQET images earlier in the chapter, the global attributes are fairly similar across cameras, though differences such as in color warmth, color balance, and tonal reproduction can be noted. The more obvious differences tend to be the local attributes, which require close inspection to observe. This "pixel peeping" process is the intent of the particular CPIQ use case of viewing 1:1 (100% magnification) on a 100 ppi computer monitor such as in Figure 9.4. Recall that the metrics with the higher quality losses were indeed those due to local attributes of sharpness, visual noise, and texture blur. An alternative CPIQ use case of viewing the resized images on a camera phone display is more similar to the visual impression of renditions in Figures 9.1 through 9.3, which mask local attributes due to resizing to the thumbnail-like sizing. Thus, based on the subjective observations of the VIQET real world images, the CPIQ metrics from lab images point to similar camera behaviors as seen in real world captures.

The real world images used for VIQET are limited in their coverage of consumer photospace. Of significance outside of the VIQET set of required captures is an important subject of consumer photography, that of people. Figure 9.13 contains a comparison of the four cameras as seen in the full image and in a cropped region of the image. If we look at the global attributes in the full images, we can observe differences such as those in skin tones and clothing colors. In the cropped images showing a closeup of an eye, we can observe differences in local attributes such as sharpness, texture, and noise. Not surprisingly, based on the results from the objective metrics, we can notice more noise in the flip phone crop. Note that the noise level is accentuated by the apparent pixelization due to the necessary upsampling to match the size of the other cameras.

9.5.5 High-End DSLR Objective Metrics

The comparison above with the 18 MP basic DSLR model and camera phones demonstrates how camera phone technology has narrowed the previously large image quality gap between standalone cameras and early camera phones; the image quality of the smartphones is representative of the current market and has even been surpassed with a newer generation and other smartphones on the market. The DSLR used in the examples above has also been followed by several newer generations and

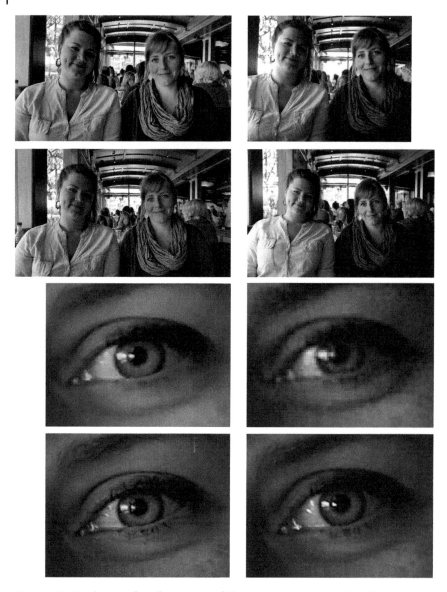

Figure 9.13 People scene from four cameras: full scene to compare global attributes and cropped to compare local attributes. Counterclockwise from upper right for each scene: (a) Flip phone; (b) Generation 5 smartphone; (c) Generation 6 smartphone; (d) Basic DSLR.

implementation of improved technology. Therefore, it is beneficial to briefly look at the advantages that CPIQ metrics can demonstrate for a more current high-end DSLR.

As mentioned previously in the chapter, the onboard "neutral" setting for the DSLR sharpness on the basic model resulted in images with low edge and texture acutance values. In fact, there was no edge sharpening apparent in either SFR curve, which was in contrast to the evidence of sharpening in the other three cameras in the set. Therefore, it was determined to select the maximum onboard sharpness setting on a higher-end 21

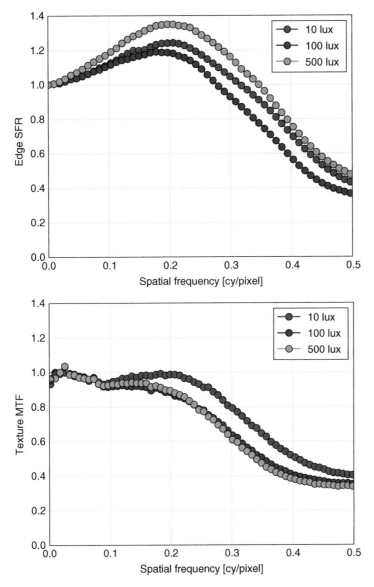

Figure 9.14 Edge SFR and texture MTF curves of high-end 21 MP DSLR at 10, 100, and 500 lux captures conditions. Note the ISO speed was fixed at 100 to achieve consistent and high acutance levels. With the onboard sharpness setting at maximum, sharpening is evident in edge SFR values greater than 1. The texture MTF shows consistent and strong texture MTF for much of the spatial frequency with minor amounts of sharpening.

MP DSLR model. This resulted in edge SFR and texture MTF curves with substantially improved behavior, including the evidence of sharpening in the edge SFR. Figure 9.14 contains the plots for edge and texture charts captured at 10, 100, and 500 lux with ISO speed fixed at 100. Compare these results to those in Figure 9.15 of the Generation 6 smartphone taken with the automatic setting under the same illuminance.

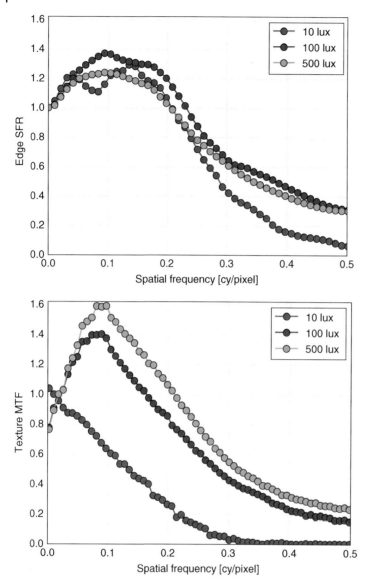

Figure 9.15 Edge SFR and texture MTF curves of Generation 6 smartphone camera at 10, 100, and 500 lux captures conditions. The image was captured in automatic mode. Note the evidence of sharpening in both plots, with SFR and MTF values reaching and surpassing 1.4.

While both show edge sharpening, the SFR peak is at 0.2 cycles/pixel for the high-end DSLR, whereas the peak is 0.1 cycles/pixel for the Generation 6 smartphone. Additionally, the Generation 6 smartphone texture MTF shows extreme sharpening for the captures at 100 and 500 lux, whereas the high-end DSLR shows consistent and strong texture MTF for much of the spatial frequency range with subtle amounts of sharpening.

Figure 9.16 compares a matched crop of the dead leaves pattern in the captured high-end DSLR and Generation 6 smartphone images at 500 lux. For comparison,

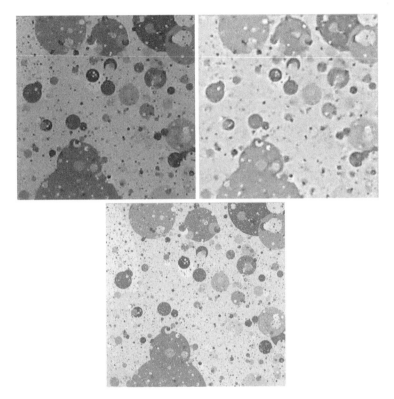

Figure 9.16 Comparison of cropped dead leaves chart captured under D65 500 lux. Top left, high-end 21 MP DSLR; top right, Generation 6 smartphone; bottom, reference dead leaves pattern. Note how the captured images both diverge from the reference, but to differing degrees.

a matched crop of the dead leaves reference pattern is included in the figure. Two key differences in the high-end DSLR images can be noted: the presence of higher frequencies in the texture MTF as seen in the smallest-diameter discs (which are not seen in the Generation 6 pattern) and the lack of ringing around edges of larger discs (which are perceivable in the Generation 6 phone, depending on the diameter and contrast). While visually the high-end DSLR dead leaves pattern appears closer to the reference pattern than that of the Generation 6 smartphone, both exhibit shortcomings versus the reference pattern.

Tables 9.11 to 9.13 contain the CPIQ objective metric data for both the edge and texture acutance values. The three use cases are 100% magnification on a computer monitor (100 ppi, 86 cm viewing distance), 4.7-inch cell phone display (326 ppi, 25 cm viewing distance), and 60-inch UHDTV (73.43 ppi, 200 cm viewing). Note how both the absolute values and the relationship between the Generation 6 smartphone and high-end DSLR differ. For example, note that the acutance values are highest for the 60-inch UHDTV condition because the angular resolution is highest for this use case. If this were a full comparison of comprehensive image quality metrics, comparing the results for these uses cases (and others) would allow the benchmarking practitioner to fill in an image quality matrix such as the one presented in Figure 1.15 in Chapter 1.

Table 9.11 Comparison of SFR and texture acutance values for the use case of 100% magnification on a 100 ppi monitor viewed at 86 cm.

	Generation 6 Smartphone			High-End DSLR		
Objective Metric	10 lux	100 lux	500 lux	10 lux	100 lux	500 lux
SFR Acutance	98.5	109	100	104	100	110
Texture Acutance	49.1	96.9	112	87.5	82.8	83.0

Table 9.12 Comparison of SFR and texture acutance values for the use case of a 4.7-inch 326 ppi cell phone display viewed at 25 cm.

	Generation 6 Smartphone			High-End DSLR		
Objective Metric	10 lux	100 lux	500 lux	10 lux	100 lux	500 lux
SFR Acutance	106	107	104	95.0	94.9	96.4
Texture Acutance	82.4	104	108	91.3	92.1	91.6

Table 9.13 Comparison of SFR and texture acutance values for the use case of a 60-inch 73.43 ppi UHDTV viewed at 200 cm.

	Generation 6 Smartphone			High-End DSLR		
Objective Metric	10 lux	100 lux	500 lux	10 lux	100 lux	500 lux
SFR Acutance	113	119	113	103	102	105
Texture Acutance	76.7	115	125	95.3	95.4	95.2

As another example, note how the texture acutance is always lower for the smartphone versus the high-end DSLR at the 10 lux capture condition, and strongly so for the 100% viewing versus the less-critical conditions of the cell phone display and UHDTV. But, in all other cases, the Generation 6 smartphone has higher texture acutance. Certainly, this advantage can be attributed to the extreme sharpening apparent in the dead leaves pattern for the smartphone and may not represent accurate texture reproduction numbers. This may be related to inadequacies of the CPIQ texture blur objective metric itself described earlier in the chapter. Referring again to Figure 9.16, note that the accuracy of the high-end DSLR texture reproduction seems much higher, that is, it has better fidelity in relation to the reference dead leaves pattern than that of the smartphone, even though the peak texture MTF values of the smartphone far exceed the high-end DSLR.

As a final note on the benchmarking examples, consider the subjective inspection of a crop from a real world photo taken between 10 and 100 lux. The image in Figure 9.17 is a closeup of hardware on an antique bellows camera and represents the first use case of 100% magnification on a monitor. Notice that the details are softer in the smartphone capture compared to those in the high-end DSLR. Note that the edges are sharp in both images, but the details on the screw dissipate for the smartphone as is typical of the loss

Figure 9.17 Comparison of cropped regions of images taken with Generation 6 smartphone, left, and high-end DSLR, right. Note the differences in texture and sharpness in the hardware pieces of an antique bellows camera. Images represent the use case of 100% magnification on a computer monitor.

in texture with image processing. The superiority of the high-end DSLR capture is tied to setting the ISO sensitivity to 100. This, in turn, increased the exposure time to 1 second compared to the auto settings on the smartphone, which resulted in an exposure time of 1/10 second and ISO 320. However, a comparison of the cropped images is indicative of the types of visual differences in the texture reproduction at lower illuminance levels, with current smartphone images typically exhibiting artifacts due to heavy noise processing compared to high-end DSLRs having more texture detail and more accuracy of the spatially related image quality attributes.

9.6 Benchmarking Validation

To close out this chapter on benchmarking, one must consider what constitutes a validation of a given approach. The process of determining how well objective metrics, even perceptually correlated metrics, estimate subjective impression of real world images is not a simple one; limitations abound in how well image quality metrics obtained from a finite set of charts predict the subjective quality in what is seemingly an endless number of scenes and objects for which a camera might be used. Thus, it is important to limit and prioritize the elements of the benchmark to be validated. For example, returning to the topic of photospace, one can determine which areas in the SCS are fundamental to the expected camera performance such as macro, low light, and landscape.

The IEEE P1858 Working Group has carried out an extensive validation study that consists of objective and subjective components using nine smartphones (Jin *et al.*, 2017). Camera test charts in the lab and real world images were captured with the nine phones. The charts were captured under numerous illumination conditions and subsequently analyzed to obtain CPIQ metric values and total quality loss predictions. For validation of the predicted image quality results, the set of real world images was captured in a way to compare matched scene content at 100% magnification viewing conditions on a calibrated computer monitor, that is, the cameras with various pixel resolutions were placed at respective distances to capture the scene subject matter such that 1:1 pixel viewing on the monitor contained equivalent subject content for each camera. The subjective evaluation of this set was performed with two psychophysical approaches, paired comparison and the ISO 20462-3 quality ruler by 18 observers, to obtain image quality assessment in JNDs. The paired comparison results were compared to the quality ruler results to ensure that the latter were indeed providing valid subjective image quality results. A Pearson correlation of 0.89 for these two subjective methods provided a validation of the quality ruler method. These quality ruler results from the subjective evaluations were subsequently compared to the predicted total quality losses from the CPIQ metrics, noting that the output of both was in SQS_2 JND units. A Pearson correlation of 0.83 was obtained for the actual and predicted total quality loss, indeed indicating that the CPIQ methodology with its Minkowski metric provided a solid camera image quality benchmarking approach despite lacking metrics in its present state for some key attributes, for example, white balance and tonal reproduction.

In addition, each selected phone has a published DxOMark Mobile score (DxO, 2016b), which can be compared to the CPIQ total quality loss and quality ruler subjective results. Figure 9.18 contains a plot comparing the predicted overall quality loss and the DxOMark Mobile Photo scores for the nine phones in the CPIQ validation study. The data is sorted by DxOMark Mobile Photo score, which is inversely related to the CPIQ quality loss because higher DxOMark Mobile scores indicate higher quality, while CPIQ results are in quality loss units, that is, higher loss is lower quality. From left to right, the general trend is that the total CPIQ quality loss decreases, that is, the predicted overall image quality increases, as the DxOMark Mobile Photo score increases. While the Pearson correlation is not high (−0.54), this could be due to

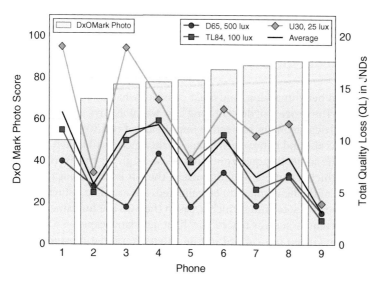

Figure 9.18 Comparison of CPIQ total quality loss predictions and DxOMark Mobile Photo scores for the 9 CPIQ validation phones (Jin *et al.*, 2017). Note the general trend that total quality loss decreases as the DxOMark Photo score increases, an expected outcome for correlated results.

many factors. For the CPIQ validation study, the cameras were all set to have the HDR algorithm turned off (if present) to minimize differences in image quality between lab captures and real world captures. This proved to be effective based on the Pearson correlation of 0.83 between these captures as mentioned earlier. The DxOMark Mobile score is always done with default mode camera setting. This means that HDR may or may not have been turned on for the tested devices for DxOMark scores, whereas, with the CPIQ results, HDR is always turned off. Also, the CPIQ validation study applied a critical use case of 100% magnification for monitor viewing to the CPIQ objective metrics because this was the condition of their formal subjective studies. DxOMark Mobile averages results from three viewing conditions (web, 8 MP and HDTV), while the CPIQ study considers the 100% magnification use case only. This is another significant difference in protocols that could help explain the overall score differences between the two systems. Additionally, DxOMark Mobile protocols consider a broader set of image quality attributes than CPIQ, including attributes such as color accuracy, white balance, exposure, flash quality, and image defects. For perspective, the Pearson correlation of the CPIQ total quality loss prediction with the megapixel count of the cameras is only −0.17.

Additional work not included in the Jin *et al.* (2017) reference was the capturing of a second set of real world images with eight of the nine phones in order to perform a VIQET analysis for the subset of cameras. As is the VIQET procedure, five scenes each of outdoor day landscape, indoor arrangement, indoor wall hanging, and outdoor night landmark were captured. Table 9.14 contains the VIQET results for the eight smartphones (Jin, 2017). Phone 2 was not part of the VIQET set.

Recall that the VIQET process uses a 1080 resize and the CPIQ quality loss predictions pointed to spatial metrics as being critical. Therefore, the VIQET resizing de-emphasizes the potential differences of interest. For example, many of the VIQET results for Outdoor Day and Indoor sets were within the known variability of ±0.1, while the CPIQ predictions ranged by 5.7 and 9.6 JNDs, respectively, for the sets captured under corresponding illuminance levels, that is, D65 at 500 lux and TL84 at 100 lux. This means that the VIQET results indicate many of the cameras produce similar image quality in the context of the use case of 1080P, whereas the CPIQ quality losses predict significant differences for the more critical 100% magnification use case. In fact, the Pearson correlations were only −0.35 and −0.40, respectively, for these Outdoor Day and Indoor comparisons (Again, the correlation is negative because the CPIQ quality loss is inversely related to the VIQET mean opinion scores.). However,

Table 9.14 VIQET results for the phones used in the CPIQ validation study (Jin, 2017). The variability for each predicted VIQET MOS value is ±0.1 standard error. Many of the results between cameras in each category are statistically the same. Note that Phone 2 was not part of the VIQET study.

Category/Phone	1	3	4	5	6	7	8	9
Overall	3.1	3.8	3.5	4.1	3.9	3.9	3.9	4.2
Outdoor Day	3.9	4.0	4.1	4.2	4.3	4.2	4.2	4.3
Indoor	3.5	4.1	4.0	4.1	4.0	4.0	4.0	4.1
Outdoor Night	1.9	3.4	2.5	3.9	3.5	3.5	3.5	4.3

the Pearson correlation was −0.78 for the Outdoor Night compared to U30 25 lux. These differences in correlation imply that the VIQET and CPIQ predictions should be considered in different ways by those using the processes. It may be that the spatial degradations become significant enough under nighttime/dim conditions that the VIQET resizing process no longer masks the quality loss differences detected by the CPIQ metrics for the use case of 100% magnification which, for the VIQET Indoor and Outdoor Day, are not apparent. Regardless, note that when the three CPIQ lighting conditions are averaged and considered collectively for these eight phones with the VIQET Overall scores (the average of the Outdoor Day, Indoor, and Outdoor Night scores), the Pearson correlation is −0.84. That this correlation is higher for the VIQET and CPIQ quality loss compared to the Pearson correlation of −0.54 for the DxOMark Photo and CPIQ quality loss is understandable because both the smartphone camera hardware and the camera settings were identical for the former comparison, whereas the latter comparison was with different hardware and included other differences as described earlier.

One final note of comparison to consider is that, despite these differences in hardware and settings, the Pearson correlation for the comparison of the VIQET Overall scores and the DxOMark Photo scores for this set of eight smartphone cameras is 0.86. What this could point to is that the VIQET use case of 1080P bridges the use case of 100% magnification for the CPIQ quality loss predictions and the combined use cases of Web, 8MP and HDTV for DxOMark Mobile Photo scores. Certainly, additional studies with both objective and subjective components and increased application, usage, and development of these three benchmarking approaches as well as other benchmarking approaches will provide more insights to the image quality community at large with respect to benchmarking validity. Thus, the results from various benchmarking approaches show strengths and weaknesses of cameras of interest. But, the results and conclusions from different approaches may or may not correlate highly. Many reasons can contribute to the differences, particularly for complex systems such as camera benchmarking, which include variations in selection of camera settings, variations in lab setup and capture, the comprehensive nature of the captured image and video set, choice of use cases, and incorporation of subjective data sets. Ultimately, the benchmarking practitioner must remember that image quality benchmarking is a prediction, made from a limited subset of images and videos, of how cameras will perform for users in the market where applications are seemingly endless. Thus, selecting items such as photospace and use cases relevant to the questions needing to be answered by one's benchmarking process will provide a solid foundation and increase the value of the findings and conclusions.

Summary of this Chapter

- Utilizing sensor megapixel count as a single image quality benchmark is not comprehensive and can often be uncorrelated with subjective evaluation.
- Benchmarking components include objective metrics for global and local image quality attributes as well as psychophysical methods to obtain subjective metrics.
- Important aspects to consider when designing a benchmarking scope include photospace, use cases, and observer impact.

- Methods of combining metrics include options from simple summation to weighted averaging to Minkowski summation.
- Existing benchmarking systems include those limited to subjective or objective metrics as well as those that have means of incorporating both subjective and objective components.
- Benchmarks limited to or weighted toward objective metrics are only as good as the comprehensive nature of the metrics in the set.
- Benchmarks limited to or weighted toward subjective metrics may not be able to provide specific enough information for responding to imaging development and engineering needs.
- Benchmarking methods should be tailored to answer the intended questions and incorporate the expected photospace and use cases for the cameras of interest.
- Benchmarking approaches vary. As such, benchmarking results may not be highly correlated between approaches, but should answer the intended questions the practitioner has set out to test.
- As imaging systems continue to evolve, the list of metrics in benchmarking needs to continue expanding, and subjective evaluation becomes more important in order to assess which objective elements of image quality need to be measured.
- Fundamentally, benchmarking is predicting image quality for broad applications from, in perspective, a very limited set of tests.

References

Artmann, U. and Wueller, D. (2012) Improving texture loss measurement: Spatial frequency response based on a colored target. *Proc. SPIE*, **8293**, 829305.

Bartleson, C.J. and Grum, F. (1984) Visual measurements, in *Optical Radiation Measurements*, vol. **5**, Academic Press, Cambridge, MA, USA.

Baxter, D., Phillips, J.B., and Denman, H. (2014) The subjective importance of noise spectral content. *Proc. SPIE*, **9016**, 901603.

Baxter, D.J. and Murray, A. (2012) Calibration and adaptation of ISO visual noise for I3A's camera phone image quality initiative. *Proc. SPIE*, **8293**, 829303.

Cárdenas, L., Hinks, D., Shamey, R.,Kuehni, R., Jasper, W., and Gunay, M. (2006) Comparison of naïve and expert observers in the assessment of small color differences between textile samples. *Conference on Colour in Graphics, Imaging, and Vision* (CGIV), Leeds, UK, pp. 341–344(4).

DxO (2012a) DxOMark for DSCs. URL https://www.dxomark.com/Mobiles/Smartphones-beat-5-year-old-DSCs, (accessed 29 May 2017).

DxO (2012b) DxOMark launch. URL http://www.dxomark.com/Mobiles/DxOMark-Goes-Mobile, (accessed 29 May 2017).

DxO (2012c) DxOMark mobile report example. URL https://www.dxomark.com/Mobiles/How-we-test-smartphones-The-DxOMark-Mobile-protocol, (accessed 29 May 2017).

DxO (2016a) DxOMark lens scores. URL http://www.dxomark.com/About/Lens-scores, (accessed 29 May 2017).

DxO (2016b) DxOMark Mobile scores. URL http://www.dxomark.com/Mobiles, (accessed 29 May 2017).

DxO (2017a) DxOMark lens with camera scores. URL https://www.dxomark.com/Reviews/DxOMark-Score, (accessed 29 May 2017).

DxO (2017b) What is DxOMark? URL https://www.dxomark.com/About/What-is-DxOMark/Test-result-reliability, (accessed 29 May 2017).

Engeldrum, P.G. (2000) *Psychometric Scaling: A Toolkit for Imaging Systems Development*, Imcotek Press, Winchester, MA, USA.

Farnand, S., Jang, Y., Han, C., and Hwang, H. (2016) A methodology for perceptual image quality assessment of smartphone cameras. *Electronic Imaging: Image Quality and System Performance*, **2016**, 1–5.

Fernandez, S.R. and Fairchild, M.D. (2001) Preferred color reproduction of images with unknown colorimetry, in *Proceedings of the IS&T 9th Color and Imaging Conference*, Scottsdale, AZ, USA, pp. 274–279.

Fernandez, S.R. and Fairchild, M.D. (2002) Observer preferences and cultural differences in color reproduction of scenic images, in *Proceedings of the IS&T 10th Color and Imaging Conference*, Scottsdale, AZ, USA, pp. 66–72.

Fernandez, S.R., Fairchild, M.D., and Braun, K. (2005) Analysis of observer and cultural variability while generating "preferred" color reproductions of pictorial images. *J. Imaging Sci. Technol.*, **49**, 96–104.

Fleming, P.J. and Wallace, J.J. (1986) How not to lie with statistics: the correct way to summarize benchmark results. *Commun. ACM*, **29**, 218–221.

FNAC (2016) FNAC methodology. URL http://www.fnac.com/labofnac/test-photo/, (accessed 29 May 2017).

GSMArena (2016a) GSMArena compare website. URL http://www.gsmarena.com/compare.php3, (accessed 29 May 2017).

GSMArena (2016b) GSMArena website. URL http://www.gsmarena.com, (accessed 29 May 2017).

He, Z., Jin, E.W., and Ni, Y. (2016) Development of a perceptually calibrated objective metric for exposure. *Electronic Imaging: Image Quality and System Performance*, **2016**, 1–4.

Hultgren, B.O. and Hertel, D.W. (2008) Megapixel mythology and photospace: estimating photospace for camera phones from large image sets. *Proc. SPIE*, **6808**, 680818.

I3A (2007) Camera Phone Image Quality – Phase 1 – Fundamentals and review of considered test methods. IEEE.

I3A (2009a) Camera Phone Image Quality – Phase 2 – Acutance – Spatial Frequency Response. IEEE.

I3A (2009b) Camera Phone Image Quality – Phase 2 – Color Uniformity. IEEE.

I3A (2009c) Camera Phone Image Quality – Phase 2 – Color Uniformity. IEEE.

I3A (2009d) Camera Phone Image Quality – Phase 2 – Initial Work on Texture Metric. IEEE.

I3A (2009e) Camera Phone Image Quality – Phase 2 – Lateral Chromatic Aberration. IEEE.

I3A (2009f) Camera Phone Image Quality – Phase 2 – Lens Geometric Distortion (LGD). IEEE.

I3A (2009g) Camera Phone Image Quality – Phase 2 – Subjective Evaluation Methodology. IEEE.

IEEE (2017) IEEE 1858-2016, IEEE Standard for Camera Phone Image Quality. IEEE.

Image Engineering (2016) About VCX. URL http://valuedcameraexperience.com, (accessed 29 May 2017).

Imatest (2016) Imatest master software for macOS. URL http://www.imatest.com/support/download/, (accessed 29 May 2017).

ISO (2013) ISO 15739:2013 Photography – Electronic Still Picture Imaging – Noise Measurements. ISO.

Jin, E., Phillips, J.B., Farnand, S., Belska, M., Tran, V., Chang, E., Wang, Y., and Tseng, B. (2017) Towards the development of the IEEE P1858 CPIQ standard - a validation study. *Electronic Imaging: Image Quality and System Performance*, **2017**, 88–94.

Jin, E.W. (2017) Personal communication.

Keelan, B.W. (2002a) *Handbook of Image Quality – Characterization and Prediction*, Marcel Dekker, New York, USA.

Keelan, B.W. (2002b) Predicting multivariate image quality from individual perceptual attributes. *Proc. IS&T PICS Conference*, pp. 82–87.

Keelan, B.W., Jenkin, R.B., and Jin, E.W. (2012) Quality versus color saturation and noise. *Proc. SPIE*, **8299**, 82 990F.

Keelan, B.W., Jin, E.W., and Prokushkin, S. (2011) Development of a perceptually calibrated objective metric of noise. *Proc. SPIE*, **7867**, 786 708.

Kirk, L., Herzer, P., Artmann, U., and Kunz, D. (2014) Description of texture loss using the dead leaves target: Current issues and a new intrinsic approach. *Proc. SPIE*, **9023**, 90 230C.

Lilja, D.J. (2000) *Measuring Computer Performance: A Practitioner's Guide*, Cambridge University Press, Cambridge, UK.

McElvain, J., Campbell, S.P., Miller, J., and Jin, E.W. (2010) Texture-based measurement of spatial frequency response using the dead leaves target: Extensions, and application to real camera systems. *Proc. SPIE*, **7537**, 75 370D.

Microsoft (2016a) Skype and Lync test specifications for USB peripherals, PCs, and Lync room systems. URL https://technet.microsoft.com/en-us/office/dn788953.aspx, (accessed 29 May 2017).

Microsoft (2016b) *Skype for Business Video Capture Specification for conferencing devices*.

Microsoft (2016c) *Skype for Business Video Capture Specification for personal solutions*.

Parr, W.V., White, K.G., and Heatherbell, D.A. (2003) Exploring the nature of wine expertise: what underlies wine experts' olfactory recognition memory advantage? *Food Qual. Preference*, **5**, 411–420.

Peltoketo, V. (2014) Evaluation of mobile phone camera benchmarking using objective camera speed and image quality metrics. *J. Electron. Imaging*, **23**, 061 102–1–7.

Phillips, J.B., Coppola, S.M., Jin, E.W., Chen, Y., Clark, J.H., and Mauer, T.A. (2009) Correlating objective and subjective evaluation of texture appearance with applications to camera phone imaging. *Proc. SPIE*, **7242**, 724 207.

Rice, T.M. and Faulkner, T.W. (1983) The use of photographic space in the development of the disc photographic system. *J. Appl. Photogr. Eng.*, **9**, 52–57.

Saad, M.A. (2016) Personal communication.

Saad, M.A., Corriveau, P., and Jaladi, R. (2015a) Consumer content framework for blind photo quality evaluation. *Ninth Int. Workshop on Video Processing and Quality Metrics for Consumer Electronics*.

Saad, M.A., McKnight, P., Quartuccio, J., Nicholas, D., Jaladi, R., and Corriveau, P. (2016) Online subjective testing for consumer-photo quality evaluation. *J. Electron. Imaging*, p. 043009.

Saad, M.A., Pinson, M.H., Nicholas, D.G., Kets, N.V., Wallendael, G.V., Silva, R.D., Jaladi, R.V., and Corriveau, P.J. (2015b) Impact of camera pixel count and monitor resolution perceptual image quality. *Colour and Visual Computing Symposium (CVCS)*, pp. 1–6.

Segur, R. (2000) Using photographic space to improve the evaluation of consumer cameras. *Proc. IS&T PICS Conference*, **3**, 221–224.

Smith, J.E. (1988) Characterizing computer performance with a single number. *Commun. ACM*, **31**, 1202–1206.

Töpfer, K., Jin, E., O'Dell, S., and Ribeiro, J. (2006) Regional preference for the rendition of people. *International Congress of Imaging Science*.

VIQET (2016) VIQET desktop tool installer. URL https://github.com/VIQET/VIQET-Desktop/releases, (accessed 29 May 2017).

Vodafone and Image Engineering (2016) VCX - Valued Camera eXperience Version 1.5.

Wang, Z. and Bovik, A.C. (2006) *Modern Image Quality Assessment*, Morgan & Claypool Publishers, USA.

10

Summary and Conclusions

Predicting the quality of an image produced by a particular camera is no trivial task. Even so, it is not impossible, and it has been the purpose of this book to show that with careful experimental design and well-established, robust objective metrics together with a sound scientific approach to subjective testing, repeatable and accurate results can be obtained.

One often hears that image quality is subjective and therefore cannot be reliably assessed by measurements and experiments. What is neglected in this statement is the fact that most people tend to agree on what is good and bad image quality. Furthermore, when asking a person what makes an image good or bad in terms of quality, statements such as "sharp," "noisy," and "low contrast" are often expressed. This tells us that firstly, if people on average agree on the quality of an image, it should be possible to employ statistical methods to quantify image quality. Secondly, the overall experience of image quality is made up of a set of identifiable categories, or attributes, that together form the impression of the quality of a certain image. The first observation also tells us that if an average observer can be established for which a representative quantitative assessment of image quality can be made, it should in principle also be possible to exchange human observers with objective measurements. From the second observation, this notion can be taken further and we may also assume that it is possible to measure how the individual attributes are affected by a certain piece of imaging equipment. Several such measurements can then be combined in order to form a comprehensive judgment of image quality. In order for this approach to be successful, it is imperative that a connection can be made between objective measurement results and perceived image quality.

There are numerous variants of image quality metrics in existence, many of which are unfortunately neither well-tested nor have a solid theoretical underpinning. The purpose of this book has been to provide the reader with the necessary background to develop their own metrics as well as to be able to critically scrutinize existing metrics for their suitability for a particular task. In order to do this, a basic understanding of the various elements affecting the impression of an image needs to be acquired. Consequently, the book started out in Chapter 1 by discussing the relationship between image content and image quality. It was shown that the elemental information of objects in a scene can be defined by a limited and select number of simple visual cues. Furthermore,

Camera Image Quality Benchmarking, First Edition. Jonathan B. Phillips and Henrik Eliasson.
© 2018 John Wiley & Sons Ltd. Published 2018 by John Wiley & Sons Ltd.
Companion website: www.wiley.com/go/benchak

in order to obtain a realistic image of a scene, a set of low-level characteristics needs to be rendered as realistically as possible. The most important such characteristics are:

- Color
- Shape
- Texture
- Depth
- Luminance range
- Motion

Attempting to quantify the imaging capabilities of an image capture device requires taking all of these aspects into account. However, as they are quite general, they need to be broken up into more specific parts. This leads to a new classification of characteristics, the *image quality attributes*, among which the most common are:

- Exposure and tone
- Dynamic range
- Color
- Shading
- Geometric distortion
- Stray light
- Sharpness and resolution
- Texture blur
- Noise
- Color fringing
- Image defects

The correspondence between the low-level characteristics and image quality attributes is evidently not one-to-one. For example, the sense of depth in an image can be dependent on sharpness as well as color. Being able to accurately characterize the main image quality attributes listed above will provide the first steps in obtaining a reasonably correct assessment of the overall image quality. However, without a precise definition of what we actually mean by image quality, it will not be possible to combine the results for individual attributes into a meaningful overall assessment. For this reason, Chapter 2 discussed the different requirements that may arise for different use cases and concludes with a relevant definition for the purposes of this book, stating that image quality is:

> *the perceived quality of an image under a particular viewing condition as determined by the settings and properties of the input and output imaging systems that ultimately influences a person's value judgment of the image.*

An important categorization of image quality attributes was made in Chapter 2 in terms of local and global attributes. These categories derive from the fact that there is a clear dependency between the conditions under which an image is viewed with respect to viewing distance and image size. This is also reflected in the image quality definition.

In Chapter 3, the concept of image quality attributes was elaborated further and several examples were provided in order to illustrate their effect on images.

Having a firm understanding of the most important aspects of image quality and its attributes, Chapter 4 then began the treatment of camera benchmarking by describing and explaining the various subcomponents of a camera and elaborated on how they affect the various attributes.

In order for measurements of the individual attributes to be effective, not only do use cases need to be established, but also a connection with perceived image quality is required. In Chapter 5, a rundown of subjective experimentation techniques was provided and the concept of just noticeable differences was introduced. This enables an absolute scale of subjective measurement results, necessary in order to correlate objective measurements with subjective results in a manner such that attributes can be directly compared and combined.

Continuing to elaborate the benchmarking approach, Chapter 6 provided descriptions of objective metrics for each of the attributes described in previous chapters. With establishment of the principles behind individual attributes and the methods to measure them, Chapter 7 presented additional methods that enable links to be made between the objective metrics and subjective results.

Chapters 1 to 7 provided the theoretical background and metrics needed in order to design and use a successful benchmarking process. In Chapters 8 and 9, more practical aspects and specifics of camera image quality benchmarking were presented. In order to produce reliable and repeatable results, the measurement protocols must have been determined and formalized. Chapter 8 described the steps necessary for building up a camera lab and also explained the procedures for measuring individual attributes. In Chapter 9, this knowledge, together with what was learned in previous chapters, was used to demonstrate a full benchmark of a number of cameras. Examples of other benchmarking processes were also presented for comparison. This ties all components together and should provide the reader with a sufficiently well-equipped toolbox to enable the design and application of image quality measurement methods for a wide range of situations.

This book has been emphasizing camera characterization and benchmarking methods, therefore the example in Chapter 9 did not include image quality assessment methods typically used for video quality assessment. Such methods have traditionally mostly been applied to the assessment of video codec quality, and therefore full-reference metrics such as PSNR and SSIM have dominated. The difference between these types of metrics and the image quality attribute metrics described in this book is that while the former try to quantify the overall quality of a particular *image*, the latter attempt to *predict* the image quality of images produced by the equipment currently under test. As such, it has not been an easy task to correlate the two methodologies. This is further complicated by the fact that image quality metrics typically require a reference, which is not readily available in a benchmarking situation. However, active research on no-reference metrics is ongoing (Shahid *et al.*, 2014). An especially stimulating recent development is to use deep learning methodologies to implement no-reference metrics (Hou *et al.*, 2014; Kang *et al.*, 2014; Xu *et al.*, 2015; Bosse *et al.*, 2016). Applications in this field go beyond image quality as defined in this book. For example, deep learning methods are being used to quantify the aesthetic aspects of photographs (Zhou *et al.*, 2015).

The natural link between image attribute metrics (predicting the quality of an image) and image quality metrics (assessing the overall quality of a particular image) is the JND. If both approaches could transform their results into JNDs, it might be possible to harmonize them, and this would be especially interesting if reliable no-reference metrics were available. Since both approaches have their shortcomings, being able to combine them in a meaningful way would be even more useful. In the case of predicting the image quality capabilities of a certain camera, the assessment of image quality attributes has a clear disadvantage in that it can only determine how a limited set of attributes affect the end result. If other artifacts are present, the result may be an overestimation of overall quality. In the case of the overall image quality metrics, it is in most cases implied that all visible artifacts in the image are taken into account. This distinction may become even more important as new imaging technologies develop, which will introduce their own specific artifacts not anticipated today. For instance, computational imaging, which has become increasingly intriguing during recent years with multi-camera systems and plenoptic cameras, will inevitably introduce new characteristics due to, for example, difficulties in compensating for parallax errors. Also, more nontraditional processing of, for example, faces in images, may require completely different approaches. Other new metrics that can be envisioned are, for instance, to measure bokeh, that is, the quality of out of focus objects in, for example, portraits. For camera image quality measurements and benchmarking, all of this will be a challenge, especially since new technology is developing at a much faster pace than the development of image quality standards, which typically take 5 to 10 years to develop into something that can be used reliably by a wider group of people.

Another issue worth raising regarding image quality attribute measurements is the increasing complexity in the image processing chain of the camera. The algorithms employed tend to be highly adaptive and therefore very scene dependent. This is evident in algorithms for HDR processing, but also in noise reduction and sharpening algorithms, to mention a few. This can make it difficult to perform accurate measurements, since an attribute that is measured via a test chart may be very different from how the attribute appears in a real scene. This may encourage new approaches to image quality measurements, such as incorporating test charts into more realistic scenes, and maybe predictive approaches that do not even necessitate shooting images for measurement in the lab.

Just as objective testing methodologies are envisioned to undergo changes as camera technology evolves, so also should one anticipate subjective testing methodology to change as well. One development may be toward crowdsourcing approaches, where a large number of people are judging images in front of their own computer screens (Burns *et al.*, 2013). This approach necessarily reduces accuracy due to the limited control over experimental conditions. However, by using a very large number of observers, estimations of the population response can be made reasonably reliable despite the larger variability inevitable in this technique. Other concepts, such as obtaining subjective evaluation directly from observations on a given smartphone model or generating observer screening protocols, may be utilized, however, to lead to subjective data with tighter variability.

Since images have become such an integral part of our lives, with a wide range of applications, the need to be able to assess image quality will continue to be very important. It is certain that the ever ongoing emergence of new imaging technology will instigate the development of new benchmarking metrics, but also the refinement of existing ones. Therefore, camera image quality characterization methods will certainly continue to evolve and become even more important for providing guidance for consumers, making the work of the imaging engineer more efficient, providing a common language for customers and suppliers of imaging equipment, and more.

References

Bosse, S., Maniry, D., Wiegand, T., and Samek, W. (2016) A deep neural network for image quality assessment, in *IEEE International Conference on Image Processing (ICIP)*, pp. 3773–3777.

Burns, P.D., Phillips, J.B., and Williams, D. (2013) Adapting the ISO 20462 softcopy ruler method for online image quality studies. *Proc. SPIE*, **8653**, 86530E.

Hou, W., Gao, X., Tao, D., and Liu, W. (2014) Blind image quality assessment via deep learning. *IEEE Trans. Neural Netw. Learn. Syst.*, **26**, 1275–1286.

Kang, L., Ye, P., Li, Y., and Doermann, D. (2014) Convolutional neural networks for no-reference image quality assessment, in *IEEE Conference on Computer Vision and Pattern Recognition (CVPR)*, pp. 1733–1740.

Shahid, M., Rossholm, A., Löfström, B., and Zepernick, H.J. (2014) No-reference image and video quality assessment: a classification and review of recent approaches. *EURASIP J. Image Video Process.*, pp.1–32.

Xu, L., Lin, W., and Kuo, C.C.J. (2015) *Visual Quality Assessment by Machine Learning*, Springer, Singapore.

Zhou, Y., Li, G., and Tan, Y. (2015) Computational aesthetics of photos quality assessment and classification based on artificial neural network with deep learning methods. *Int. J. Signal Process. Image Process. Pattern Recog.*, **8**, 273–282.

Index

ΔE 180
ΔE_{94} 180
z-score   139, 158
2AFC   125
35 mm film   19
3A algorithms   101
3AFC, *see* three-alternative forced
      choice

a

aberrations   65
   chromatic   51, 65, 67, 213
   lens   45
   monochromatic   65
   third-order   65
absolute category rating   157, 161, 162
absolute threshold   118
absorbance   174
acceptability rating   117
acceptability scaling   122–125, 302, 312
ACF, *see* autocorrelation function
ACR, *see* absolute category rating
acutance   235–240, 284, 287
ADC, *see* analog to digital converter
additive white Gaussian noise   257
adjacency effects   191, 203
Airy disk   73
akinetopsia   16
albedo   3
aliasing   46, 52, 96, 194
   temporal   58, 305
analog to digital converter   76, 78, 79, 89
analysis of variance   158
anchor images   132

anchored scaling   122, 123, 125, 135, 142,
      144, 302, 312
angular subtense   313
aperture   68, 70, 73, 101, 289
APS, *see* active pixel sensor
arithmetic coding   101
aspect ratio   325
astigmatism   66
audio-video synchronization   59, 293,
      295, 311
auto exposure algorithm   216
auto white balance   216
autocorrelation function   209–211
autofocus   293, 302, 311, 318
autofocus responsiveness   216
AWB, *see* auto white balance
AWGN, *see* additive white Gaussian
      noise

b

B-frames   110
backlighting   280
backside illumination   80
baffle   277
Bartleson–Brennan effect   245
Bayer pattern   81, 82
Bessel function   72, 188
beta movement   15, 16
bidirectional prediction   110
bidirectional reflectance distribution
      function   176
bilateral filter   96
binocular disparity   11
black body   171

Camera Image Quality Benchmarking, First Edition. Jonathan B. Phillips and Henrik Eliasson.
© 2018 John Wiley & Sons Ltd. Published 2018 by John Wiley & Sons Ltd.
Companion website: www.wiley.com/go/benchak